DRIFTING CONTINENTS & COLLIDING PARADIGMS

DRIFTING CONTINENTS

SCIENCE, TECHNOLOGY, AND SOCIETY

Ronald N. Giere and Thomas F. Gieryn, General Editors

& COLLIDING PARADIGMS

Perspectives on the Geoscience Revolution

John A. Stewart

INDIANA UNIVERSITY PRESS
Bloomington & Indianapolis

Manufactured in the United States of America

Library of Congress Cataloging-in-Publication Data

Stewart, John A.

Drifting continents and colliding paradigms : perspectives on the geoscience revolution / John A. Stewart.

p. cm. — (Science, technology, and society)

Includes bibliographical references.

ISBN 0-253-35405-6

1. Continental drift. 2. Science—Philosophy. I. Title. II. Series: Science, technology, and society (Bloomington, Ind.)

QE511.5.S86 1990

500—dc20 89-45195

CIP

1 2 3 4 5 94 93 92 91 90

Contents

LIST OF FIGURES

LIST OF TABLES

Preface

In the late 1960s and early 1970s most geoscientists experienced a revolutionary transformation in their conceptions of the geological processes forming the Earth's surface. They moved from a general belief in the stability of the continents and the permanence of ocean basins to a dynamic view involving the movement and collision of continents and the birth and death of whole ocean basins. This book includes a description of the historical development of plate tectonics theory—from the rise and fall of continental drift theory over fifty years ago to the acceptance of plate tectonics by the early 1970s. The historical account is only the first part of this book. In the rest of the book this history becomes "data" as we examine how science works. We will consider different philosophical, historical, and sociological "theories" about how scientific ideas develop and compare these theories to both the history of plate tectonics and some comments made by many of the geoscientists who developed plate tectonics or who continued to oppose it. In other words, this book is more than a history of plate tectonics because it uses this specific scientific revolution to illustrate *theories about* the development of scientific knowledge.

I hope that a variety of readers will find something of interest to them in this book. Geoscientists who "know" most of the history of plate tectonics and continental drift might be interested both in how some perspectives on the development of science illuminate aspects of this history *and* in the comments made by their colleagues. Philosophers and historians of science may value the brief history of continental drift and plate tectonics, the discussion and illustrations of sociological perspectives, and the personal comments made by the geoscientists about this revolution. Sociologists of science may find some value in my efforts to illustrate how some aspects of cognitive science and organizational studies can be applied to the study of science and to develop a quantitative analysis that combines features of both the "functionalist" and the "constructivist" approaches to science. Finally, it is hoped that lay people will find an understandable history of this revolution, a simple introduction to a variety of perspectives on science, and how these different perspectives illuminate aspects of this history.

Whereas I *hope* that these different readers will find something of interest in this book, I am *certain* they will find something to criticize. A single individual cannot master the diverse knowledge and skills needed to understand and describe the nature and process of the scientific endeavor—even within a limited domain. Geoscientists may object to the emphasis on social factors. Historians of science will deplore the heavy reliance upon secondary sources. Historians and philosophers may feel that the applications of the different perspectives are too simplified or that their favored perspective is not even mentioned. All of the above will certainly regard some of the quantitative applications as too "crude" to be useful, even though these analyses are more "relevant" to their interests than previous quantitative studies in the sociology of science. Despite these certain objections, I hope that this book encourages others to study this particular episode in the history of science so that we will gain a fuller understanding of the intellectual *and* cultural processes leading to the theory of plate tectonics and its widespread acceptance.

Acknowledgments

Numerous individuals have played an important role in the ideas presented in this book, which grew out of my doctoral dissertation in sociology at the University of Wisconsin-Madison (Stewart, 1979). My thesis advisor, Warren Hagstrom, not only coached me—a recent chemistry graduate—through the graduate program in sociology, but also conceived of this project, obtained funding from the National Science Foundation, and patiently let me pursue the study of plate tectonics in my own manner. His broad conception of sociology, willingness to consider diverse viewpoints, and concern for his graduate students have provided an important role model.

Scott Long in the Sociology Department at Indiana University read an early draft and provided valuable comments that greatly improved the content, organization, and style of writing. Richard Thiessen in the Department of Geology at Washington State University provided very helpful comments and corrections to major sections of the manuscript. Their efforts are greatly appreciated. Both Washington State University and the University of Hartford provided greatly appreciated financial support.

Others providing valuable comments, encouragement, or influences on my conception of science include Victor Hilts, Herbert Wang, Robert Dott, Ron Giere, Hal Winsborough, Aage Sørensen, and my wife, Susan Randolph, whose endless support made this book possible.

I am especially grateful to the many geoscientists who granted me interviews and shared many of their thoughts and recollections about the development of plate tectonics. This information helped me develop my ideas about how scientists think, act, and interact with each other and it provides the "glue" that holds together the diverse subjects discussed in this book. Although they may not share the perspectives on science discussed in this book, it is hoped that they will find it interesting and challenging.

DRIFTING CONTINENTS
& COLLIDING PARADIGMS

CHAPTER ONE

Perspectives on Scientific Revolutions

In the course of just one decade, from 1960 to 1970, most geoscientists experienced a major revolution in how they thought about the earth. In 1960 the vast majority believed that the continents and ocean basins were permanent and geographically fixed features of the earth's surface. Most thought the earlier ideas about continental drift were ridiculous and had been given a proper burial before World War II. Yet by the early 1970s the same geoscientists were using the new theory of plate tectonics to calculate the rates of continental movements over the last 200 million years, which included the formation of the Atlantic Ocean.

This remarkable change in beliefs raises a number of interesting questions. How did this sudden transformation occur? What evidence and which scientists caused this change? Since some of the evidence for plate tectonics is similar to that presented for "continental drift" theory fifty years earlier, why didn't geoscientists accept "drift" theory earlier? If scientific knowledge can change so rapidly, why should we believe it? More generally, what distinguishes scientific "knowledge" from other "beliefs"? Is scientific knowledge based upon a special or "rational" process? What do we mean by "rational"? How do we study science so that we can answer such questions as these?

These questions illustrate some of the topics covered in this book, which not only describes the history of this revolution, but compares this history to a variety of perspectives on the development of scientific ideas. It will become clear that just as plate tectonics theory cannot explain all the known details or "facts" about the earth, neither can any of these different perspectives on science explain all of the details of this history. This limitation is obvious if we simply sketch the numbers of "actors," products, and processes involved.

Plate tectonics theory resulted from over sixty years of research by thousands of geoscientists who expended millions of dollars and even more days of human labor, thought, and interaction with each other and the physical environment. This massive effort was coordinated by a variety of individuals thinking and working within a diverse set of informal and formal organizations (geoscience specialties, research teams, universities and departments, journal publishers, funding agencies, international research projects, etc.), who produced and used numerous communications, including thousands of scientific publications. These components—the physical environment, scientific instruments, empirical data, people, organizations, time, labor, thought, social interaction, publications, etc.—and their interactions constitute the "reality" of the geoscience research that produced plate tectonics theory.

None of the perspectives considered in this book can explain all the details of this history, but they are not meant to do so. Instead their purpose is to identify the fundamental "processes" underlying this and other scientific accomplishments in the same way that plate tectonics theory specifies the fundamental processes shaping the surface of the earth. The next section briefly describes some of the differences among the perspectives considered in this book. Following that are brief descriptions of the data sources used in this book and an overview of the different chapters.

Diverse Perspectives on the Nature of Scientific Knowledge

Although we will consider only a few perspectives on science, they illustrate a range of viewpoints. The chosen perspectives were either (a) widely assumed to characterize science, such as "logical empiricism," (b) explicitly mentioned by various geoscientists, such as Popper's (1959) "falsificationism" or Kuhn's (1962) "paradigm" perspective, (c) applied to this revolution by historians, such as the perspectives developed by Lakatos (1970) and Laudan (1977), (d) of special interest to me, such as the various sociological perspectives, or (e) those that seemed particularly important for specific aspects of this revolution, such as Harré's (1976, 1986) suggestions about the role of "models and analogies" in scientific thought.

A brief discussion of these perspectives and their differences provides a general context for the history presented in the next three chapters. It will emphasize what these perspectives imply about how scientists make "decisions," especially those concerned with the choice between competing theories or "paradigms" in the same discipline. The stress here is on the more conscious decisions of scientists, but later chapters will generalize the decision-making approach to include "unconscious" cognitive processes.

We can start by examining the logical empiricist perspective, which is a view about scientific reasoning that is widely held among both the public and scientists themselves, even though few philosophers of science now accept it. The logical empiricists accepted two fundamental assumptions in their attempts to account for scientific practice. First, they assumed that empirical data were "objective" and provided primitive truths about the natural world that were shared by all scientists. Second, they regarded formal deductive logic and mathematics as the only correct or "rational" means of relating theoretical propositions to each other and to empirical data.

Logical empiricists separated the decisions made by scientists into two realms: those concerned with the context of discovery and those concerned with the context of justification of scientific theories. They had little to say about the context of discovery. Their major interest was in showing that scientific theories were rationally justified because deductive logic connected theoretical terms to empirical data. Since logic and mathematics were *rule*-based activities, there was little room for "decisions" by scientists as they sought to evaluate scientific theories. Correct predictions provided support for theories and incorrect predictions decreased support, so one simply accepted the theory that gave on balance the best fit to the known facts. The discovery of new facts might prompt theory change, but most logical empiricists believed these changes always increased the total number of explained facts. Thus they tended to be "realists" because they believed that over time scientific theories more accurately described an external reality.

This basic perspective evolved as logical empiricists found internal inconsistencies in their views and inconsistent behavior by scientists (Brown, 1977). For example, the "problem of induction" is always present in science because its theories are supposed to be *universally* true, but this belief is always induced from a *finite* number of empirical confirmations. It cannot be "rational" because deductive logic cannot justify such inductive generalizations. In addition to such internal problems, the rules of logic did not specify adequately what practicing scientists considered confirmations of theoretical propositions. For example, the statement that all "island arcs" are next to ocean "trenches" is *logically* consistent with empirical observations on the Aleutian trench/island arc system *and* the lack of such a combination in Ohio. Yet scientists do not regard the latter as confirming the association between trenches and island arcs. As a result of these and other considerations (e.g., Brown, 1977; Suppe, 1977; Ziman, 1968, 1978; and Weimer, 1979), philosophers developed various modifications or alternatives to logical empiricism.

The critical rationalism of Karl Popper (1959) was one of the most influential modifications of logical empiricism and was occasionally cited by geoscientists as the proper method of scientific research (Sengör and Burke, 1979). This alternative was primarily a response to some of the logical problems inherent in logical empiricism, but it was also an effort to make it more realistic with respect to actual scientific practice. Critical rationalism shares with logical empiricism both its belief that "rational" thought was equivalent to the use of deductive logic and its relegation of the discovery of scientific theories to an irrational or non-rational realm. To avoid the problem of induction, Popper discounted the importance of confirmations and suggested that scientists accept theories and laws as universal givens, which they try to *disprove or falsify*. If a theory made an incorrect prediction, then it must be rejected for the same reasons that one would reject the truthfulness of a set of logical axioms whenever logical deductions gave false propositions. In other words, in a strictly logical sense we could know for certain *only* that a theory was false, not whether it was true. Popper argued that good theories had implications for a wide range of empirical phenomena and stood up to repeated attempts to falsify them. The history of science was the history of "conjectures and refutations"—of learning by our mistakes so that our knowledge gradually approached a description of the real world. Good science should be in a state of constant competition between alternative theories with attempts to falsify them. Hence, his theory is often called "falsificationism."

Popper, however, recognized that the empirical data or "basic statements" used to test theories were also scientific statements and hence falsifiable. That is, empirical data were not "theory-free" and could not be unambiguously accepted as confirmed "facts" used to test or falsify a theory. Thus any falsifying empirical result could be a consequence of comparing a false theory to "true facts" or of using "false facts" to test a possibly true theory. The acceptance of these basic statements depended on the methodological "decisions" of scientists, and when agreement was reached, then the theory could be rationally, that is, logically, tested. As he put it (Popper, 1968: 108): "From a logical point of view, the testing of a theory depends upon basic statements whose acceptance or rejection, in its turn, depends upon our decisions. Thus, it is decisions which settle the fate of theories."

Hence Popper required more "decisions" on the part of scientists. As with the logical empiricists, the "decision" about the consistency between a theory and empirical data merited little discussion, but he added a critical decision: which empirical data would be accepted as "true." Popper did not describe how this decision was made and apparently believed that there would be a widespread consensus on what were the important "facts." Yet we will see that scientists frequently disagree on which are the important "facts" and even on whether accepted "facts" are consistent with a given theory. More serious for Popper's perspective was the willingness of many scientists to accept a theory even when it conflicted with widely accepted "facts." For Popper the failure of scientists to follow his methodology only meant that they illustrated poor scientific practice: His methodology had strong normative implications for how scientists should behave. For others such examples indicated that the philosophy of science needed a new system of fundamental assumptions.

One of the most influential of these alternative perspectives on science is Thomas Kuhn's (1962, 1970*a*) *The Structure of Scientific Revolutions*. Kuhn suggested that a mature scientific field was usually in one of two fairly distinct stages: "normal science" with research directed by a shared "paradigm" or "revolutionary science" with competition between paradigms.

A paradigm is a constellation of beliefs, values, procedures, and past scientific achievements that is shared within a community of scientists, directs their research activity, and is learned during their training or in their common research experiences. These shared group commitments provide a framework that highlights specific scientific problems, restricts possible solution tactics, and establishes criteria for the evaluation of solutions. Hence, normal science has a "puzzle solving" character, which resulted in rapid accumulation of scientific knowledge because the community of researchers work within the framework of the accepted paradigm instead of trying to overthrow it.

A key component of a paradigm is the "exemplar," which is a widely recognized, specific research accomplishment that provides several important functions for the research community. First, exemplars concretely illustrate how the more abstract components of a paradigm, such as metaphysical models and assumptions, can be related to the results of specific instruments. Second, they establish standards for how well theoretical predictions must correspond to empirical observations. Finally, these functions of the exemplar are learned *tacitly* during scientific training by repeating the research in the paradigm's exemplars. In other words, it is the tacit knowledge provided by exemplars that connects together theory and data instead of the rules of logic emphasized by the philosophies of science discussed above (Kuhn, 1970c: 16). Kuhn (1977: 308–316) argues that this tacit knowledge is broader than that contained in a system of verbal rules. Rules may capture selected aspects of this tacit knowledge, but only by distorting the essential nature of this method of learning. For example, only a period of training with actual fossils provides the tacit knowledge allowing a paleontologist to identify specimens. Verbal descriptions or rules cannot summarize this knowledge.

Although normal science is constrained science, Kuhn argues that it inevitably produces some inconsistent or unexpected results, which become "anomalies" for the paradigm. In fact, a paradigm always has anomalies. Some of these provide the "puzzles" to be solved, but scientists may ignore more resistant anomalies, hoping that later developments of the paradigm will solve them. However, a "crisis" may develop if accepted anomalies resist repeated efforts by the best researchers, accumulate in number, or are particularly severe threats to the basic assumptions of the paradigm. An extended crisis encourages researchers to question and modify various paradigm elements in attempts to account for the anomalies. Eventually different groups of researchers will develop their own competing paradigms with different assumptions, models, and exemplars. Thus, a crisis can lead to a period of "revolutionary science" marked by competition between two or more communities with different paradigms for similar, but not identical, empirical results. A crisis could, of course, end with the resolution of the important anomalies by acceptable modifications of the old paradigm.

Kuhn's suggestions about the processes ending revolutionary science and reestablishing a new period of normal science cause the strongest reactions from philosophers of science. Since the competing paradigms in revolutionary science develop from different solutions to different anomalies and vary in their metaphysical models, assumptions, and exemplars, Kuhn argues that the researchers following different paradigms will experience communication problems as they argue about which paradigm should provide the basis for future research. There will be a lack of agreement on the "important" facts, on which anomalies have been adequately "solved," and even on the meaning of some of the "same" concepts used by proponents of different paradigms. These factors make the different paradigms "incommensurable." For these reasons the choice between different paradigms cannot be reduced to a matter of logical proofs, rules, or algorithms because the different paradigms do not share the basic premises required for these procedures.

Persuasion becomes the key process and it is based on the abstract values shared by all the different groups. During revolutionary science proponents of different paradigms compare alternative paradigms in terms of such values as their ability to explain past puzzles, generate new and interesting puzzles, make quantitative or more accurate predictions, and in their relative scope and simplicity. Kuhn emphasizes the distinction between appealing to shared rules and appealing to shared values. The use of shared rules can lead only to one conclusion, whereas the use of shared values or ideology may result in considerable individual variation. "Group behavior will be affected decisively by the shared commitments [values], but individual choice will be a function also of personality, education, and the prior pattern of professional research" (Kuhn, 1970*b*: 241). These differences result from the differences that individuals place on the relative importance of these shared values (Kuhn, 1970*b*: 262).

The "choice" of a paradigm is not the result of a period of neutral weighing of the relative merits of each paradigm. Rather, the "decision" is often in the form of a dramatic "gestalt switch" as the scientist changes world views, language habits, and significant others. Older or established scientists, who contributed to the older paradigm, are most likely to experience this type of shift to the new paradigm. Whereas younger scientists or those entering a field from a new area are more likely to help develop new paradigms, they also find it easier to adopt them. The revolutionary period ends and normal science begins when a new paradigm succeeds in gathering the majority of members in the specific scientific community—by conversion of most members or at least of the younger members, who eventually form the community.

One of the major misunderstandings of Kuhn's model of scientific development concerns the sense in which the new paradigm is an "improvement" on the older paradigm (e.g., Lakatos and Musgrave, 1970). Kuhn asserts that his is a "relativist" position in the sense that one cannot say that the new paradigm is a better approximation to "reality" or to "what is really out there." Furthermore, Kuhn does not accept that each revolution necessarily increases the number of "explained facts." Although scientific revolutions are usually portrayed as progressive increases in the explained facts, Kuhn argues that much of this apparent "progress" occurs because textbooks in the new paradigm emphasize the old anomalies solved by the new paradigm and ignore the old paradigm's successes. This results in the concealment of many of the non-progressive aspects of revolutions and explains why some scientists can still question the adequacy of the new paradigm.

Instead of emphasizing the explained facts, Kuhn argues that there is progress between revolutions in the satisfaction of such values as simplicity, scope, accuracy, and especially quantitative predictions. With each new paradigm there is an increase in one or more of these values, but which values are increased depends on the revolutions considered. Kuhn's emphasis on persuasion processes using shared values and his deemphasis of explained facts prompted some philosophers of science to accuse him of arguing that change in scientific theories involves an irrational element, even "mob psychology." Yet Kuhn never suggests that this is an "irrational" process. Brown (1977) suggests that Kuhn has given a *new definition* for what is meant by "rational belief." Instead of beliefs based upon deductive logic, rational beliefs are those based upon the decisions made in a specially trained community. For example, Kuhn suggests the following procedure for making theory choice in science.

> . . . take a *group* of the ablest available people with the most appropriate motivation; train them in some science and in the specialties relevant to the choice at hand; imbue them with the value system, the ideology, current in their discipline (and to a great extent in other scientific fields as well); and, finally, *let them make the choice.* If that technique does not account for scientific development as we know it, then no other will. There can be no set of rules of choice adequate to dictate desired individual behavior in the concrete cases that scientists will meet in the course of their careers. Whatever scientific progress may be, we must account for it by examining the nature of the scientific group, discovering what it values, what it tolerates, and what it disdains. (Kuhn, 1970*b*: 237–238)

Kuhn argues that his position is "intrinsically sociological" and it is clear that the "decisions" required of scientists are numerous and critical for the success of science. However, Kuhn offers few suggestions for how these decisions might be studied, except to cite Merton's sociological perspectives mentioned below. Naturally, philosophers of science dislike Kuhn's emphasis and his failure to specify in more detail how theory-choice is actually made. Just as naturally, sociologists are attracted to these "failures" in Kuhn's analysis because Kuhn's "paradigms" are essentially "cultural traditions" (Barnes, 1982) and sociologists have the "tools" for the analysis of such traditions. However, they have pushed the sociological analysis of science in directions beyond those that even Kuhn would probably accept.

As a consequence of Kuhn's book, the formulation of general perspectives on science has taken two divergent routes, but both incorporate some of his insights. First, philosophers of science have tried to establish better "rules" that define the "rationality" of scientific decisions and protect the "progressiveness" of scientific knowledge. Sociologists (and a few philosophers and historians of science) have taken a second route by extending some of the social processes implicit in Kuhn's view. Although those following the different routes share the same concern with explaining the belief systems of scientists, they differ so greatly in their basic models, concepts, assumptions, and methods that one can say that the study of science is itself in the midst of a "revolution." We will briefly consider the perspectives of Lakatos (1970) and Laudan (1977) as illustrations of the route taken by some philosophers and then examine some of the different sociological perspectives on science.

The "sophisticated falsificationism" of Lakatos (1970) modified and extended Popper's methodology by incorporating some of the elements of Kuhn's perspective. He proposed, as did Popper, that "rational thought" was equivalent to the use of deductive logic, and that theories could not be empirically tested without accepting a chain of related theories, assumptions, boundary conditions, and agreed upon facts. Lakatos extended Popper's perspective by suggesting that the tested theory and the bundle of other necessary assumptions and supporting theories were related so closely that they should be considered an evolving unit, which he called a "scientific research programme."

According to Lakatos, the history of a science was understood best as a succession of competing research programmes, where the dominant programme in any given period was the one explaining the most facts. Lakatos, therefore, believed that confirmations played an important role in the acceptance of research programmes. He accepted Popper's observation that apparent falsifications of a programme could be removed by suitable modifications in the supporting theories and assumptions, but tried to describe when these adjustments were "rational" and when a research programme should be abandoned even though it was always possible to "patch it up."

Lakatos proposed that scientific research programmes could be divided into two parts: the "hard core," which contained the essential parts of a set of related theories, and the "auxiliary belt," which contained the assumptions and other theories needed to relate the hard core to observational data. Scientists did not question the hard core and protected it from falsifications by modifying the auxiliary belt. Thus, most changes in a research programme occurred in the auxiliary belt, whereas the core was relatively stable, but could undergo theoretical elaboration somewhat independently of empirical tests. For example, Newton's laws of motion constituted the hard core of the "Newtonian" research programme, which was applied to such subjects as planetary motions and the temperature and pressure properties of confined gases.

Successive tests of a research programme resulted in protective modifications of its auxiliary belt, but these modifications were not "rational" unless they allowed the hard core to be consistent with the new evidence *and* implied new empirical tests ("novel facts"). In Lakatos's terms these modifications must be "progressive problem shifts" and further tests must verify the implied "novel facts." It was rational, at least for a while, to continue work in a research programme even if it failed some empirical tests, but if the new modifications did not generate novel facts and only served to protect the hard core from refutation, then the programme is in a "degenerate phase" and should be abandoned. Ideally, good scientific research would be marked by competition between two or more research programmes. Scientists should tolerate new research programmes as long as they developed through progressive problem shifts, but given two mature programmes it was possible to rationally compare their explanatory abilities.

Lakatos believed the history of science showed that the competition between scientific research programmes produced rational and objective advances in scientific knowledge and a closer approximation to "reality." The historian of science should provide a "rational reconstruction" of episodes in science. Only if the actual history of the episode did not match the rational reconstruction, then the historian might use social or other non-rational factors to explain any deviations.

An important part of Lakatos's methodology was the necessity of certain decisions.

> We cannot avoid the decision [about] which sort of propositions should be "observational" ones and the "theoretical" ones. . . . We cannot avoid either the decision about the truth-value of some "observational propositions." These decisions are vital for the decision whether a problem-shift is empirically progressive or degenerating. (Lakatos, 1970: 127)

Kuhn (1970*b*: 256) notes that Lakatos's view shares many aspects with his own view, e.g., Lakatos's "hard core," work in the "protective belt," and "degenerative phase" of a research programme correspond with Kuhn's paradigm, normal science, and crisis. However, Kuhn suggests that Lakatos has failed to specify precise "rules" for making the key decisions mentioned above and that, in fact, his criteria are based upon more general values that individual scientists will apply in different ways. For example, if we substitute "values" for "tastes" in the following quotation from a geoscientist, we can see that indeed individual geoscientists differed in their value assessment of plate tectonics theory.

> Thus it is a fact that scientific truth is ultimately a matter of subjective taste. Accordingly the best science is done by those individuals who have the best taste. I do not share the tastes of those who set the geotectonic fashions during the '60s. . . . my taste finds the terrestrial disharmonies of the New Global Tectonics to be unbalanced, unconnected with the planetary motions, unmotivated, awkward, uneconomical of energy: in a word, ugly; and therefore false.
>
> We may permit outsiders to suppose that scientific judgements are strictly objective, but when we begin to believe this ourselves we begin to make fools of ourselves. (Fowler, 1972: 12)

In subsequent chapters we will see that such value differences help us understand differences in who accepted plate tectonics and when they did so, but equally critical for Lakatos will be the differences in the way geoscientists made decisions about *empirical issues*, such as what were the "important facts" and whether a theory adequately predicted accepted facts. Lakatos's methodology is premised upon uniformity in such decisions, otherwise it is impossible for scientists to rationally evaluate research programmes in terms of the number of explained facts or in the verification of "novel facts." These problems also apply to the following perspective on science.

Laudan (1977, 1981*a*) combines several of the features in the above perspectives, but is critical of both. For Laudan the essential nature of scientific research is "problem-solving activity." Scientists make "rational" decisions in the choice between competing theories to the extent that they chose the theory that solves the most "problems." This involves *weighing* the relative importance of different types of "empirical" *and* "conceptual" problems. The weights assigned to problems depend on two other factors: the competing theories and the larger belief structures ("research traditions") in which theories are embedded. Research traditions include broad metaphysical assumptions and methodological norms or rules about how to do research. Thus, an empirical problem has more weight in a scientist's decision-making if it is "anomalous" in that only some of several competing theories can solve it. A conceptual problem can arise, for example, because of conflicts between a theory and some of the metaphysical elements in its research tradition.

Rational decisions between theories require a type of "cost-benefit" analysis. "[O]ur principle of progress tells us to prefer that theory which comes closest to solving the largest number of important empirical problems while generating the smallest number of significant anomalies and conceptual problems" (Laudan, 1981*a*: 149). Laudan argues that scientific progress occurs because scientists rationally choose theories based upon their general problem-solving capacity. However, he use "rational" in a new sense. Now it means maximizing some "utilities" in a decision-making situation, rather than using a logical or mathematical method of connecting facts and theories. Furthermore, he denies that the changes in scientific beliefs increasingly approximate some external "reality" because we can never know this and because changes do not always involve increased *empirical* content. Sometimes a theory with less empirical content is preferred because it has less serious conceptual problems.

Laudan's research tradition shares some of the elements of both Lakatos's scientific research programme and Kuhn's paradigm. His emphasis upon problem-solving is similar to Kuhn's puzzle-solving in "normal science" and he seems to agree with Kuhn's view (in contrast to Lakatos) that we cannot say that science is approaching a better description of "reality." However, he agrees with Lakatos in that he regards the dominance of one paradigm during Kuhn's normal science as neither desirable nor a common feature in the history of science. Instead, he argues that scientists are usually faced with several competing research *traditions* for the same empirical phenomena and that they use the same problem-solving assessment procedures in deciding between traditions as they use for theories within the same tradition. Furthermore, Laudan suggests that scientists will adopt a variety of attitudes toward competing theories or research traditions, instead of the dramatic dichotomy of oppose/accept described by Kuhn. Research traditions will be assessed both on their overall problem-solving achievements and the *rate* of problem-solving progress. Thus, it is rational for scientists to "pursue," but not "accept," a research tradition or theory making rapid progress, even though another has solved more problems.

A final aspect of Laudan's approach is extremely important. Is he, like Popper and Lakatos, *prescribing* what scientists must do to be "rational" or is he attempting, like Kuhn, to *describe* what scientists actually do, but suggesting that scientists actually follow a "rule"-based definition of rationality that contrasts with Kuhn's group "value" satisfaction definition? Laudan adopts the latter position and draws from it a number of implications for the study of science.

Since his perspective or theory about scientists' rationality is simply one of many, he suggests it has to be "tested" against the history of science. This is done by first identifying episodes in the history of science which are widely seen as "rational" periods using our "pre-analytic intuitions" about rational actions rather than some abstract model of rationality. The intuitions are seen as providing more trustworthy assessments of "rational" action than the abstract models. Once some "rational" episodes in science are identified, then the actions of scientists in these episodes are compared to the various abstract models of rational action, such as Laudan's or Lakatos's models. The abstract model providing the best fit to actual behavior is the best model. Once the best model is identified, then historians of science can use it to examine other episodes in the history of science and identify choices of scientists that are "rational" and those that are not. The "rational" choices need no further explanation in terms of other "causes." The irrational decisions, however, need "causal" explanations, which might include social factors. Thus, the historian of science is dependent upon the philosopher of science for developing the best model of rationality and the sociologist of science must wait until both of these groups have finished their work before attempting to study the actual decisions of scientists.

Both philosophical and sociological criticisms have been directed at Laudan's view. Doppelt (1983) has noted that Laudan's method might work for theory choice *within* the same research tradition, but it is more problematic for choices between traditions *themselves* because traditions differ in their standards for identifying and assessing the importance of problems. This makes it impossible to compare the problem-solving ability of different traditions in the same manner as theories within the same tradition. Barnes (1979) suggests that the "preanalytic intuitions" are essentially social "prejudices" based upon current science and conventional beliefs about the past. Furthermore, Laudan's "rules" cannot be applied to the actual, day-to-day decisions facing the practicing scientists as they make choices about, say, what facts are important or how well a theory fits the empirical data. For example, we will see in later chapters that three scientists can look at the same data and conclude that theory gives a "lousy" fit, one as "good as can be expected," or one that is "too perfect" for the theory to be correct. These different assessments arise because of the different assumptions made by these three scientists, not because of differences in their ability to reason rationally. Only later, when some of these assumptions become so widely accepted that they become unquestioned conventions or "tacit" knowledge, will some of these scientists be seen in *retrospect* as "rational" or "irrational."

Although Laudan and Lakatos severely limit the possibility of a sociological analysis of scientific decisions and beliefs, this has not stopped sociologists from developing alternative perspectives on science. The first perspective considered, the Mertonian model, predates the philosophical debate created by Kuhn's *The Structure of Scientific Revolutions* and is broadly compatible with any of the above perspectives on science. The other perspectives developed later and were, in part, stimulated by Kuhn's writings. Thus they directly challenge the philosophical approaches to the study of scientific knowledge.

Merton's (1942) approach to the study of the scientific community is a "functionalist" approach. The functionalist perspective in sociology argues that the different subsystems of society provide functions essential for societal survival.

For example, the family system provides reproduction and socialization functions and the economic system provides a means for the distribution of goods and services. Merton suggested that the societal function of the scientific community is the production of "certified knowledge" by the scientific method. His major concern was not how the "scientific method" worked, but how it would operate most effectively in social environments where specific social norms or rules prescribed the proper behavior of scientists toward each other.

Merton (1942) and others (Barber, 1952; Hagstrom, 1965) proposed that the following norms facilitated the operation of the scientific method. The "universalism" norm required that scientists evaluate the merits of a knowledge claim solely on the basis of intellectual criteria and ignore such irrelevant criteria as the race, religion, personality, or nationality of the person making the knowledge claim. "Communality" directs scientists to place their findings in the public domain so that others may either criticize it or use it for their own work. "Independence" and "individualism" demand that scientists should be free to chose their own research topics and be personally responsible for checking the validity of the results upon which they base their own work. "Disinterestedness" restricts the acceptable range of motivations for research: Science should only be done for "science's sake," not for personal profit or prestige among nonscientists. Finally, "emotional neutrality" requires that scientists avoid emotional involvement in their own or other's beliefs, so that they can adopt a new perspective when the evidence warrants it.

Merton (1957) added an important modification with his suggestion that scientists are not only motivated by the joys of scientific discovery, but also by their desire for recognition from their colleagues. This desire had several important implications. First, it suggested the existence of an additional norm: scientists must acknowledge or publicly recognize the contributions of others to their own research. Without such a norm scientists' motivation for recognition would always be frustrated. Second, Hagstrom (1965) suggested that the desire for recognition provided a means of social control in science because scientists may deny recognition to those who violate the norms or fail to adopt new theories or techniques. Finally, the autonomy of the scientific community is protected by the intervening role of *peer* recognition in the distribution of other social rewards. For example, academic scientists' access to such rewards as rank, tenure, research funds, or higher salaries is based to a large extent on their contributions to the scientific community and the recognition earned by these contributions. This arrangement helps to protect scientists from outside political and social pressures.

Since these norms pertain more to decisions regarding the *relationships between scientists* than to decisions made *in the research process itself*, Merton's perspective is fairly compatible with any of the above viewpoints. However, the evidence for the existence and importance of these norms is mixed. A few studies attempt to measure directly how much scientists express agreement with statements representing the different norms and find little support (West, 1960; Mitroff, 1974). Most studies focus on the implications of the key norm of universalism, which requires that scientists ignore intellectually irrelevant characteristics of others when evaluating their contributions. These studies have focused on specific decisions, such as those related to article acceptance by journal editors and referees (Zuckerman and Merton, 1971), the grant review process (Cole, Cole, and Simon, 1981), the recognition given scientists for their contributions (e.g., Cole and Cole, 1973; Reskin, 1977; S. Cole, 1978), and academic appointments and mobility (e.g., Reskin, 1976; Long, 1978; J. Cole, 1979).

Although the evidence in these studies of the functionalist norms has been mixed and open to alternative interpretations, the recent "constructivist" approaches in the sociology of science are not so much a reaction to the Mertonian approach as an attempt to develop the sociological implications of various perspectives in philosophy and sociology, including Kuhn's perspective. Constructivists inquire about knowledge *in general* and how it is constructed through interactions among individuals as they interpret their experiences. Rather than viewing science as an institution with a special reasoning method or special norms, they emphasize the study of science because knowledge construction is more visible than in other social institutions (Collins, 1981; 1985).

There are several varieties of constructivist approaches, but most of them share a set of four assumptions or tenets that have been formalized as the "strong programme" in the sociology of knowledge (Bloor, 1976; Law, 1975). The first tenet of the strong programme specifies that all beliefs have *causes*. Second, analysis of beliefs should be *impartial* with respect to whether the beliefs are considered "true," "false," "rational," or "irrational" in any sense. That is, all beliefs have causes, even those currently accepted as "true." Third, the *symmetry* tenet holds that the same types of explanations should be used for "true" *and* "false" beliefs. That is, it is not acceptable to ascribe "mistaken" beliefs to "social" factors, while giving other explanations, such as the weight of empirical evidence, for beliefs currently accepted as correct. Even if scientific beliefs are "true" descriptions of reality, they still have causes that can include social factors. Finally, the *reflexivity* tenet suggests that the three previous tenets apply as well to the explanations given by those following the strong programme. The contrast with the views of Lakatos and Laudan are obvious and have prompted some debates (Brown, 1984).

Beyond these programmatic aims, constructivist studies share two methodological preferences (e.g., Latour, 1987; Latour and Woolgar, 1979; H. Collins, 1975, 1981, 1985; Knorr-Cetina, 1981; Knorr-Cetina and Mulkay, 1983; Barnes, 1982). First, they emphasize the detailed study of how scientists actually do research and reach conclusions about the meaning of their results. Some constructivists actually live in the lab with the scientists so that they can observe the scientists' experiments, interactions, and writing activities. Second, they emphasize the social factors that shape the actual beliefs that scientists have about the natural world. We shall consider two of these constructivist approaches: the "relativist" and "interests" approaches.

We can illustrate the issues addressed by these two perspectives with a quotation describing some of the problems faced by a geologist trying to apply plate tectonics theory. During a 1974 interview, Ron Oxburgh mentioned the difficulties in relating the abstract plate tectonics theory to his specific field research.

Oxburgh: Plate tectonics provides a superb paradigm for the understanding of present day [earthquake patterns], of the distribution of the gross tectonic features of the earth's surface today, or current volcanicity. It also provides a good explanation for the magnetic strips on the ocean floor. But it has . . . two aspects that ought to be emphasized more. The first is that it does describe what happens today [as opposed to the past] and the second is that it describes what happens on a large scale—on a scale of thousands of kilometers. Now most of the problems which confront geologists—and this is a distinction between geology and geophysics—are much smaller scale than plate tectonics in its purest sense has anything to say about. You are concerned with the field area . . . a hand specimen . . . or something of this kind. Now if you want to apply plate tectonics to problems . . . which the geologist in the [field] is working on, [you've] got to make an extrapolation back in time and an extrapolation downward in scale. . . . The more serious problem, in terms of us using this for geology, is the drop in scale. There it involves making assumptions, physical extrapolations, imagining processes . . . that are tied into the logical extensions of the larger scale, and here at the moment, there is very little control. This is why we are almost getting back into the sort of situation we were in ten to fifteen years ago—of geologists inventing their own physics to explain their own level of problems. They're doing it now within this broad framework, but it isn't really solving any problems in an exclusive or analytical way.

JAS: So some people are claiming that what they say follows from plate tectonics, when it really doesn't?

Oxburgh: In fact, [they] could have probably stated the exact reverse [conclusions]. We don't have the rigid guidelines of the paradigm at these lower scales.

This statement illustrates that paradigms are abstract and cannot be connected to empirical observations without additional interpretative activity or construction of connecting links. This process is a characteristic of all scientific research. Even the connections that are now accepted without question were constructed previously. For example, Oxburgh accepts the ability of plate tectonics to explain the magnetic strips on the ocean floor, but, as we will see, this connection required the "invention" of a model of the seafloor with certain properties that allow it to record the magnetic reversals of the earth's field. Once this model is invented then it can be used to generate predicted patterns that can be compared to the observed patterns. Yet scientists contested the model of the seafloor and differed in their assessments of the quality of the fit between the observed and predicted patterns. However, a consensus eventually develops on both the proper assumptions and what is an "adequate" fit between theory and the "facts." The two constructivist approaches outline some of the processes that might cause this consensus.

The "relativist" approach (Collins, 1981, 1985) adds to Kuhn's basic image that the consensus in a scientific community results from persuasion based on shared values. This imagery is extended by emphasizing the interaction aspect between scientists so that they are seen as "negotiating" what assumptions, beliefs, and interpretations will be accepted. Furthermore, the negotiating resources include more than Kuhn's shared values or other "scientific" resources, such as finding predicted results. A scientist's previous scientific recognition, institutional prestige, style of presentation, and personality are possible "social" resources that help convince others to accept his or her ideas.

Relativists tend to study areas of scientific controversy where these processes are most apparent, such as gravity wave research (H. Collins, 1975), learning in flatworms (Travis, 1981), and solar neutrino studies (Pinch, 1981). In areas of controversy scientists have not yet developed a consensus on the proper assumptions, which is required before challenges to these assumptions are defined as "irrational" by other scientists. Although relativists would accept that "scientific" negotiating resources, such as finding expected results, influence the negotiation process, they would emphasize that this does not free scientists from social influences because social factors helped establish these shared expectations and values in earlier negotiations. Consequently, some relativists argue that the logical gap between theory and data is ultimately bridged only by social factors and they adopt the *methodological* assumption that the "natural world has a small or non-existent role in the construction of scientific knowledge" (Collins, 1981: 3). This assumption is applied in the following manner: explain scientific beliefs as much as possible with social factors and allow "Nature" to explain the residual. After describing the "interests" perspective, we can consider some of their common problems.

Complementing the relativists' focus on communications and interaction dynamics, the "interests" proponents focus on the psychological dynamics of why scientists adopt specific theories when logic cannot bridge the gap between data and theory. The answer is found in the "interests" that scientists have in different theories. This position has been developed in a general sense by Barnes (1982) as some of the implications of Kuhn's (1962) analysis of scientific revolutions and has been applied in a variety of case studies from mathematics (Bloor, 1978), taxonomy (Dean, 1979), to high energy physics (Pickering, 1980), many of which are reviewed by Shapin (1982). The concept of "interests" is very general and can include a "scientific" interest in using quantitative theories to more "social" interests, such as the "interests" of the scientific elite in maintaining the theoretical foundations of their previous research or the compatibility of specific theories with political ideologies (MacKenzie, 1978).

Both of these constructivists argue that the scientists' sense of what is really in the external world is a social construction that becomes "reified"—taken as a description of the real world. That is, when a socially constructed model becomes widely accepted among scientists, its constructed nature is forgotten and it is viewed as a description of what is really present in the external world. Some constructivists emphasize that the social construction of scientific beliefs does *not* imply that the beliefs must be "false" in any sense of the word (Barnes, 1974). Since they believe that scientific beliefs develop through the same processes that cause beliefs in the typical human, they are less concerned with the "truth" status of scientific beliefs. Naturally this view conflicts with traditional philosophical concerns with demarcating scientific from "everyday" reasoning by postulating some special characteristics of scientific reasoning, such as the use of deductive logic or choosing theories with increased problem-solving capacity. For example, Laudan (1981*b*) attacks the "symmetry" thesis of the strong programme and argues that "rational" beliefs should not be seen as caused by the same factors that cause "irrational" beliefs. In response, constructivists emphasize that they are only interested in the *empirical* study of the belief-formation process and are unwilling to let philosophers dictate some beliefs as "off-limits" (Bloor, 1981).

As a sociologist, I think the basic constructivist viewpoint provides a good starting point. This is not surprising in light of the "reflexivity" thesis of the strong programme: I have lots of "interests" in adopting this position instead of those proposed by some philosophers. However, my "interests" do not match all of those held by the constructivists, so I see some "problems" with their perspective. One that philosophers might agree with is that relativists and interests proponents do not describe systematically the "negotiation" process or how "interests" impact on beliefs. For example, these perspectives need to be refined into "theories" stating when and under what conditions social and scientific resources (or interests) will vary in their impact on scientific decision-making. Not only do these processes need better description, but the description should include statements about how the results of empirical evidence ("Nature") combine with these other factors as causes of beliefs. An empirical analysis must consider these factors as they operate *together* rather than adopting the relativist's one-sided emphasis on social resources. Similarly, the interests proponents are faced typically with complex intercorrelations between the beliefs and the social and scientific interests of scientists, but *subjectively* assess which interests are the most important determinants of beliefs. Neither of these ways of determining the relative importance of intercorrelated social, scientific, and empirical factors seem adequate. Given my training ("interest") in quantitative methods, I think that the empirical study of scientists' beliefs can be aided by the *quantitative* assessment of the *relative* importance of different intercorrelated causes, even if it does so in a rather crude manner (Stewart, 1986, 1987).

A quantitative procedure that could empirically measure the relative importance of "social" and "scientific" resources would promote the development of better "theories" about when these different types of resources or interests will influence scientific decision-making. Before describing briefly this procedure or measuring "instrument," we need to consider one other perspective gaining increasing attention in the philosophy of science: the use of models and analogies in scientific reasoning. This perspective will illuminate some aspects of the history of plate tectonics *and* suggest how we might proceed in our own study of this revolution.

For some philosophers of science the analysis of models, analogies, and metaphors in scientific reasoning provides an alternative to the traditional emphasis on scientists' use of logical and mathematical reasoning processes (e.g.,

Kuhn, 1962; Harré, 1976, 1986; Hesse, 1966, 1974, 1976); Ziman, 1978; Giere, 1988). The same processes are identified in human reasoning in general (e.g., Minsky, 1985), so this view is compatible with the constructivists' assumptions that scientific reasoning is similar to everyday reasoning *and* it implies that reasoning by social scientists involves the same basic process (Harré, 1976; Hesse, 1976).

According to this perspective the really essential or creative aspect of scientific thought involves the use of some "model" from which "analogies" are drawn to explain observed phenomena. For example, Bohr used Rutherford's simple "planetary" model of the structure of an atom to obtain analogies for the analysis of hydrogen's spectra. Similarly, Harré (1976) suggests that the virus explanation for some diseases was based upon the bacterial model for diseases. Much of scientific reasoning involves borrowing models from other areas of science and even outside of science, as in Darwin's development of "natural selection" using "domestic selection" of plant and animal breeders. However, Harré emphasizes that only selected aspects of the original "source" models are borrowed as scientists create their "iconic" models for the analysis of their particular subjects. For example, in the "billiard ball" model for the pressure, temperature, and volume properties of a contained gas, the concept of "color" is not carried over to the iconic model that a gas is composed of very tiny "balls" interacting according to Newton's laws of motion.

This book uses models and analogies in several ways. First, we will see how geoscientists developed some of their theories by borrowing models from other areas of the geosciences. Second, we will use models from other social science disciplines to analyze how scientists think. In particular, a general "decision-making" model provides the basis for relating cognitive psychology and studies of organizations to our analysis of the history of plate tectonics. Finally, the history of plate tectonics itself provides a model of what might be needed before the *study of science* can reach the same level of theoretical development.

A major factor contributing to the development of plate tectonics theory was the development of instruments producing quantitative measurements. On the assumption that a similar development might help the study of science, I will propose a quantitative procedure for measuring the relative importance of social and scientific resources in the negotiations among scientists. This procedure starts with a very simple "iconic" model of the decisions involved in generating a citation to a scientific article. The quantitative procedure derived from this model can be related to both the functionalist and constructivist perspectives on science, as well as previous studies of social organizations in general. Chapter 7 will illustrate how this "instrument" can measure important aspects of the history of plate tectonics.

Data and "Theory Testing" in the History of Science

The previous section has outlined several perspectives about scientific reasoning that will be "tested" by comparing them to the history of continental drift and plate tectonics theories. Given the diverse assumptions among these perspectives (one is tempted to call them "paradigms"), it is not surprising that they vary in what aspects of the historical "reality" are most important to them. For example, the constructivists, especially the relativists, would prefer access to more of the full day-to-day details. Indeed, Latour's (1987) first "rule of method" in the study of science is to "follow scientists in action." This rule is easiest to follow with current science, but this option is not available to us and we are left with a fairly limited set of potential data.

The most obvious sources are the scientific publications, but these have a number of problems associated with testing ideas about how scientists reason and reach a consensus in their beliefs. Various observers of science (Medawar, 1964; Knorr-Cetina, 1981; Gilbert and Mulkay, 1980; Yearley, 1981) have suggested that scientific articles do not represent the actual scientific thoughts and practices behind the reported results.

Despite this problem the published record does include a number of debates, conference comments, and letters to the editors of geoscience journals that will help us understand how scientists reach a consensus. Furthermore, numerous books and articles have been written on this history by a diverse set of authors, which includes historians and philosophers of science, science writers, and especially geoscientists themselves, many of whom directly participated in this revolution.[1] Menard's accounts of the day-to-day problems of oceanographic research (Menard, 1969) and his candid history of plate tectonics are especially rich sources of correspondence and interview material (Menard, 1986). Glen (1982) also provides extensive quotations from interviews with participants in the revolution.

This is one of the advantages of studying a recent revolution: many of the participants are still alive and can be interviewed. This important source of information will be used throughout this book, using both interviews conducted by others and myself. Between 1974 and 1980 I interviewed about forty geoscientists, who either lived in or were visiting the United States or Canada. I selected most of these informants on the basis of their contributions to the new theory or their continued opposition to it. Others were interviewed because they were located at the same institutions as the original sample and were mentioned as good information sources or continuing "skeptics."

An initial list of preselected questions was based on Kuhn's perspective and how one might apply it to the development of plate tectonics, but the interviews were unstructured and started with questions about the relevance of the geoscientist's current work to plate tectonics. The preselected questions were asked when they seemed appropriate, when the conversation needed some direction, or at the end if some of the questions remained unanswered, but the most interesting and relevant information was obtained by pursuing unanticipated topics mentioned during the interviews.

All of the informants allowed their interviews to be recorded, and each was sent approval forms for the interview excerpts used in this book. The approval form provided the informants with the context of the excerpt and allowed them to specify the manner in which they were identified with the quotation—either by name or anonymously.[2] Whenever interview material is included in the text, it is indented or set off by quotation marks. A "*JAS*:" denotes my questions, and the scientist's name or "*A*:" denotes the responses.

There are some obvious problems with this type of information source. First, the informant may not remember previous events accurately, especially when these events occurred more than forty years earlier. This problem may be reduced by asking a number of informants about the same period and by consulting historical documents. Both of these corrective procedures may help reduce the second problem of unconscious distortion in recollections. Latour and Woolgar (1979) noted this problem in their study of laboratory practice: after some time their respondents often changed their accounts of why they tried certain procedures or held certain beliefs. Finally, all humans have expectations and models about reality—including their own and other people's behavior. These models affect what they perceive and recall (Gilbert and Mulkay, 1982). For example, Meyerhoff sees the continued dominance of plate tectonics theory as due in part to its endorsement by major funding agencies. "If you . . . take a look at people who like drift, you will find that by and large they are theoreticians, very few field geologists. . . . They are people who live on grants, partly or entirely, and, therefore, they know entirely what side of the bread their butter is on." In contrast, proponents see Meyerhoff's and others' continued resistance as due to personality factors:

> *A*: I think that his [Meyerhoff's] pride is mixed up a little in it now. It's like he couldn't back off if he wanted to.
>
> *JAS*: Do you think this applies to Beloussov?
>
> *A*: Yes, it seems similar. He was the brilliant young man in Russia and had established himself with vertical tectonics and then found his work more or less wiped off the map by plate tectonics.

These different perceptions arise from fundamental differences in the models these scientists have about their own and others' behavior. This model problem is characteristic of all human perception and of course it applies to the present analysis. For example, there is the influence of my own model in what questions I ask, in my selection of interview excerpts to use in this book, and in my interpretation of the excerpts. In other words, these problems are related to the objectivity and rationality of my own and other's methods of observation and reasoning. These issues cannot be resolved here for they are among the subjects considered in this book. At the end we will be in a better position to answer them.

Overview of the Chapters

Chapters 2 through 4 describe the development of continental drift and plate tectonics theories. This history provides essential information about the major researchers and their contributions to the development of plate tectonics, as well as illustrating how geoscientists think about the earth. It differs from other histories by including (a) both the earlier debate about continental drift and later developments in plate tectonics theory, (b) more discussion of the opposition to plate tectonics, (c) more information on the geoscientists involved in this history, and (d) quantitative information on the growth of this literature and the influence of selected articles and geoscientists.

In particular, Chapter 2 describes the background, development, and reaction to Alfred Wegener's theory of continental drift. Chapter 3 covers selected events from the end of World War II to about 1959, during which few geoscientists accepted drift theory even though they unknowingly developed the evidence for plate tectonics. The major focus of Chapter 4 is the 1960s, which saw the development of seafloor spreading and plate tectonics theories. By 1970 the revolution was nearly complete, even though a vocal minority of geoscientists opposed this theory. Chapter 4 also briefly covers this opposition and some of the subsequent developments in plate tectonics theory.

By the end of Chapter 4 the reader will have some basic "data" about the mixture of ideas, people, and research results that make up this history, so Chapter 5 begins the contrasts between the different philosophical and historical perspectives outlined above. With some modifications Kuhn's perspective seems to fit the history best, but we will need to elaborate some of the key processes he identified. One of these is the use of models and analogies in science, so the end of Chapter 5 deals with this topic and how it can be applied to this history.

Chapter 6 illustrates how a "decision-making" model helps us understand scientists' reasoning both in a "typical" research project and in their cognitive processes in general. A review of decision-making in social organizations introduces the discussion of the sociological perspectives on science. Each of the sociological perspectives has implications for how scientists write articles and who they decide to cite in their articles. Consequently, we will develop a simple model of the decision about whom to cite and use it to derive a quantitative procedure to measure the relative importance of "social" and "scientific" factors in predicting the importance of an article in the generation of new knowledge. The final section of Chapter 6 describes how this procedure relates to the study of social organizations in general.

Chapter 7 applies this quantitative approach to several topics in the history of plate tectonics. The first analysis demonstrates that the technique of "cocitation analysis" can identify the publications containing the "exemplars" of plate tectonics research. The second analysis shows that "scientific" resources are the most important predictors of citations to a sample of geoscience articles published in 1968. A final analysis examines the predictors of published opinions on continental drift in the 1900 to 1950 geoscience literature. The finding that more productive scientists tended to oppose continental drift illustrates how quantitative analyses can provide some support for the "interests" perspective.

The final chapter, Chapter 8, reviews each chapter in light of the major themes and results of later chapters and describes some of the implications of these results for the study of science. It concludes with some reflective illustrations of how the perspectives discussed in this book might apply to its own creation and its reception by others.

CHAPTER TWO

The Rise and Fall of Continental Drift Theory

Since scientists develop and respond to theories within the context of their intellectual and social environments, we first examine selected aspects of geological thought present at the beginning of this century, when continental drift theories first gained some recognition. The second section emphasizes Alfred Wegener's drift theory and his supporting arguments. The third section examines some of the responses to Wegener's theory, especially those expressed at a 1926 symposium in North America. Finally, we consider some of the modifications of drift theory proposed in the 1930s, even though few geologists, especially those in North America, seriously considered drift theory after this symposium.

Geological Thought in the Early 1900s

By the early 1900s a number of the earlier debates among geologists had reached tentative resolutions. Debates about the relative importance of "water" and "fire" in the formation of rocks had yielded the conclusion that both processes were important: rocks on the earth's surface had formed both from the deposition and compaction of the sediments in oceans and the cooling of molten rocks (lavas or "magmas"), as well as other processes. A similar compromise resulted from the debates between proponents of "uniformitarianism" and "catastrophism." The principle of uniformitarianism held that the geological processes operating in the past were the same as those we see operating today ("actualism"), that these processes worked at the same rate as that observed today, and that there was no "progressive" change in the overall characteristics of the earth's surface. The opposing catastrophism view held that most of the earth's features were the result of such catastrophes as huge floods or the sudden uplift of mountains. By the end of the 1800s, geologists tended to accept "actualism," but allowed long-term changes in the rates of today's observed processes (Greene, 1982; Gould, 1977). Geologists believed the earth's surface showed progressive development from its origin, which was believed to be several hundred million years ago. These general beliefs were part of the foundations of geological research during the late 1800s when geology made rapid advances in its knowledge about the earth.[1]

THE SEDIMENTARY RECORD AND THE AGE OF THE EARTH

Much of the geological research in the 1800s emphasized the interpretation of the layers (strata) of sedimentary rocks exposed by mines, canals, and natural erosion. Geologists believed these strata formed when the continents were below sea level and they used several principles to interpret or reconstruct the formation of these strata. These principles included (a) the principle of superposition—in any succession of strata not severely deformed, younger strata are above older strata, (b) the principle of original lateral continuity—strata originally extended in all directions until they thinned to zero or terminated at the edge of their basin of deposition, and (c) the principle of fossil assemblages—strata containing the same assemblage of fossils are of the same age.[2]

Throughout the 1800s geologists combined these principles with careful observation of exposed strata and classification of fossils to develop a *relative* time scale for past geological events. Geological time was divided into "eras" and "periods" based on the type of life preserved in the fossil record. Widespread extinctions of plant and animal species provided the bases for separating the eras. Dispersed throughout the geological record was evidence for periods of intense geological activity, called "orogenies," which included the transgressions of the seas onto the continents and the extensive formation of new mountain belts. Column 1 in Table 2.1 presents one of the early attempts in the 1800s to classify three eras in the history of the earth. Column 2 presents the time scale widely accepted at the beginning of this century. Each period within an era is represented by distinctive assemblages of fossils and named after representative sedimentary sequences exposed in specific locations.

This time scale had two problems. First, it only gave the *relative* sequence of events and did not give the *absolute* ages or time between events. The absolute ages of these eras and periods were estimated only in this century with the development of radiometric methods of age dating. Column 3 in Table 2.1 provides the absolute ages accepted today. The second problem with the relative time scale arose from its use of fossils for dating rocks. Thus it could not date igneous and metamorphic rocks or sedimentary rocks formed before the evolution of life. Despite these limitations to the time scale, geologist used their knowledge of sedimentary strata to estimate the absolute age of the earth, to correlate similar fossils and geological structures on different continents, and to produce models for the geological processes that produced mountain chains.

During the last half of the nineteenth century geologists began to estimate the absolute age of the earth. This effort arose because Lord Kelvin, an eminent British physicist, mounted a series of attacks on the geologists' assumption that the age of the earth was in the hundreds of millions of years. Kelvin had helped develop the laws of thermodynamics and he applied these laws to estimate the age of the earth by assuming the heat released by the sun and planets came from the potential energy released as meteorites aggregated to form the solar system. Initially Kelvin estimated the earth to be about a hundred million years old, but by 1900 he had reduced this to thirty million years or less.

Burchfield (1975) and Hallam (1983) describe how geologists initially accepted these decreasing limits and even produced their own estimates using the time needed to form the exposed sedimentary record or to produce the salt concentration in the oceans. Burchfield suggests that geologists were intimidated by Kelvin's scientific stature and the quantitative nature of his arguments, but as his estimates grew shorter they began to question his assumptions. This movement was aided by the discovery in 1903 that radioactive decay released a definite amount of heat. This provided an alternative source of heat for the sun and the earth, which was adopted quickly by geologists as another reason to doubt Kelvin's results. Burchfield (1975) suggests that geologists tended to distrust quantitative, geophysical arguments in later years because of this experience with Lord Kelvin. Their distrust helps explain some of the developments discussed later.[3]

During the nineteenth century geologists used the fossil assemblages to identify rock strata of the same age on different continents and identify fossils unique to specific areas and geological ages. Since the same unique fossils were found on some continents that are presently separated by oceans, geologists concluded that these continents must have been connected in the past and offered several competing explanations for the continuity of land fossils across vast oceans. First, the early form of the principle of uniformitarianism included an assumption of equilibrium of continental masses: if an area of land subsided below sea level, then an equal area became elevated above sea level. Thus identical land fossils on each side of the Atlantic could be explained by assuming that the crust of the Atlantic Ocean was above sea level during the specific geological age when the fossilized animals or plants were alive. This explanation was favored in Britain, whereas European geologists tended to believe that some oceans were just collapsed continental areas that did not rise again. Another explanation, which was accepted more in North America, assumed that continents and oceans were permanent features of the earth and that "land bridges" had connected the continents at various times in the past, but subsequently subsided into the ocean basins.

MOUNTAINS AND GLOBAL THEORIES ABOUT THE EARTH

The relative time scale and the ability to identify different sedimentary strata were also important in deciphering the internal structure of mountain ranges and in developing plausible theories for their formation. Greene (1982) describes the numerous debates about mountain formation that occurred in the 1800s and how a consensus had been reached that mountains, such as the Alps and the Appalachians, were produced by the folding and compression of originally level sedimentary strata laid down in ocean basins. This explanation required the identification of horizontal forces that could shove the sedimentary strata into folds. Since geologists believed that the earth was cooling slowly, the needed horizontal forces could result from the buckling of the cool, rigid crust of the earth as it collapsed onto a cooling and contracting mantle.

During the last quarter of the nineteenth century Eduard Suess, an Austrian

TABLE 2.1

Evolution of the Geological Time Scale from Relative to the Absolute Ages and Durations (in millions of years: M.Y.) of Geological Eras and Periods.

	Lyell, 1833		Chamberlin and Salisbury, 1905		Present Usage	Age at Base of Period, M.Y.	Duration M.Y.
Recent	Newer Pliocene Older Pliocene Miocene Eocene	Cenozoic	Present Pleistocene	Cenozoic	Quaternary: Present, Pleistocene	2	2
Tertiary Period			Pliocene Miocene Oligocene Eocene		Tertiary: Pliocene, Miocene, Oligocene, Eocene	65	63
Secondary Period	Cretaceous wealden	Mesozoic	Cretaceous Comanchean	Mesozoic	Cretaceous	144	79
	Oolite or Jura limestone group		Jurassic		Jurassic	213	69
	New red sandstone group		Triassic		Triassic	248	35
		Paleozoic	Permian	Paleozoic	Permian	286	38
	Carboniferous Group: Coal measures		Coal measures or Pennsylvanian		Pennsylvanian (U. Carboniferous)	360	74
	Carboniferous Group: Mountain limestone		Subcarboniferous or Mississippian		Mississippian (L. Carboniferous)		
	Carboniferous Group: Old red sandstone		Devonian		Devonian	408	48
	Carboniferous Group: Graywacke and transition limestone		Silurian		Silurian	438	30
			Ordovician		Ordovician	505	67
Primary Period			Cambrian		Cambrian	590	85
		Proterozoic	Keweenawan Animikean Huronian	Precambrian	Regionally defined systems	4600	85% of Earth's history
		Archeozoic	Archean complex				

SOURCES: Adapted from Wyllie, The Dynamic Earth, copyright 1971, by permission of John Wylie and Sons. Ages and durations from Harland et al. (1982).

geologist, synthesized most of the ideas and results mentioned above into a global perspective in his four volume work, *The Face of the Earth*. Suess held that earth contraction was the major force producing the major features of the earth's surface, especially the formation of oceans and the location of mountain chains. He proposed that the continents had been larger in the past to explain the known distribution of fossils across current oceans, but as the earth cooled some of these continental areas subsided below sea level and the sea level regressed from the remaining continental areas. As the ocean basins filled with sediments the seas transgressed back onto the continental areas until the next collapse. Accompanying the collapse of continental areas were horizontal forces that raised up mountain chains in different parts of the world. Thus Suess could explain similarities in structures and fossils across continents and the apparent cyclic nature of the regression and transgressions of the seas and the formation of mountain chains. Greene (1982) suggests that Suess's synthesis had gained an increasing number of followers during the last quarter of the nineteenth century, but it was progressively challenged by the results of new studies. Before considering these studies, we should briefly consider an alternative view that was more common in North America.

Since the middle of the nineteenth century, many American geologists used the "geosyncline" model to explain how flat sedimentary strata could be compressed, folded over, and elevated above sea level to produce mountains. This model proposed several stages for the formation of mountains. Initially, sediments were deposited in a "geosyncline," a trough below sea level and located along the margin of a continent. The weight of accumulating sediments caused continued sinking or subsidence of the trough until a mile or more of sediments had been deposited. Eventually the heat of the earth at the bottom of the trough and/or an external compressive force on the sides of the trough folded and uplifted the strata into a mountain range above sea level.[4] Like the Europeans, most American geologists accepted that earth contraction played a major role in the formation and compression of the sediments in geosynclines. However, they thought that today's oceans and continents formed very early in the earth's history and were permanent features throughout the geological record.

Suess's global theory lost broad support in the early 1900s because of the implications of radioactive heating and "isostasy." During the last half of the nineteenth century geophysicists proposed the concept of isostasy to account for various geological phenomena, such as the present uplift of areas of Scandinavia that were covered by thousands of feet of ice during the last glacial period. Isostasy explained vertical movements of the earth's crust by assuming that the crust of the earth was "floating" on a "fluid" underlying mantle. Increased loads of sediments or glacial ice would cause the crust to settle a little deeper into the underlying mantle, just as a loaded ship floats deeper than an empty one. Similarly, erosion of material from the surface would cause uplifting of the lightened crust.

This concept had two important implications. First, isostasy should act to produce an uniform gravity field. Consequently, gravity surveys provided important information about underlying geological structures. For example, gravity surveys over mountain ranges and in oceans showed that they were in isostatic balance, which implied that the mass of the mountains was compensated for by less dense "roots" below the mountains and that ocean basins must be composed of rocks denser than continental rocks to compensate for the less dense oceans. Second, the concept implied that the mantle was capable of some sort of flow. These results conflicted with Suess's suggestions that the less dense continental areas could have subsided to form some of the ocean basins.

Little was known about the oceans in the early 1900s, but what was known tended to support the idea that ocean basins were not simply foundered continents. Most of this knowledge came from British oceanographic explorations in the 1870s and the laying of telegraph cables across the Atlantic. Flat sediment planes, deep ocean trenches, and a submarine mountain range in the middle of the Atlantic Ocean showed that the ocean basins did not have a simple, uniform topography. The dredging of rocks from the ocean floors indicated that most of the igneous rocks—rocks formed by the cooling of molten rock—on the ocean floor were composed of silicates of heavier elements, including manganese, which were called "sima." In contrast, most continental, igneous rocks were less dense and composed of silicates of aluminum and other light elements, which were called "sial."[5] These density differences were reasonably consistent with the gravity studies showing isostatic balance.

By the early 1900s some geologists began to express doubts about the contraction theory. The discovery of radioactive heat in 1903 and the estimated abundances of radioactive elements in the earth's crust implied that the earth may be cooling less than previously thought; it might even be heating up. Furthermore, some estimates of the amount of crustal shortening needed to form the folded sediments in the Alps indicated that the original rocks covered too large an area to be consistent with plausible shrinkage rates for the earth's surface.

Despite these doubts, earth contraction continued to be defended by disputing the estimated crustal shortening in the Alps or by modifications of the reasons for contraction. For example, Thomas Chamberlin proposed the "planetismal" theory in which contraction resulted from the rearrangement of material in the interior, which he held had never fully melted. But these attempts were unable to regain what had been lost: the direct connection of global contraction to the details being found in the study of specific features of the earth's surface. Consequently,

> [t]he career of geology in the following decades confirmed the worst fears of the theorists who had striven to maintain coherence in geology through grand geotectonic syntheses. The rapid proliferation of subspecialties and their concomitant isolation from one another made impossible the kind of syntheses enhanced by wealth of detail and breadth of mastery that had been achieved by Suess. . . . geology in 1910 was a pyramid with a rapidly expanding base but without an all-seeing eye at the top. (Greene, 1982: 275)

To summarize, in the early 1900s there were major areas of agreement and disagreement. The principles of superposition, lateral continuity, and fossil assemblages were accepted means of interpreting the structural relationships in sedimentary strata observed in field work. The relative timing of such events as transgressions and regressions of the seas onto the continents, orogenies (periods of mountain formation), ice ages, and the stages in the evolution of life were generally accepted or were being worked out in more detail. The age of the earth was believed to be in the hundreds of millions of years. Finally, there was an implicit assumption that the continents did not move horizontally on the surface of the earth.

Some of the areas of disagreement included the general characteristics and permanence of the ocean basins, the explanation of fossil continuities across ocean basins, and the implications of isostasy and the heat of radioactive decay. These disagreements were not randomly dispersed, but tended to be correlated with nationality and favored methodological approaches to the study of geological processes. For example, Americans tended to favor permanence of the oceans and continents, to use the geosyncline model to explain how mountain ranges are added to the edges of continents, and to favor the collection of geological "facts" before trying to achieve a global theory. Europeans tended to favor subsidence of continents to form oceans and account for fossil distributions, to neglect geosyncline concepts, and to prefer the development of broad theories before all the "facts" were in.

Compounding these differences was a flood of new specialist studies. Consequently, by the early 1900s there were a number of attempts to modify old theories or create new ones that could both increase the areas of agreement and incorporate the new findings. Although not widely discussed, various forms of continental drift theory were among the alternatives.

Continental Drift Theories

The essential idea behind the various continental drift theories is that in the geological past the present day continents were united in a single land mass that subsequently split apart to form new oceans. The basic idea had been presented in 1858 by Snider-Pellegrini, a layman, who argued that the Noachian flood was the result of a splitting apart of the continents. His major evidence was the apparent match between the coast lines on the two sides of the Atlantic Ocean. Several other presentations of drift were given later by Osmond Fisher, William Pickering, and Howard Baker, but they suggested the cause of drift was the catastrophic event of the moon being torn from the earth by a passing planetary body.[6] In 1910 Frank Taylor (1910) used drift theory to explain the location and youth of the Tertiary (see Table 2.1) mountain ranges as the result of a slow creep of the continents toward the equator, which in turn was caused by the disruptive tidal forces acting on the earth when the moon was captured, rather than lost. These suggestions of drift theory received little consideration by other geologists because little systematic evidence for drift was provided and such catastrophes were not compatible with the "uniformitarian" beliefs of most geologists (Hallam, 1973: 3). Drift theory received little attention until it was combined with better evidence, given an "actualist" cause, and promoted by established scientists. The first person to combine these characteristics was Alfred Wegener.

ALFRED WEGENER'S THEORY OF CONTINENTAL DRIFT

Alfred Wegener was born in 1880 in Berlin and was educated in Germany and Austria in astrophysics and meteorology. He made several meteorological expeditions to Greenland and died during one of them in 1930. Georgi (1962), a student and colleague of Wegener, described him as an effective leader, unpretentious for such a prominent scientist,[7] and a persuasive teacher. Wegener's early professional career involved a lectureship at the University of Marburg, after which he became director of the meteorological division of the German Marine Observatory in Hamburg. Although an accomplished scientist, he could not obtain available professorships in meteorology at German universities because of his additional strong interest in continental drift. In 1924 he accepted a professorship of meteorology and geophysics at the University of Graz in Austria—a position that was created specifically for him.

Apparently it was the correspondence in the coasts of South America and Africa that gave Wegener the initial idea that these two continents may have been united in the past (Wegener, 1924: 5). He did not pursue this idea until 1911, when he came across an article citing paleontological evidence for a land bridge between these two continents. Subsequent study of the literature convinced him that drift theory was compatible with the geological evidence. In 1912 he gave a public presentation on this theory and published two articles in German scientific journals (Wegener, 1912*a*, 1912*b*). In 1915 he published a book version, *Die Entstehung der Kontinente und Ozeane*; revised editions appeared in 1920, 1922, and 1929, with a 1936 revision edited by his brother.

The 1922 edition was translated into four languages, including a 1924 English version, *The Origin of Continents and Oceans*. The 1924 English edition is divided into three parts: (1) a general summary of the theory and its relationship to other global theories and concepts prevalent at that time, (2) various lines of evidence supporting the theory, and (3) a consideration of the evidence for some of the underlying assumptions and implications of the theory. A brief summary of these three sections of *The Origin of Continents and Oceans* will help us understand the nature of Wegener's reasoning and some of the reactions to his theory.[8]

In the first part summarizing his theory, Wegener suggested that during the Mesozoic Era[9] the continents were united in a supercontinent, which he called Pangaea. During the Jurassic and Tertiary periods Pangaea split apart and the separated continents began to move westward, toward the equator, or in both directions. The Atlantic and Indian Oceans were produced as the continents separated from each other at different times. Some mountain belts were created on the western edges of North and South America as these continents plowed their way through the resisting sea floor. Other mountain belts, such as the Alps and the Himalayas, were created when two continents collided as each moved toward the equator. Figure 2.1 depicts Wegener's estimates of the positions of the continents and oceans at various times since the beginning of the Mesozoic Era.[10]

Wegener also considered the evidence for the contraction theory, land bridges, and the permanence of the locations of the oceans and continents, as well as the their relationships to his theory. Although he acknowledged that the contraction theory had provided an admirable synthesis of earlier evidence, he suggested that an alternative was needed for a number of reasons. These included the discovery of radioactive heating and the large crustal shortening represented in the compressed sediments of mountain ranges. Furthermore, he noted that if the earth was contracting, it should produce a uniform distribution of mountain ranges on the surface of the earth, whereas mountain belts had a very distinct distribution of linear chains, which were often on the edges of continents. He accepted that continents and oceans were permanent features of the earth's surface *only* in relative size, but not in relative locations. While accepting the paleontological evidence for past land connections, he rejected the subsiding land bridges explanation for fossil continuities across deep oceans. Land bridges between continents separated by shallow seas, such as the Bering Straits, were reasonable, but only continental drift could explain connections across deep oceans.

In the second part of his book Wegener presented supporting evidence from the various areas of the geosciences. This evidence was from five general areas: geophysics, geology, paleontology and biology, paleoclimatology, and geodesy. Wegener used geophysical arguments to attack the contraction theory of the earth, particularly the European version associated with the subsidence of continents. He argued that geophysical data indicated two basic types of surface crust. The continental crust was on average just above sea level and composed mostly of sial rocks, whereas the oceanic basins were composed of the denser sima rocks. These were physically and chemically different and not interchangeable, hence, isostasy would not allow the less dense material of continents to subside into the denser material of the ocean basins. Other geophysical evidence supported this interpretation. For example, gravity measurements showed that oceans and continents were in isostatic balance, seismic studies indicated continental and oceanic rocks had different properties, and ocean dredge hauls often contained sima types of rocks. Thus, he held that continents were large blocks of less dense sial floating in a "sea" of denser sima. He stated that the geophysical data fit this conception so well that "these ideas have been accepted by most geophysicists" (Wegener, 1924: 41).

Wegener developed his geological evidence for drift by first matching the continental shelves to reconstruct Pangaea. He then showed that many of the geological features, presumably formed before separation, had matching configurations on each side of the initial rift. These geological features included

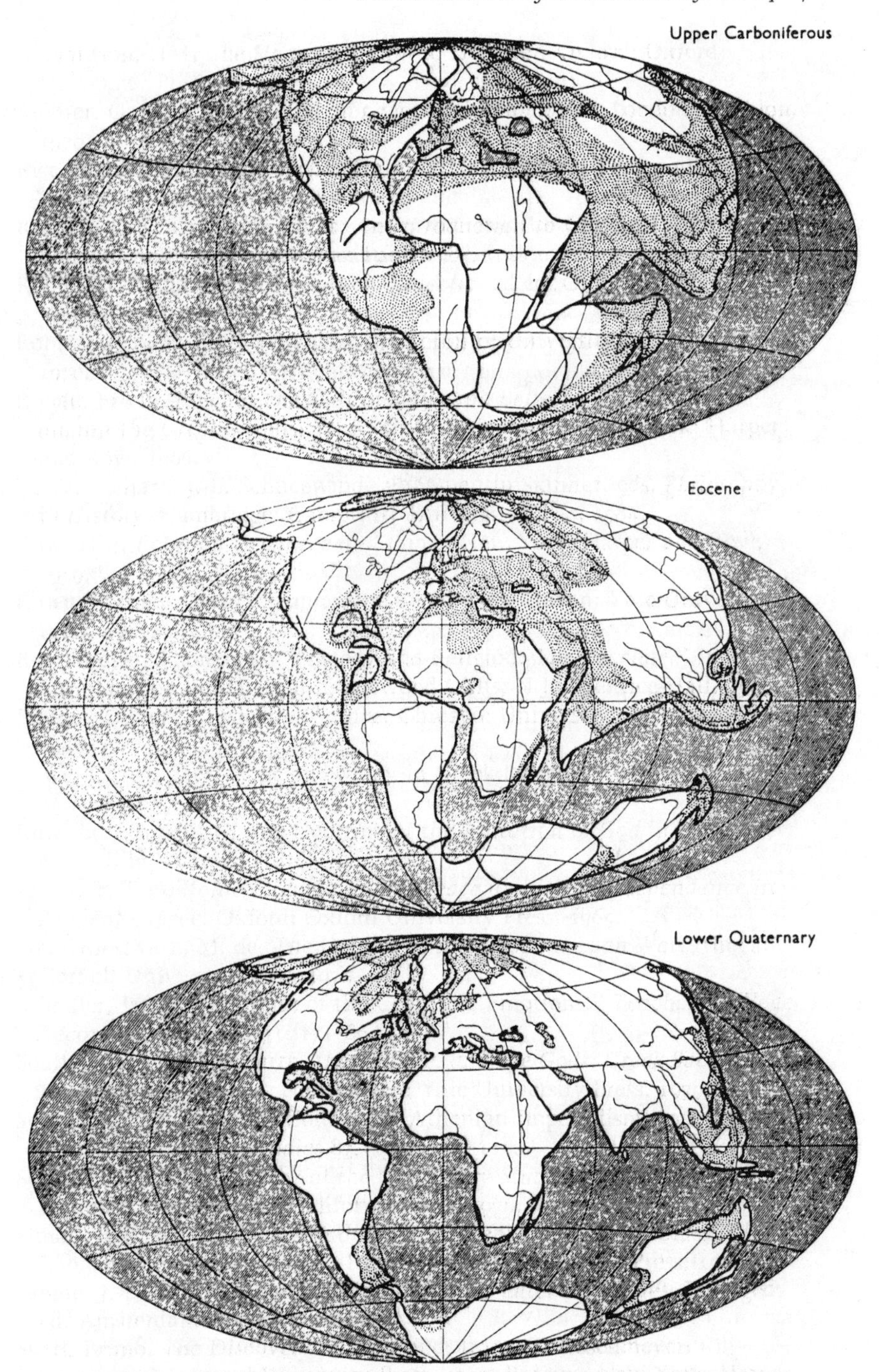

Figure 2.1 Wegener's reconstruction of continental positions in three geological periods. The dotted areas indicate shallow seas. (From Wegener [1929], with permission, Friedr. Vieweg & Sohn.)

rocks with similar ages, composition, and strata sequences, as well as continuations of older mountain ranges from one continent to another.

The state of paleontological evidence at that time is illustrated by how Wegener used this evidence. Such studies tended to concentrate on specific geological time periods or on specific fossil types. To account for the observed distributions of fossils, some researchers had postulated the existence of land bridges between various continents because some organisms could not have dispersed across oceans. The results were such that in any given geological period some fossils indicated the existence of land bridges between two continents and other fossils did not, or at least similar fossils had not yet been found on other continents. He noted the number of opponents and proponents for land bridges for each geological age and whenever there was a majority proposing land bridges, Wegener assumed that the relevant continents had been in direct contact. For the other geological periods when the opponents of land bridges between two continents were in the majority, he assumed the continents had drifted apart. In this manner Wegener estimated the timing and pattern of drift for different continents.

Wegener was coauthor of one of the first textbooks on paleoclimatology (Koppen and Wegener, 1924) and he used this type of evidence in all editions of his book. He noted that climate is influenced by many factors, but latitude is one of the primary factors: some types of rocks and geological events can only occur at specific latitudes. For example, continental glaciers can only occur in the polar regions, whereas large reptiles cannot live in these regions. Consequently, the rocks and fossils in a specific area contain information about the climate, and hence the latitude, for the area in different geological periods. His most impressive use of this type of evidence was his interpretations of the evidence for the widespread locations of known Permo-Carboniferous glaciations. Evidence for these glaciers was present in the southern regions of South America, Africa, and India; the latter is presently located near the equator. Wegener pointed out that if Pangaea is reconstructed, all three of these continental areas are united next to each other in the southern hemisphere, as shown in Figure 2.1. He suggested that the Permo-Carboniferous glaciers could have formed on these diverse areas when they were united and this area of Pangaea was located near the South Pole.[11]

Geodesy, the study of the shape and dimensions of the earth and its features, also seemed to provide evidence for continental drift. Wegener suggested that if the continents were still drifting apart, then the measured distance between them should be increasing over time. He found published evidence for the movement of Greenland away from Scandinavia. However, the evidence indicated a movement of two to twenty meters per year, which Wegener felt was rather high and probably contained some errors because it was based upon triangulations with the position of the moon. Determinations based on radio waves were just being implemented and Wegener hoped they would give more accurate results.[12]

The last section of Wegener's book considered the various assumptions needed in his theory, some of the theory's implications, and evidence relevant to both of these issues. Since his theory implied that the continents were like "icebergs" floating in a "sea" of sima, Wegener had to support his belief that the sima could "flow." He suggested that the vertical, isostatic adjustments of the continents and the equatorial bulge of the rotating earth implied the mantle was capable of viscous flow over a long time period. Thus horizontal movements must also be possible and the compressional forces creating mountain ranges indicated that horizontal movements had occurred. Although earthquake data implied that the mantle was rigid, Wegener argued that this rigidity was only true for short-term forces. He used the analogy that pine pitch was hard and brittle to rapidly applied forces, but given a sufficient time it would flow into a puddle as a result of the weak gravitational force. In a chapter on oceanographic data Wegener suggested that the known trenches around the Pacific Ocean were due either to the continents overriding the ocean floor, as in South America, or to the rifting of the crust as the Asian continent moved west. Wegener's other chapters in this section dealt with the structure of the sial continents, continental shelves, and the folding and rifting of continental rocks.

Wegener's final chapter examined what he recognized as a major problem with his theory: What forces could cause the continents to move through a resisting sea floor?[13] He felt the evidence for continental drift was so strong that it could be accepted without knowing the actual causes, just as geologists accepted the existence of past ice ages without agreement on their causes. Yet he tentatively offered two weak forces to explain why the continents might move toward the west or toward the equator. These were the pole-fleeing and the tidal forces.

Since the earth is a rotating body, the centrifugal force at the surface of the earth is zero at the poles and maximum at the equator. This force was used by geophysicists to explain the equatorial bulge of the earth, and Wegener noted that it also would cause any objects on the surface of the earth, specifically continents, to move toward the equator. Although this force was considered quite weak, it operated in the right direction and Wegener suggested that the sima of the ocean floors would yield slowly to it over the long term. The forces producing the westward drift of the continents were the familiar tidal forces of the sun and the moon, which produce ocean tides. Since the earth rotates to the east, the sun and the moon should exert a slight, westward, retarding drag on the continents. Again, Wegener recognized that this was a very weak force, and suggested in the 1929 edition of his book that the "Newton of drift theory has not yet appeared."

Although Wegener's theory was the predominant and most criticized "mobilist" theory, there were other such theories appearing from 1920 to 1940. In the 1920s and 1930s Taylor and Baker were still presenting their theories of drift. Reginald Daly, a Harvard geologist, accepted some of the evidence for drift, but proposed a different mechanism based on gravitational sliding of the continents from bulges in the mantle (Daly, 1926). Several well-known European geologists held "mobilist" viewpoints that included continental drift as a mechanism for the formation of mountains. Emile Argand and Rudolf Staub were the most prominent among a group of Swiss alpine geologists accepting drift as an explanation for the Alps (Brunnschweiler, 1983; Carozzi, 1985). Edward Baily, a prominent British structural geologist, supported drift theory, but few British geologists accepted drift, although many were fairly sympathetic toward it (Wood, 1985; Marvin, 1985). Spanish and Dutch geologists also tended to be supportive of drift (Carozzi, 1985). Despite several "mobilist" viewpoints and a number of widely scattered supporters, Wegener's writings were given the most criticism, especially in North America.

The Response to Wegener: 1920–1945

Even before Wegener's theory was published in the English language, it had received generally negative reviews in America and Britain. In Britain the 1923 meeting of the Royal Geographical Society considered Wegener's theory and, according to Marvin (1973: 86), the overall tone of the meeting was "essentially hostile." A central opponent present at this meeting was Harold Jeffreys, who was soon to be Britain's most prominent geophysicist. Jeffreys remained opposed to the theory into the 1970s and his arguments against drift were echoed in the later responses of geologists. The most important public discussion of drift theory in North America occurred in a 1926 symposium sponsored by the American Association of Petroleum Geologists. The papers at this meeting were published (Waterschoot van der Gracht, 1928) and included strong opposition from prominent geologists. This symposium marked the end of serious consideration of drift theory in North America.

There were, however, some geologists who continued to advocate drift theory and offered additional evidence and modifications. The most significant and outspoken of these was the South African geologist Alexander du Toit (1937). Other geoscientists suggested some modifications to drift theory, but did not pursue it strongly. Among these was Arthur Holmes, the British geologist who made major contributions to determining the absolute age of the earth. Although Holmes was skeptical about drift, he proposed a mechanism for drift that was an early anticipation of "seafloor spreading," which is described in Chapter 4. The negative responses to Wegener's drift theory can be illustrated by considering the contributions of Harold Jeffreys and those contained in the 1926 symposium. The contributions of du Toit and Holmes will illustrate the positive responses to drift theory.

HAROLD JEFFREYS

In 1924 Jeffreys published *The Earth: Its Origins, History, and Physical Constitution*, which exemplified how mathematics could be applied to geophysical problems. One of the major contributions of this book was its defense of contraction theory. Jeffreys overcame many of the previous objections to this theory by proposing a highly quantified model for the evolution of the earth that was consistent with contraction. In a section on continental drift theory he emphasized the weakest part of Wegener's theory: his causal mechanisms. Jeffreys quantified Wegener's tidal and pole-fleeing forces and argued they were too weak to force continents through the cold, rigid sima of the ocean basins. He suggested several lines of evidence indicating that the sima did not flow. For example, if the sima in the ocean basins could flow, then it should not be able to support the known mountain chains in the ocean basins and data from earthquakes indicated a rigid mantle and seafloor. The strongest support for Wegener's theory, the geological evidence, was considered only briefly and dismissed. In particular, he argued that the fit of the continents was actually quite poor. Jeffreys maintained this position through all editions of *The Earth*, including the fifth edition appearing in 1970.

The book was an instant success among geophysicists and caused many of them to doubt continental drift theory. No longer could Wegener assert that most geophysicists accepted his ideas (Hallam, 1975). Even geologists, who were wary of quantitative arguments by physicists after their experience with Lord Kelvin's estimates of the age of the earth, quickly accepted Jeffreys's arguments and repeated them in their attacks on drift theory. Furthermore, many of them challenged the geological evidence presented by Wegener. The nature of some of these attacks can be seen in the articles presented at the 1926 symposium.

THE 1926 AAPG SYMPOSIUM

This symposium was sponsored by the American Association of Petroleum Geologists and held in New York during the evening of November 15, 1926. Participants included many of the eminent men in North American geology. By this time Wegener's theory had received considerable public and scientific attention and many of the arguments for and against it were well known. Consequently, much of the discussion consisted of organizing and repeating the arguments in favor of one's initial beliefs.

There was, however, one new element in this debate. The proponents of drift theory cited John Joly's recently proposed theory of mantle convection currents as a possible force for continental drift. Joly, an Irish geologist, had calculated that the present heat flow from earth's surface was sufficient to remove only the heat generated by radioactive elements in the crust. Hence, the mantle below the crust must be heating up and would eventually be hot enough to form convection currents, which would remove the excess heat. Joly related this periodic heating and cooling of the mantle to major geological processes, especially the cycle of mountain building periods—the orogenic cycle. When the mantle heated up, it would become less dense and allow the continental crust to sink a little deeper into the interior, thereby causing the transgressions of the oceans onto the lower continental areas. The hotter, convecting mantle would cause the increased volcanism of orogenies and remove the excess heat. When the interior cooled, it would become more dense, which would uplift the continents, cause the regression of the oceans, and compress and uplift geosynclines into mountain ranges.

Joly used his theory to explain the orogenic cycle, but proponents of drift theories realized that convection currents might be one of the causes of drift or that the forces suggested by Wegener might be sufficient for continental drift when the mantle was hotter and more fluid. Initially, Joly did not accept drift theory, but he was present at the symposium and gave drift theory a limited endorsement. This made little difference to the opponents of drift and they mounted a rather hostile attack. It is worthwhile to consider in some detail the papers from the symposium because they illustrate how scientists can "talk past" each other when there is little consensus on the "facts," assumptions, and even the ground rules for how to talk about a topic. These same problems reappear in the 1960s and 1970s when the opponents of drift are the minority.

The opening statement at the symposium was given by the chairman, Willem van Waterschoot van der Gracht. He asked that the participants approach the subject with an "open and unprejudiced mind," to not "lose ourselves in minor details," and to recognize that many of the recent global theories all have a "considerable amount of truth in them." He then repeated many of the arguments for and against drift, described various forms of drift theory (including his own version), and concluded that the evidence for drift was persuasive. Nevertheless, he conceded that since no theory is entirely correct, the discussion should focus on whether some form of drift theory provides a possible "working hypothesis."

The next article by Bailey Willis, an emeritus professor at Stanford University, claimed to consider continental drift with "avowed impartiality"—then rejected it. Willis argued that drift was self-contradictory. He accepted that there was a reasonably good fit in the coast lines of Africa and South America, but argued that if drift had occurred, then the coast lines would be deformed. Hence, the good fit showed that drift had *not* occurred. Although he accepted the paleontological evidence for a connection between these two continents, he accepted the permanentist assumption and suggested that land bridges provided a better explanation. He concluded:

> When we consider the manner in which the theory is presented we find: that the author offers no direct proof of its verity; that the indirect proofs assembled from geology, paleontology, and geophysics prove nothing in regard to drift unless the original postulate of drifting continents be true; that the fields of related sciences have been searched for arguments that would lend color to the adopted theory, whereas facts and principles opposed to it have been ignored. Thus the book leaves the impression that it has been written by an advocate rather than an impartial investigator. (p. 82)[14]

Rollin Chamberlin, a University of Chicago professor, didn't even pretend to consider drift seriously. He asked, "Can we call geology a science when there exists such difference of opinion on fundamental matters as to make it possible for such a theory as this to run wild?" (p. 83). He proceeded to outline objections to the theory without qualification, defending his strategy with "Wegener's own dogmatism . . . makes categorical comments somewhat less objectionable than would otherwise be the case" (p. 83). He argued that the rocks on the southern continents do not match, the causal forces were too weak, sima and sial rocks would not behave as Wegener proposed, and that drift was not a global theory since it did not account for mountains formed before the Mesozoic, which is when Pangaea was supposed to have rifted apart. He concluded that Wegener's hypothesis is "of the foot-loose type. . . . The best characterization of the hypothesis [is] 'If we are to believe Wegener's hypothesis we must forget everything which has been learned in the last 70 years and start all over again' " (p. 87). Chamberlin believed that the "planetesimal" version of the contraction theory was still viable.[15]

The next two articles were favorable. John Joly gave a very short comment and concluded that drift was a logical possibility within his theory of periodic convection currents. G. Molengraff, a professor at the Institute of Technology in Holland, argued that eastward drift was also possible and cited his 1916 article that suggested the mid-Atlantic Ridge marked the location from which Africa had drifted east and South America had drifted west.

The short article by J. W. Gregory, a professor at Glasgow University, suggested that the evidence for a sima composition of the ocean basins was still inconclusive and that subsidence of former continental areas into oceans was a better way to explain the evidence for drift. He also attacked the adequacy of the geodetic measurements that seemed to show considerable movement of Greenland.

The longest article opposed to drift was presented by Charles Schuchert, an emeritus professor of paleontology at Yale. Unlike other opponents who accepted a match between the southern continents, Schuchert argued the fit was poor, using an eight-inch globe with sliding cutouts of the continents. He considered the geological evidence for matches between these continents, but only admitted that some fossils showed remarkable "threads of affinity" between the two sides of the Atlantic. In his opinion land bridges between Brazil and Africa provided the best explanation. He admitted it was hard to dispose of bridges, but was "confident that the geophysicists will in time find the way in which this was accomplished" (p. 142). In his concluding remarks Schuchert stated that "the whole trouble in Wegener's hypothesis and in his methods is . . . that he generalizes too easily from other generalizations. . . . Facts are facts, and it is from facts that we make our generalizations, from the little to the great, and it is wrong for a stranger to the facts he handles to generalize from them to other generalizations" (p. 139).

Wegener's short and rather weak paper focused on two issues. First, he gave alternative interpretations of some evidence for glaciation of North America during the Permian period, when his own reconstructions suggested that North America should have been located at the equator. Second, he emphasized that the hypothesis of continental drift had "one advantage over all other geological theories, that it involves the possibility of checking its truth by repeated astronomical observations" (p. 100). He cited data supporting the movement of Greenland away from Europe. He thought there might be some errors in this evidence, but hoped that more accurate measurements would provide the necessary proof for drift theory.

The other presentations at the symposium did not introduce anything new. Chester Longwell, an associate professor at Yale, was quite skeptical about aspects of drift theory, but argued that much more knowledge was needed. He warned the participants of the "zeal of the advocate" and the "prejudice of the unbeliever" and suggested that if "the doctrine of continental displacement is accepted as a working hypothesis, to be tested and tried fairly along with others, it may be productive of valuable results" (p. 157). Frank Taylor was present and repeated his version of drift with the moon's capture as the causal force. William Bowie raised old geophysical objections and D. White wondered why the continents had stuck together until the Mesozoic. J. Singewald considered the idea of drift very plausible and noted the "disagreements among the observers as to the facts." E. Berry's main objection to the theory was Wegener's method of presentation, which he claimed was not scientific, but rather a "selective search through the literature for corroborative evidence . . . ending in a state of auto-intoxication in which the subjective idea comes to be considered an objective fact" (p. 194). The last article was by van Waterschoot van der Gracht, who rather optimistically suggested that "minds have become more open." He then systematically considered the critiques of drift theory and offered possible defenses.

Although the proponents of drift were represented adequately at the symposium, they had little influence on subsequent geological thought in North America. The quotations from the symposium help us understand why this occurred. Drift theory was seen as a challenge to some cherished beliefs among geologists, especially North American geologists. It challenged the permanence of the ocean basins, the contraction theory, and the geosyncline model and it seemed to be working down from a global theory to the "facts" rather than the reverse (Le Grand, 1986*b*). Furthermore, there was little agreement on many of the relevant facts or on how these related to drift theory. For example, some opponents denied the fit between the edges of the southern continents, while other opponents acknowledged the good fit, but used it to argue against drift. Even when there was agreement on some of the facts explained by drift theory, the favored theories were protected by proposing alternative mechanisms or simply stating that a solution will be found. For example, the continuity of fossils across oceans was explained by land bridges with the hope that geophysicists would be able to explain their subsidence, but Wegener was not allowed to appeal to the same "hope" for his problem of the causes of drift. Finally, the opponents were prominent North American geologists and the proponents were less established or "outsiders" in the sense of being from Europe or being non-geologists.[16] Wegener, in particular, was seen as a German climatologist who should not meddle with geological data. Consequently, the opponents even challenged the motivations and general competence of Wegener.

Given this hostile attack by prominent North American geologists it is not surprising that drift theory found few proponents in North America. Although subsequent debates on drift tended to repeat many of the same old arguments and counter-arguments for each position, Frankel (1981*a*, 1984, 1987) has shown that there was some evolution in these arguments within different specialties in the geosciences as pro-drifters fought it out to a "stalemate" with anti-drifters. By all accounts, however, Alexander du Toit and Arthur Holmes made major new contributions to the overall case for continental drift.

ALEXANDER DU TOIT AND ARTHUR HOLMES

Du Toit was a professor at the University of Johannesberg in South Africa and an esteemed member of the Geological Society of South Africa, serving as its president in the late 1920s. As early as 1921 he had noticed the apparent similarities between Africa and South America. Upon the urgings of R. Daly, the Harvard geologist who accepted a form of drift theory, du Toit carried out some South American field studies, which were sponsored by the Carnegie Institute of Washington (Marvin, 1973: 108). Du Toit found many matching features and advocated continental drift until his death in 1949. His major contribution to drift theory was his 1937 book, *Our Wandering Continents.*

Du Toit began his book with a polemical attack on the current orthodoxy by suggesting that current beliefs were so diverse and contradictory that their only common aspect was a rejection of mobile continents. Du Toit suggested that drift theory was the "antithesis" of current beliefs and represented a "holistic" viewpoint in geology that provided testable predictions. After reviewing several variations of drift theory, he noted that most of the criticism had been directed at Wegener's version, while neglecting other promising expositions, particularly those of Rudolf Staub.

Du Toit offered new evidence for similarities between the continents and offered corrections to some of Wegener's evidence. However, he differed from Wegener in dividing Pangaea into two supercontinents: "Laurasia" in the north and "Gondwanaland" in the south. His reconstruction of Gondwanaland is presented in Figure 2.2. He also differed from Wegener in proposing that all the ocean basins had a thin sial layer. The final chapter offered an eclectic set of causes for drift. Wegener's forces were mentioned, but considered too weak. Convection currents were considered likely and he suggested the mantle may have layers and circulation might occur within these layers, just as in the earth's atmosphere. He even suggested that "cyclones" might exist in the mantle and noted several circular distributions of continental fragments.

Du Toit seemed to have little influence on North American geologists for several reasons. First, the evidence for drift was not as obvious in North America. Northern geologists did not have to directly deal with the Permo-Carboniferous glacial evidence or the "Gondwana" fossil assemblage, both of which were confined to South America, Africa, and India. Second, the most outspoken opponents of drift were in the north and they kept up an insistent attack on pro-drift publications. For example, du Toit's book was reviewed by none other than R. Chamberlin. Third, there seems to have been a distrust of the works of geologists in the southern hemisphere. Evidence for this will be considered

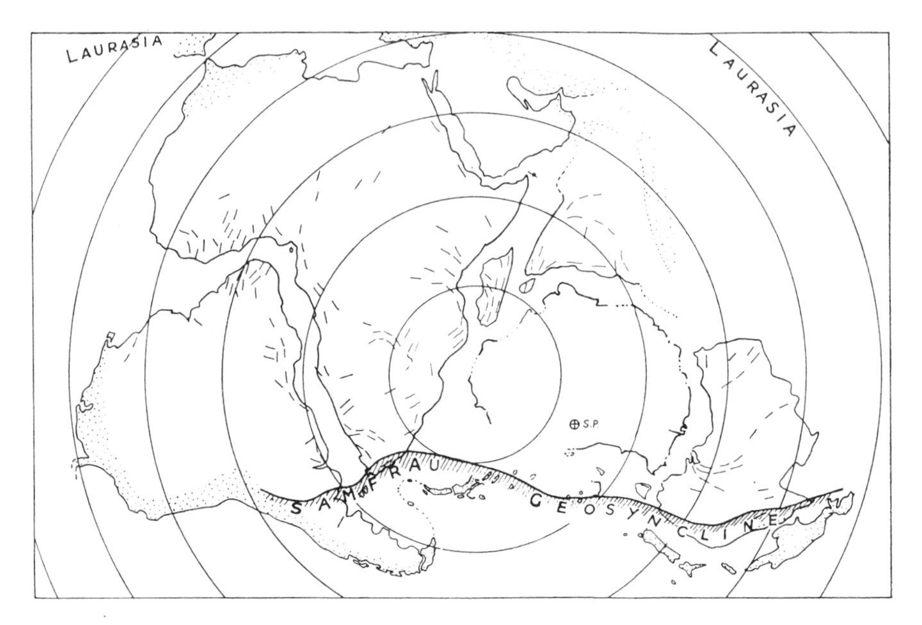

Figure 2.2 Du Toit's reconstruction of the southern continents' positions in the Paleozoic Era. (From du Toit [1937], with permission, Oliver & Boyd.)

later. Fourth, Hallam (1973) suggests that du Toit's style was too much that of a religious advocate to be persuasive to skeptics. Finally, the outbreak of World War II followed the publication of his book and may have adversely affected the ability of a novel and fledgling theory to win new followers.

During the time that du Toit was working in South America, Arthur Holmes was becoming the most prominent British geologist of the twentieth century. More than any other geologist, Holmes was responsible for developing the implications of radioactivity for the earth sciences. His fame was based on using the radioactive decay of elements to give *absolute* dates to igneous rocks associated with the sedimentary rocks containing fossil assemblages upon which the relative time scale had been constructed. These results were given in *The Age of the Earth* (Holmes, 1913), which was published when he was twenty-three years old. He gave geologists an absolute time scale, but not without meeting some resistance because the estimated age of the earth was in the billions, not millions of years (Burchfield, 1975).

Following *The Age of the Earth*, Holmes published a series of five papers from 1915 to 1925 dealing with the implications of radioactive heating for the contraction theory of the earth (Frankel, 1978). By 1930 he had arrived at the belief that convection currents must occur in the mantle and that continental drift was a possible consequence (Holmes, 1931). Thus he was led to drift theory from a completely different line of evidence than previous proponents. In 1944 he published his extremely popular book, *Principles of Physical Geology*, which included Figure 2.3. In contrast to Wegener, who had continents moving through the sima or basalt of the ocean floors, this figure shows that the continents were held in an outer rigid layer. Upwelling convection currents in the mantle below this layer would split the continent apart and form a new ocean basin between the continental fragments. The areas where the currents descended would form the ocean trenches with associated mountain ranges due to the compression of the sediments on the continental shelves. The continents themselves could not return to the mantle because they were less dense than the sima or basalt of the mantle. As we shall see, Holmes' mechanism is a remarkable anticipation of Harry Hess's (1962) concept of seafloor spreading.

Holmes had several confrontations with Jeffreys over the convection current model and possible continental drift. Frankel (1978: 147) suggests that Holmes's drift ideas were not pursued in Britain largely because of Jeffreys's opposition and the popularity of Jeffreys's quantitative model of a contracting earth. However, there is another possible reason: Holmes did not firmly believe in continental drift. He was only committed to drift as a possible working hypothesis. In an article published in 1953 Holmes stated that "despite appearances to the contrary, I have never succeeded in freeing myself from a nagging prejudice against continental drift; in my geological bones, so to speak, I feel the hypothesis to be a fantastic one" (Holmes, 1953: 671). This is hardly the attitude

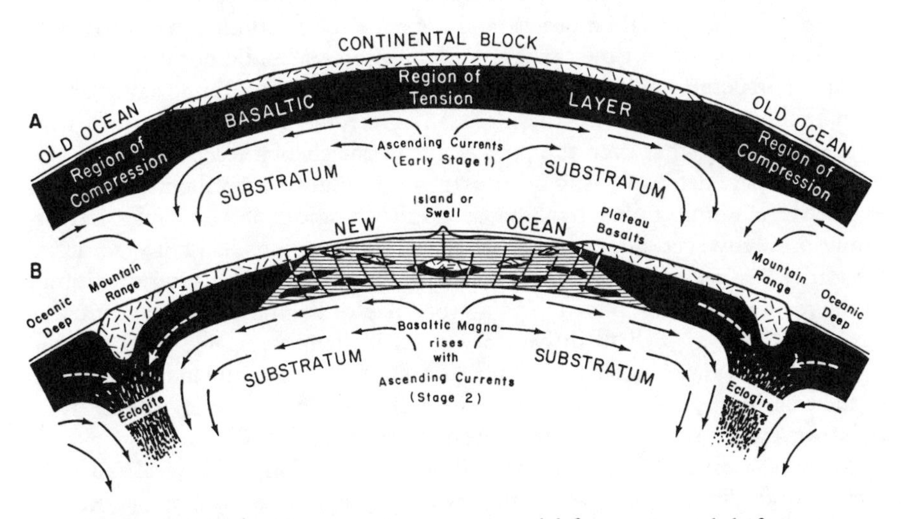

Figure 2.3 Holmes's convection current model for continental drift. Part A shows an early stage before the continental block is rifted apart by the ascending currents in the mantle. Part B shows the new ocean basin created after the continent is rifted apart. Mountain ranges are formed where the less dense continental block resists descent into the trenches. (Reproduced from Holmes: *Principles of Physical Geology* [1944], with permission from the publishers Van Nostrand Reinhold [International], London, U.K.)

of a strong supporter who will encourage his colleagues and students to advocate a novel and widely opposed theory.[17]

Discussion and Conclusions

Despite the general contributions of Holmes, du Toit, and others to the theory of continental drift, the number of geoscientists expressing *any* opinion was a very small percentage of the geoscience community. Most geoscientists just ignored the whole issue and, presumably, continued their research within a "fixist" framework. The few who did support drift tended to be localized in rather specific groups. Among these were Swiss alpine geologists (Brunnschweiler, 1983), Spanish and Dutch geologists (Carozzi, 1985), and many South African geologists. In other nations the reaction was more mixed (Le Grand, 1986*a*; Frankel, 1981*a*, 1984, 1987), but the predominant North American attitude was quite negative (Marvin, 1973; Wood, 1985; Oreskes, 1988). Bailey Willis (1944) may have expressed the typical American opinion when he stated that drift theory was a "fairy tale" which actually hindered the development of geological knowledge.

These are the empirically observed patterns of acceptance of drift theory, but what are the underlying "causes" of this pattern? What do such responses indicate about the nature of (geo)scientific reasoning? Were the opponents or proponents being "irrational"? Was the general rejection of drift theory "fair"? Does this episode fit any of the perspectives outlined in Chapter 1?

Certainly the logical empiricist and Popper's falsificationist perspectives do not seem to fit. At that time reasoning in the geosciences did not have a formal deductive structure, nor was there much agreement on the important facts or even methods of research. Furthermore, any given theory was inconsistent with numerous "facts" that even the proponents of the theory accepted. These two perspectives are defeated easily because they presume a "scientific method" that is relatively uninfluenced by broader methodological and metaphysical assumptions. However, one might argue that the geosciences had not yet reached "maturity," so we will examine these perspectives in Chapter 5 after the history of the revolution is completed. The perspectives of Kuhn, Lakatos, and Laudan have each been applied to this history (e.g., Hallam, 1973, 1983; R. Laudan, 1980; Ruse, 1981, Frankel, 1979*a*, 1979*b*, 1981*b*), but these include discussions of the transition to plate tectonics, so we will save most of them until Chapter 5. However, we can make some relevant observations here.

Although drift theory had few followers, it was one of several *competing* "research traditions" or "scientific research programmes" (Frankel, 1979*a*, 1979*b*). This structural aspect is more compatible with the perspectives of Lakatos or Laudan than it is with Kuhn's suggestion of a "paradigm" monopoly during normal science. Although Kuhn might argue that the geosciences were in a "revolutionary" state with competing paradigms during this time, this state existed for over fifty years and included significant scientific progress by any definition. Laudan's emphasis on both empirical *and* conceptual problems seems more appropriate than Lakatos's emphasis on empirical solutions because many of the opponents stressed the *conceptual* incompatibility between drift and a rigid mantle and seafloor. Nor do we see the tolerance that Lakatos suggested should be extended toward new "research programmes," at least not among North Americans. However, all three of these viewpoints predict the observed pattern of adjusting peripheral aspects of one's favored theories to make them compatible with the empirical "facts." These observations are peripheral to the major difference among these perspectives, which is concerned with how scientists reason and evaluate alternative theories or "paradigms." Although Chapters 5 and 6 consider this aspect in more detail, we will consider here several explanations for the limited level of interest in drift theory and its geographic pattern of acceptance.

First, the vitriolic attacks by esteemed geologists, when compounded with the "religious zeal" of some proponents, may have discouraged many geologists from seriously considering the new theory. In reference to continental drift, Raup suggests that even in science there is a tendency to regard a new theory as "guilty until proved innocent, and the pre-existing theory is innocent until proved guilty" (Raup, 1986: 18). Second, Le Grand (1986*a*, 1986*b*) and Frankel (1984) suggest that "localism" and methodological preferences help explain the diverse reactions to drift theory. Geologists tended to study their own locality or region and to focus on very specific problems, some of which were more easily analyzed within the context of drift theory. In general, the evidence for drift was more ambiguous in the northern hemisphere and open to a number of alternative and more "acceptable" interpretations. Third, there was a movement toward abandoning global theories and emphasizing more detailed regional and specialized studies (Dott, 1981; Bullard, 1975*a*). Fourth, Frankel (1976) suggests that the narrow specialization present in the geosciences caused opponents of drift to fail to see how it solved problems in other specialties or how the results in other specialties conflicted with their favored alternatives to drift. Fifth, the lack of knowledge about the ocean basins permitted geologists to save their favored hypotheses by assuming the ocean basins had certain properties. Finally, a number of factors help explain the stronger opposition of North American geologists: (a) their lack of knowledge of the views of non-English speaking geologists made them less aware of the mobilist viewpoints on other continents (Brunnschweiler, 1983), (b) they distrusted the reported field evidence from southern hemisphere geologists (Craw, 1984; Frankel, 1984); (c) they emphasized the exploration and interpretation of North American geology (Dott, 1981); and (d) they tended to favor an "inductivist" methodology of working from the "facts" to the "theory" (Le Grand, 1986*b*; Oreskes, 1988). Many of these detriments to drift theory would change during the two decades following World War II.

Drift theory did not die entirely. It remained in use where the proper organizational resources and geological evidence were available: the Swiss Alps, Holland, and South Africa. In these locations there were esteemed geologists who adopted the theory (e.g., Argand, Staub, and du Toit) and researchers were confronted with evidence that was interpreted easily in terms of continental drift, such as the Alps, the East Indies, and South Africa. However, these pockets of acceptance played little role in the acceptance of plate tectonics in the 1960s. The major contributions to the rise of plate tectonics were the result of geoscience research conducted in different specialties during the twenty years following World War II. The next chapter covers the period from 1945 to about 1960. By 1960 continental drift theory had a new group of defenders: geophysicists, some of whom were in Harold Jeffreys's own department at Cambridge.

CHAPTER THREE

Specialization without Integrating Theory: 1945–1959

Following World War II geoscience researchers continued their emphasis on specialized research without much concern with the development and testing of global theories. The older contraction beliefs provided an implicit assumption in much of this research, but some researchers occasionally would invoke convection currents in the mantle. However, even advocates of the latter process seldom thought that drift was a necessary result. The vast majority of geoscientists felt that continental drift was not a plausible component of any global theory. This conclusion is supported by counts of the publications related to continental drift from 1920 to 1970 (see Figure 3.1) and the number of citations to Wegener's and du Toit's "classics" on continental drift theory (see Table 3.1). These data show a small peak of interest in the late 1950s, but sustained interest did not start until about 1965.

Even though drift theory was seldom considered in the 1945–1959 period, it is important to summarize selected developments because they provided much of the evidence cited in support of seafloor spreading and plate tectonics in the 1960s. This chapter summarizes these developments by geoscience specialty or research area and provides occasional data on selected scientists and citation counts to specific papers.[1] The first section describes the major developments in oceanographic research, which provided some of the most important evidence for the revival of drift theories. The second, third, and fourth sections briefly cover developments in the study of earthquakes (seismology), geochronology, and the geosyncline model. In the fifth section we find that studies of the earth's past magnetic field (paleomagnetism) produced new evidence for drift. The final section describes two global theories that attempted to synthesize the postwar developments. One theory assumed an expanding earth with drift of the continents, whereas the other assumed a special version of a contracting earth without drift. This information will prepare us for the revolution described in the next chapter.

Oceanography

Before World War II very little was known about the ocean floor, yet it covered about sixty percent of the earth's surface. A few major oceanographic institutes produced the most important postwar research. The U.S. institutions included

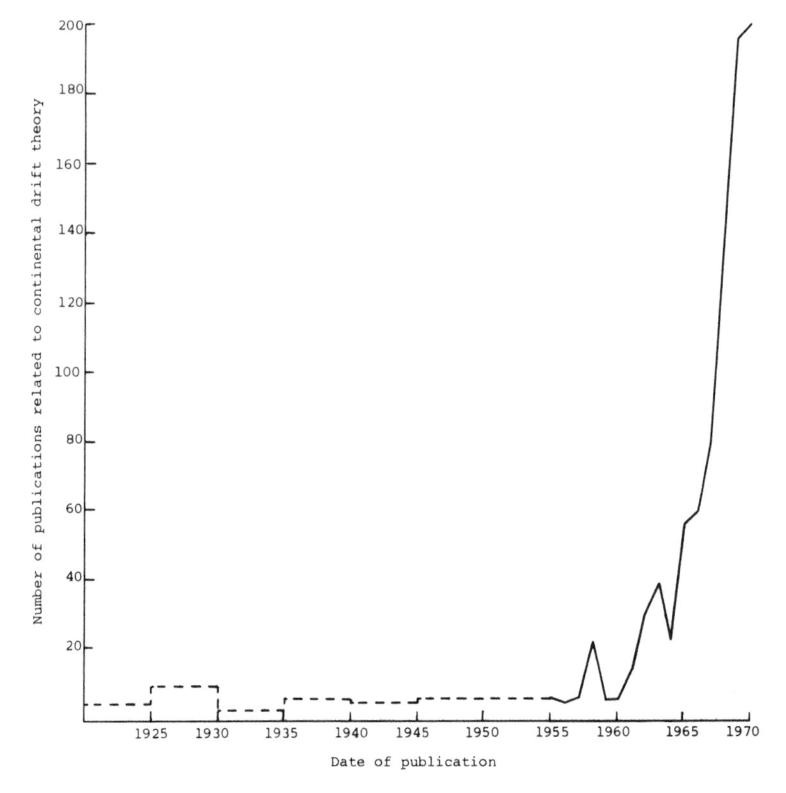

Figure 3.1 Number of publications related to the topic of continental drift from 1920 to 1970, from a personally compiled bibliography. The dashed lines are five year averages. Peaks before 1968 are due to publications from symposiums.

Lamont Geological Observatory (associated with Columbia University), Scripps Institute of Oceanography (associated with the University of California, San Diego), and Woods Hole Oceanographic Institute in Rhode Island. Later research at Cambridge University was very important for the development of plate tectonics.

Oceanographic research experienced rapid growth after World War II for several reasons. First, the military increased its interest in and funding of oceanographic research. Federal support of just academic research on the oceans was over 25 million per year in 1965—an increase of more than a hundredfold since 1941 (Menard, 1986: 38). Second, some of the scientific research groups that were organized for war-related problems remained fairly intact and took up more basic research questions on the oceans (Bullard, 1975*b*). Third, there was greater intellectual interest in this subject as geologists realized that global geology could not be complete without better knowledge of the ocean basins.

TABLE 3.1 Citation Counts to the Two Major Books on Continental Drift before 1940

	Science Citation Index citation counts for:																
	1955	'56	'57	'58	'59	'60	'61	'62	'63	'64	'65	'66	'67	'68	'69	'70	'71
The Origin of Continents and Oceans (Wegener, any edition)	1	1	5	1	3	3	4	2	6	3	8	8	7	9	17	18	16
Our Wandering Continents (du Toit, 1937)	0	3	4	1	1	4	3	3	4	7	10	5	5	10	16	13	13

SOURCE: Institute for Scientific Information (1984, 1972).

Finally, instrumental techniques had been improved and new ones developed, partly because of research efforts associated with the war. These instrumental advances are so important that they merit at least brief descriptions.

By the early 1950s most oceanographic expeditions employed many of the following instruments.

1. The sonic depth recorder or "sonar" had been developed in the 1920s, but by 1950 it was possible to measure ocean depth to six miles.
2. The magnetometer was developed during the war for submarine detection and used by many postwar expeditions, especially those from Lamont Geological Observatory.
3. In the 1930s Vening-Meinesz, a Dutch geophysicist, made the first gravity measurements at sea from submarines, but by the early fifties researchers had developed gravitometers capable of continuous measurements from surface ships.
4. In the late 1930s Maurice Ewing adapted petroleum research techniques to ocean settings. These included seismic reflection and seismic refraction techniques. The *reflection* technique involved setting off periodic explosions of TNT from a moving ship. The generated "shock" (seismic) waves were reflected from the different rock layers of the ocean floor and the time delays of the reflected waves provided information about the thickness of the different layers.
5. Seismic *refraction* studies were more difficult since they required separation of the explosion and receiving locations. When the seismic waves enter a rock layer at an angle, some are bent or "refracted" and travel along the layer itself for a distance until refracted again back to the receiving ship. Since the speed of travel was a function of the density of the rock, this technique indicated the density of the separate layers in the ocean floor, and density placed constraints on possible compositions of the layers.

6. In the early 1950s Edward Bullard, a British geophysicist, developed a sea-floor probe that could measure the rate of heat flow through the ocean crust. Because of his association with Scripps, most of the early heat flow data came from this institution (Menard, 1986).
7. Samples of the sediments in the upper thirty meters of the ocean floor were obtained by a coring device, which consisted of a weighted pipe that was released just above the floor. Gravity drove the pipe through the soft sediments, which were then returned to the ship for analysis.
8. Ever since the earliest oceanographic studies in the 1800s various dredging nets or devices were used to collect samples of rocks and bottom life.

These instruments[2] provided the basic data for the many oceanographic expeditions of the 1950s. The empirical results of these expeditions are discussed under the next two headings; they are (a) the topography of the ocean basins and (b) their geological and geophysical properties.

TOPOGRAPHY OF THE OCEAN BASINS

Information on the topography of the ocean floors was provided mostly by sonar. By the end of the fifties the sonar soundings were sufficient for Lamont scientists Bruce Heezen,[3] Marie Tharp,[4] and Maurice Ewing[5] (1959) to produce a "bird's-eye" view of what the North Atlantic Ocean basin might look like without water. Figure 3.2 is such a view and illustrates various features that were discovered before or during the fifties.[6]

The Mid-Ocean Ridge System

Segments of this ridge system were known from earlier oceanographic research and the laying of telegraph cables in the last century. In the fifties Ewing and Heezen at Lamont noticed a correlation between the location of many oceanic earthquakes and the known crests of the ridge system and predicted that the known global encircling band of earthquakes in the earth's oceans indicated that the mid-ocean ridge system extended around the globe as well. Tharp and Heezen also found that the Atlantic portion of the ridge system had a nearly continuous, deep rift valley running down its central axis. The Atlantic ridge system was generally located at the center of the ocean basin and was elevated and surrounded by jagged mountain peaks on each side.

Fracture Zones

Marine studies in the Pacific Ocean by Bill Menard[7] at Scripps had discovered large faults or fracture zones that could be traced for hundreds of miles. These fracture zones were marked by large offsets in the relative heights of ocean floor on each side of the linear fault line. In Figure 3.2 these fault zones are offsets in the ridge crest and the extensions of the offsets. However, the first fracture zones discovered in the Pacific were extensions and did not seem related to the Pacific ridges, which tended to be less pronounced in height and without a prominent rift valley. Similar fracture zones were discovered later in the Atlantic, where they tend to run in an east-west direction. Consequently, they were discovered first by Russian expeditions sailing a north-south direction because they were denied access to North American ports (Wertenbaker, 1974: 184).

Submarine Canyons and Abyssal Plains

Inspection of Figure 3.2 shows that the continents continue at a shallow depth into oceans to form the continental "shelves," such as the Grand Banks off Newfoundland. The edges of the continents are marked by the continental "slopes," where the floor descends rapidly to a great depth, eventually leveling out into the abyssal plains before rising again to the mid-ocean ridge. Deep submarine canyons periodically cut into the continental slope. In the fifties one of the more common explanations of these canyons suggested that they were eroded by rivers when the sea level was thousands of feet lower. Some support for this view was given by the presence of sand in some core samples from the abyssal plains. Sand can only be formed on land and is too dense to be transported very far after a river flows into the ocean. However, in the fifties Ewing and Heezen proposed that the sea levels had always been relatively constant and that the submarine canyons were cut by "turbidity currents," massive underwater landslides originating on the continental shelves and flowing hundreds of miles, eventually settling to form the broad abyssal plains. Thus, much of the ocean basins was covered by turbidity current deposits instead of the microfauna "ooze" deposits expected in the deep ocean basins.[8]

Ocean Trenches and Island Arcs

Just as one would not expect the shallowest parts of the oceans (the ridges) to be located in the middle of the basins, one would not expect the deepest parts (ocean trenches) to be adjacent to land. Yet earlier laying of telegraph lines and the later oceanographic surveys only found trench systems close to land, either along continental margins like the west coast of South America or along "island arcs," which are arc-shaped strips of volcanic islands, such as the Aleutian Islands.

Seamounts and Guyots

Sonar studies had located numerous underwater mountains or peaks that stood above the surrounding terrain. Many of these were isolated, but others seemed to form chains. Most of the smaller oceanic islands, except those in island arcs, were seamounts that reached above the water surface. Studies of these islands and dredge hauls from the submerged seamounts indicated that most were composed of igneous rocks and were formed by volcanic action.

Many of the Pacific seamounts were flat topped, called "guyots." Some geologists proposed the flat tops were formed by the erosion of ocean waves when the sea level was much lower. Coring studies of the guyots supported this interpretation since some cores contained corals, which can only form in shallow water. However, as we shall see in the next chapter, geoscientists

MID-OCEAN CANYON
-10500
-9510
-462
(15538)
ISLAND OF NEWFOUNDLAND
CAPE BRETON ISLAND
MIQUELON ISLANDS
-42
-330
GRAND BANKS
SABLE ISLAND
-228
-12
LAURENTIAN CONE
-168
FLEMISH CAP
-15420
-14760
MILNE SEAMOUNT
-336
(15664)
-15900
-15300
-4500
(11500)
RIFT VALLEY
AZORES
Pico
7615
(23615)
KELVIN SEAMOUNT
(10588)
-4590
SOHM ABYSSAL PLAIN
-17400
OCEANOGRAPHER FRACTURE ZONE
CORNER SEAMOUNTS
HUDSON CANYON

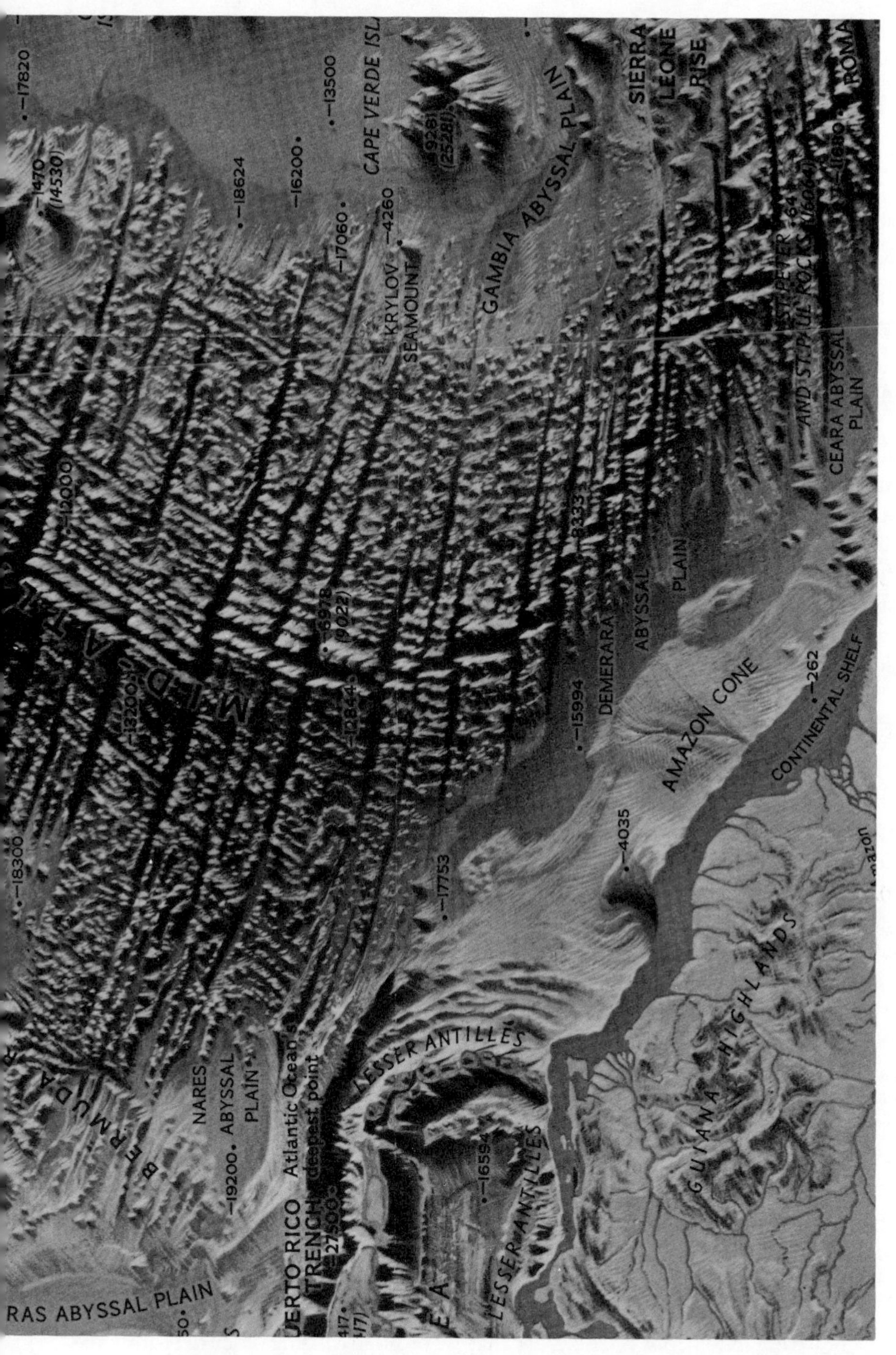

Figure 3.2 A "bird's eye" view of the Atlantic Ocean basin prepared by the National Geographic Society showing the northern part of the Mid-Atlantic Ridge. (Reproduced with permission from the National Geographic Society.)

eventually accepted that it was the seamount, not sea level, that was rising and falling over time.

THE GEOPHYSICAL PROPERTIES OF THE OCEAN BASINS

A number of general characteristics of the ocean basins were inferred from various geophysical data, including the analysis of heat flow measurements, magnetic and gravitational field studies, and seismic reflection and refraction studies. These studies produced even more surprising results.

Gravity Measurements

In general the oceans displayed normal gravity readings, which indicated they were in isostatic balance. However, the gravity fields over the trenches and the mid-ocean ridge crest were lower than expected. These fields were consistent with several structural interpretations. Since the 1930s a common interpretation of the trench structure was that they were caused by descending convection currents, which pulled down the crust. This dip in the crust was called a "tectogene." However, a splitting apart or extension of the crust was also consistent with the lower gravity field and this interpretation was used often for the anomaly at the ridges. Seismic reflection and refraction studies provided additional constraints on these models.

Seismic Studies

By the late 1950s seismic reflection and refraction studies indicated that the ocean basins tended to have three fairly distinct layers. The first layer varied considerably in its thickness and generally was thinner nearer the mid-ocean ridges. As expected, coring samples indicated that it was composed of unconsolidated sediments, but its average thickness was only about five percent of the several miles of sediments expected to have accumulated in the ocean basins, which oceanographers thought were billions of years old. Refraction studies indicated that the second and third layers were thicker and denser and were regarded generally as igneous rocks. The bottom of the third layer was defined by a distinctive change in the rate of seismic wave propagation, which was called the Mohorovicic discontinuity (the "Moho") and used to define the lower boundary of the crust of the earth.

Under continents the Moho varied from ten miles to thirty miles under mountains, but the ocean basins had a fairly uniform crustal thickness of about four miles. Under the ridges, however, seismic waves traveled slower and it was assumed the ridges were underlaid by less dense material. The refraction studies of the crust over trenches were subject to various interpretations. The Scripp researchers tended to find thickened and deeper oceanic crust and favored a convection current explanation, whereas Lamont scientists thought that trenches were extensions of the crust that were filled with less dense sediments (Menard, 1986). Despite these differences, the overall seismic results suggested that the lower layers were too dense to have the sial (granitic) composition characteristic of the continents. Thus subsidence of continents or land bridges into ocean basins became less plausible.

Heat Flow

Until 1950 the measurements of heat flow from the earth were restricted to the continents. In 1950 Edward Bullard[9] developed a temperature probe that measured the rate of heat flow through the ocean sediments. Subsequent oceanic heat flow studies showed that on the average the oceans were giving off as much heat as the continents and the flow tended to be higher near the ridge crests and lowest near the trenches (Bullard, Maxwell, and Revelle, 1956). This higher rate was unexpected because the major source of heat was assumed to be the decay of radioactive elements, which were more common in the granitic, continental material than in the ocean's basaltic rocks. Consequently, these researchers suggested that mantle convection currents might explain the heat flow pattern.

Magnetic Field Measurements

At first the study of the magnetic field at sea did not produce interesting results. Most of these studies were conducted by Lamont surveys in the Atlantic, which indicated a large magnetic anomaly over the central rift valley with minor variations on each side that were not correlated with the underlying topography. In the mid-fifties a visiting professor at Scripps, Ron Mason,[10] borrowed the Lamont magnetometer and towed it behind a Coast Guard ship that was making a detailed systematic survey along the west coast to establish the topography before installing a submarine detection system (Menard, 1986). Mason found that the minor variations in the field along a ship's course lined up with the variations recorded on adjacent courses. Figure 3.3 shows that the entire area surveyed was covered by "linear magnetic anomalies," which he illustrated by shading the peaks of magnetic intensity between parallel survey lines. Furthermore, these linear patterns appeared to be disrupted by the known fracture zones (Mason, 1958).

The wiggles in the magnetic profiles were distinctive enough that matching patterns could be found on the other side of the fracture zones, but only by assuming the zone marked a displacement of the ocean floor. Some of these displacements were over 600 miles. These results indicated that the sea floor had experienced considerable movement. Although the causes of the linear magnetic anomalies were unknown, they seemed to record large crustal displacements and geologists had to explain why these displacements did not continue onto the adjacent continental margins. Early explanations of the anomalies suggested that they were caused by a linear pattern of vertical faulting or changing composition in the ocean floor (Vacquier, 1959; Mason and Raff, 1961). Later, some geologists thought they could trace the faults onto the continental margins.

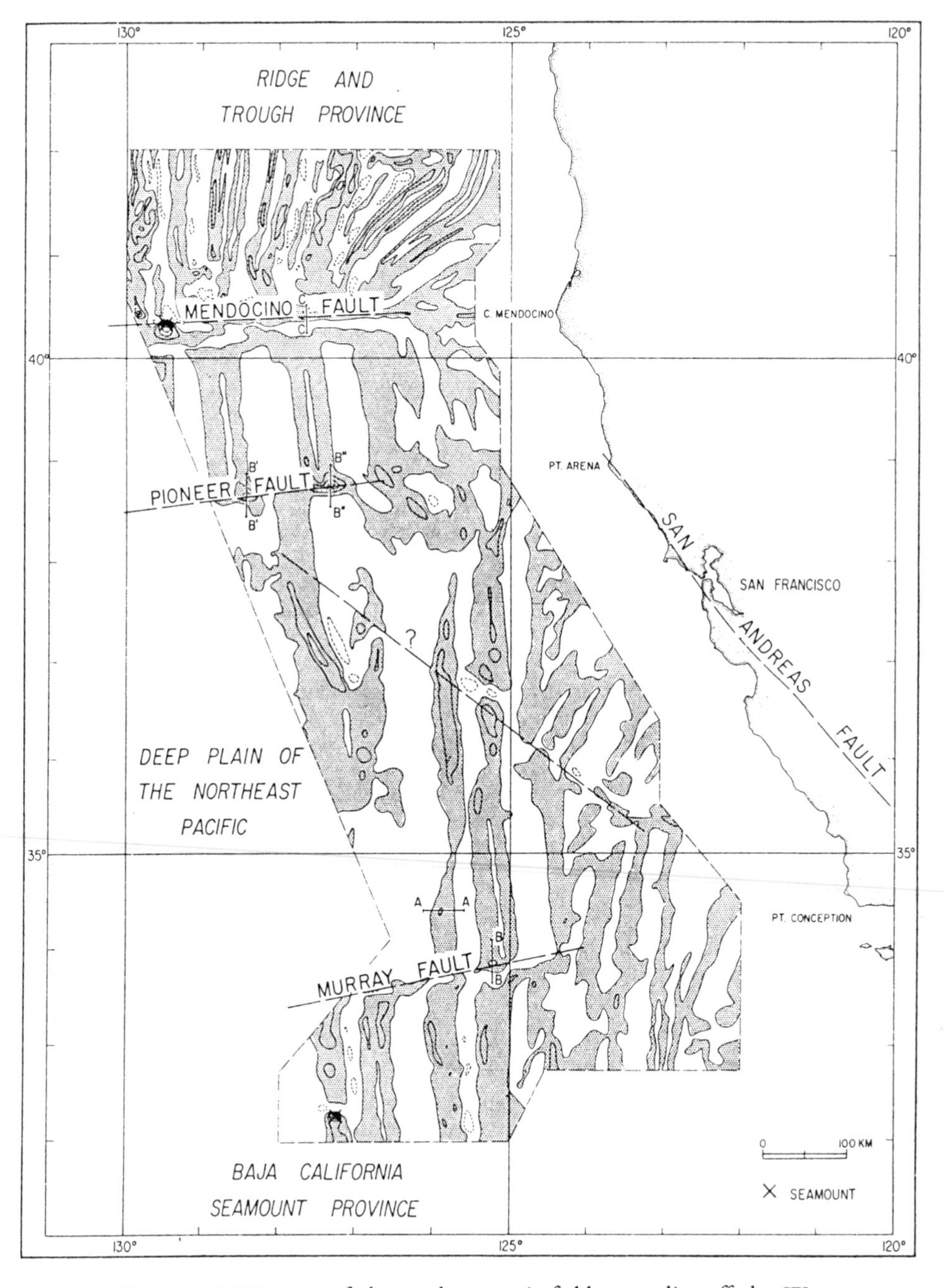

Figure 3.3 Diagram of the total magnetic-field anomalies off the West Coast of North America. Areas of positive anomaly are darkened and several fracture zones are noted. (From Mason and Raff, *GSA Bulletin* 72 [1961]: 1259–1266, with permission, Geological Society of America.)

IMPLICATIONS OF THE OCEANOGRAPHIC RESEARCH

The results of these numerous oceanographic studies challenged many of the initial expectations about the ocean basins. Less than a tenth of the expected miles of sediments were found in the ocean basins and the oldest dated fossils from dredges were only about 180 million years old. Both of these findings were inconsistent with expectations that the basins were billions of years old. The seafloor was not a flat surface, but was mostly rolling hills with clusters of larger sea mounts. Most prominent was the mid-ocean ridge system, which was the largest mountain range on earth. Furthermore, the ridge seemed to have a central rift valley, where the ocean floor might be splitting open. Unlike the continents, the crust above the Moho was fairly thin and uniform in thickness, yet the ocean's heat flow was similar to the continental rate. Finally, the linear magnetic anomalies were totally unexpected and indicated that the Pacific Ocean floor had experienced very large lateral displacements along the fracture zones.

These results challenged the expectations, but it was always possible to reconcile the differences. For example, one could suggest that the sedimentation rates were different in the past or argue that the second layer in the floor was not igneous, but highly compressed sediments. Thus it was always possible to preserve beliefs in global theories. Even the displacements along the fracture zones were interpreted within a "fixist" framework that allowed movement of ocean floor segments without continental movements. One of the few attempts to relate oceanographic results to continental drift theory occurred in the 1949 Mayr Conference on drift theory (Mayr, 1952). Ewing and Walter Bucher summarized the current marine research and concluded that both continental drift and land bridges were implausible and argued that permanence of the ocean basins was the best interpretation of the available evidence. There are important sociological aspects to this research. Menard (1986) emphasizes that the influx of funds encouraged so many expeditions that researchers rarely had time to write up their results. Furthermore, each expedition produced so many new results that older results became outdated. Thus one had to be in an informal network of researchers to keep on top of the field. Two other social aspects followed from the nature of oceanographic research. First, the months-long confinement on ships supported increased cooperation and communication among the specialists involved. These ties helped the integration of research results that later would provide the foundations for plate tectonics. Second, of necessity mostly geophysical techniques were used in oceanographic research. Thus continental geologists tended to ignore the oceanographic results because of the specialization gap between the geologists and the geophysicists and because oceanographic results were seen as largely irrelevant to understanding continental geology (Menard, 1986). Consequently, plate tectonics theory was constructed largely by geophysicists and only later was made relevant to other geological specialties. Even different geophysical specialties varied in their reactions and contributions to the new theory, but some of the key evidence in favor of plate tectonics came from the study of earthquakes.

Seismology

Prior to World War II seismologists used the study of seismic waves generated by earthquakes to refine an "average earth" model that assumed the earth was composed of onionlike layers. This was done by studying the arrival times of the "P" and "S" seismic waves generated by earthquakes. Since the P waves travel faster than the S waves, the difference in arrival times at a seismograph gives the distance to the earthquake. Comparison of the recordings of the same earthquakes at different seismographs allowed researchers to locate the centers of earthquakes on the earth. Such research provided the data used by the Lamont researchers to predict the location of the mid-ocean ridge system by using known earthquake locations. The speed of transmission of P and S waves also depended on the density of the transmitting rocks, and S waves could not pass through fluids. These latter properties were used to identify an inner, fluid core, an outer core, several density layers in the mantle itself, and the "Moho discontinuity," which marked the lower surface of the earth's crust.

After the war there were many advances in seismology arising from better communication systems, standardization of seismographs, more research funds, international geophysical projects, computer aided calculations, and theoretical developments. Some of these developments were prompted by efforts to monitor Soviet underground bomb tests. These developments allowed seismologists to use more of the information recorded by a seismograph, especially the actual pattern of "wiggles" that constituted each type of recorded wave. This information helped seismologists identify more internal structures inside the earth and study the relative movement of the earth's crust along fault zones, which was of great importance to structural geologists.

Hugo Benioff[11] (1954) used this type of information in his study of the zones of intense earthquake activity associated with ocean trenches. Benioff extended prewar Japanese work in the mapping of an inclined zone of earthquake foci associated with trenches and suggested possible geological processes occurring at trenches. Part A of Figure 3.4 shows surface locations of the earthquakes associated with a trench—the cross-hatched area in Figure 3.4-A. The cross-section in Part B shows the depth location of these earthquakes with respect to the trench. It shows that the earthquakes occur in a "Benioff zone" descending down from the trench. Part C of the figure is Benioff's (1954) interpretation of the geological structures producing this inclined zone of earthquakes. He suggested that the upper 700 kilometers of the earth's surface was rigid and trenches marked the location of sloping faults through this rigid surface. The earthquakes were caused by one side of the fault moving against the other side.

Benioff applied a theoretical model of earthquake mechanisms to infer the direction of movement along this fault plane. The model assumed that earthquakes were generated by slippage between two planes of contact and it allowed seismologists to infer from the pattern of wiggles in the seismic waves the direction of slippage along the plane of contact. These "first motion studies" yielded ambiguous results in that two possible directions of slippage would

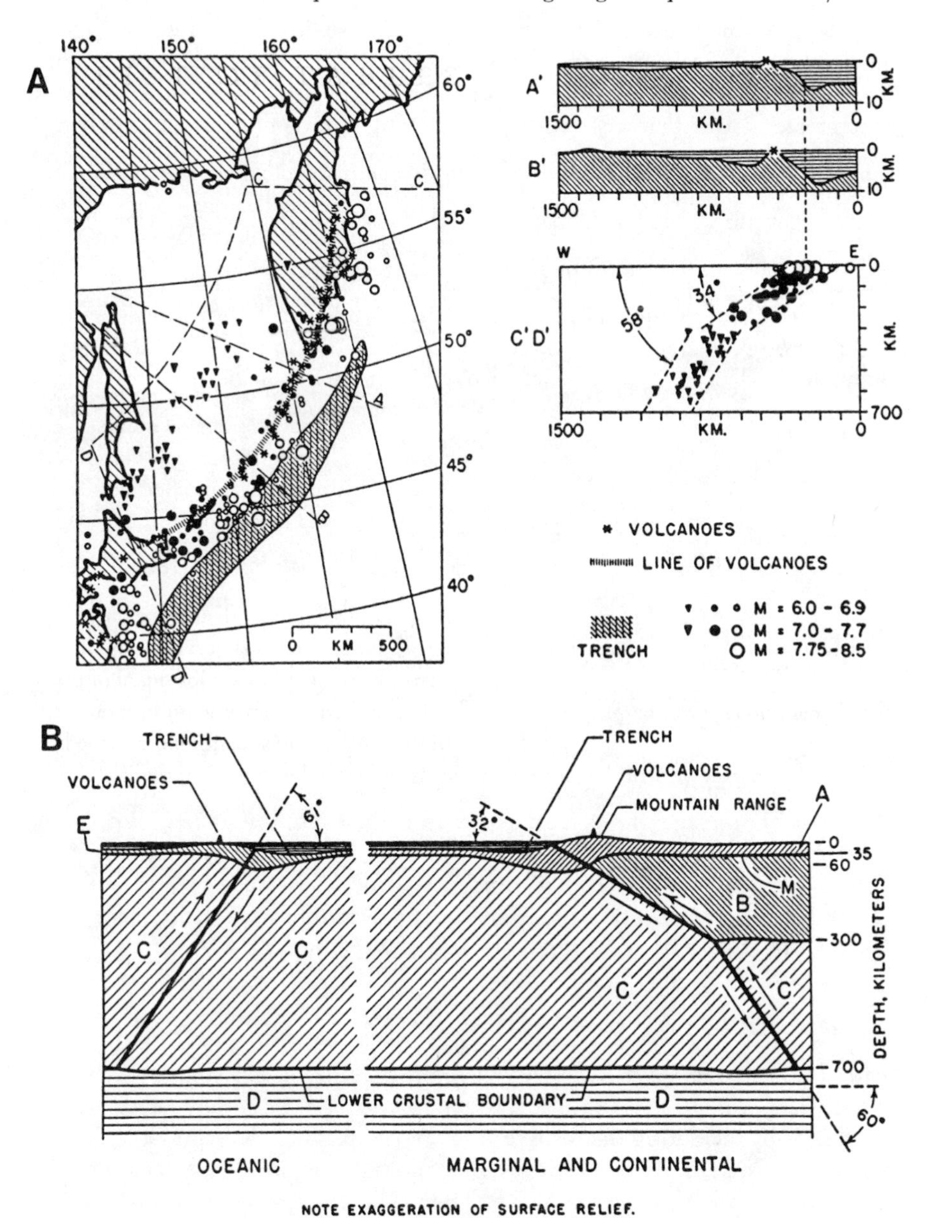

Figure 3.4 Composite diagram from Benioff (1954) showing the location and structural interpretation of earthquakes associated with the Kurile Trench. Part A shows the surface location of the earthquakes, the topography along two lines (A and B) crossing the trench, and a plot of the depths of the earthquakes as a function of the distance from the trench, which defines the "Benioff zone." Part B is Benioff's interpretation of the geological structures and processes (thrust faulting) that might produce the observed pattern of earthquake activity associated with trenches. (From Benioff, *GSA Bulletin* 65 [1954]: 385–400, with permission, Geological Society of America.)

produce the same pattern. For Benioff zones one of these planes was along the inclined plane of earthquakes and was consistent with a thrust fault, where one side of the crust was being thrust under the other side. The other possible interpretation was that a strike-slip fault existed at Benioff zones, where the two sides of the crust were moving horizontally past each other. Initially Benioff (1954) interpreted these zones as thrust faults, but later changed his mind and argued that the whole Pacific basin, which was surrounded by extensive trench systems, must be rotating with strike-slip faulting along its edges. Agreement on the thrust interpretation of the faulting in Benioff zones only occurred later with better data, theoretical models, and field studies of the earthquake areas.

Two other major developments occurred in the 1950s. Press and Ewing summarized data on "surface" waves to make inferences about the structure of the oceanic crust. Surface waves are also generated by earthquakes, but unlike the P and S waves, they only travel through the crust and were of little interest to those trying to determine the earth's internal structure. Press and Ewing showed how these waves contained information about the thickness and composition of the intervening crust. Their studies indicated that the ocean crust was uniformly thin and had a basaltic composition, which generalized the earlier results of the few, very localized seismic refraction studies conducted on oceanographic surveys. The second finding was the identification of a "low-velocity zone" in the upper mantle, where the slower speeds of P and S waves indicated a zone of lower density than those above and below it. This zone was about sixty miles below the earth's surface.

These developments in seismology played a major role in the evidence for plate tectonics in the late 1960s. Particularly important was the ability to infer direction of slippage along fault zones from first-motion studies. But again, most studies in seismology focused on limited subjects, such as delineating the depths to the different layers in the earth's interior, and were not seen as relevant to the issue of continental drift. The same lack of interest in drift theory was apparent in studies of geochronology and geosynclines.

Geochronology

After the discovery of radioactivity at the beginning of this century some geologists, especially Arthur Holmes, began to calculate the absolute ages of rock specimens by measuring the relative amounts of "parent" elements and "daughter" decay products. Each radioactive element spontaneously decays at a constant rate into specific lighter elements or isotopes. The first applications of this means of dating used the decay of certain uranium isotopes to lead isotopes and calculated the absolute ages of geological eras and even the absolute age of the earth. During the first half of this century many problems were solved. For example, the decay rates and decay products were studied and better instruments were developed. However, rocks without uranium could not be dated, nor could rocks younger than a few million years because of the long half-life of uranium.

The fifties saw the development of a technique based on the decay of potassium isotopes to argon isotopes. Since this decay had a short half-life and potassium was a very common element, the potassium-argon technique allowed the dating of a wider diversity of rocks that were less than a few million years old. One of the key developments was the improvement in the mass spectrometer for separating the minute amounts of the relevant isotopes. The establishment of the magnetic reversal chronology, which will be described shortly, relied heavily upon this particular advance in geochronology (Glen, 1982). One of the major applications of geochronology was the mapping of the ages of different parts of the continents.

Geosyncline Theory and Continental Accretion

The geosyncline theory for mountain formation was a key element in geological thought because it united several different geological specialties during its development in this century. As noted earlier, the most studied mountains—the Alps and Appalachians—were interpreted as the buckled sediments of geosynclines,[12] which were long troughs that accumulated sediments before the sediments were compressed and raised into mountains. By the 1950s this concept was widely accepted and geologists sought the present day geosynclines that would be the basis of future mountain ranges. One of the popular views argued that ocean trenches ("tectogenes") were modern geosynclines: they were depressed troughs, seismically active, associated with volcanism, and located along continental areas or island arcs, which could provide a source of sediments. However, as noted previously, Lamont researchers thought this view was inaccurate.

Despite the differences in opinions, the geosyncline concept was the most "global" of the theories influencing researchers and it united the work of a number of specialists. By the 1950s the combined work of structural geologists and petrologists—those who study the origins of various rock types—suggested that geosynclines consisted of two parallel troughs separated by a ridge. This distinction was based on the different types of rocks and the patterns of metamorphism—changes due to heat and pressure—found in the different troughs. In particular, the outer trough often contained a distinctive assemblage of rocks, the "ophiolite suite," which consisted of igneous rocks high in iron and manganese overlaid by deepsea sediments. Furthermore, geochemical and petrological studies on the origins and metamorphism of rocks occurring in mountains indicated that the compression of a geosyncline followed a certain pattern: intrusions of molten rocks were more common in the outer trough and the sediments in the outer trough were folded and compressed first.

Although the geosyncline model related structural geology with several other geoscience specialties, there were considerable disagreements on the causes of the subsiding troughs, the sources of the ophiolites, and the origins of the compressive forces, not to mention numerous elaborations of secondary concepts as the basic geosyncline model was applied to different settings (Aubouin, 1965). For example, Russian geologists emphasized vertical forces and western geologists advocated horizontal forces as the major forces producing geosynclines and the resulting mountains. Other national differences in opinions had diminished, e.g., most geologists accepted that the geosyncline model should be used to understand the formation of mountains and most thought that continents grew in size by the addition of mountain ranges on their edges.

This idea of "continental accretion" was another global aspect of the geosyncline model. From the earliest days of the geosyncline theory it had been tied to the idea that continents grew in size as geosynclines added mountain ranges to their edges. That is, continents seemed to have an older core or "craton" with progressively younger belts toward the edges. Better evidence for this general structure of continents had developed by the 1950s and was aided by the improvements in geochronology mentioned in the previous section (Wyllie, 1971; Dott, 1979; Aubouin, 1965). Yet geologists studying mountains or continental structure seldom proposed continental drift theory as the cause of the horizontal compression of geosynclines. Explanations based on contraction theory or mantle convection currents were more common, but more often the causes of horizontal compression were simply neglected. In the 1950s continental drift theory was supported most actively by a few geologists in the southern hemisphere and by researchers in the geophysical specialty of paleomagnetism, the study of the earth's past magnetic field.

Paleomagnetism

During the period from 1945 to 1959 the study of the earth's past magnetic field provided the evidence that eventually persuaded some geophysicists to accept continental drift. Before describing the development and interpretation of this evidence, several features of the earth's magnetic field must be reviewed. First, the magnetic field is quite weak—about one Gauss in intensity, whereas a toy magnet may have a strength of ten Gauss or more. Second, the earth's field is dipolar with the magnetic poles closely approximated by the geographic poles of physical rotation. Third, the directions of the lines of magnetic force at any point on the earth's surface can be described by two components: the *declination*, which measures the *horizontal deviation* of the North end of a compass needle from the North geographic pole, and an *inclination*, which measures the *vertical deviation* or dip from the horizontal. The inclination is zero at the magnetic equator and ninety degrees at the magnetic poles. Finally, the field is not uniform over the surface of the earth, nor is it stable over time. Small daily changes occur, and over hundreds of years the declination and inclination of the same spot on the surface of the earth may change as the magnetic poles fluctuate around the geographic poles.

Although geoscientists did not know the causes of the magnetic field, they were aware that some rocks could form a natural remanent magnetism (NRM) that recorded the inclination and declination of the magnetic field when the rock was formed. This occurred when molten rock cooled below the Curie Point (about 500 degrees Centigrade). Sedimentary rocks also retained a weaker

NRM. It was also known that chemical and physical alteration of rocks could induce a "secondary" NRM recording the magnetic field at the time of the alteration.

Two aspects of NRM played a major role in the evidence for continental drift and plate tectonics. The most important studies focused on the magnetic reversals in the NRM of rocks, but this area of research did not develop until the late 1950s and was not seen as relevant to continental drift until 1963. The other area of research focused on the directions of the NRM preserved in rocks. These studies developed in the early 1950s and were being related to continental drift theory by 1954.

DIRECTIONAL STUDIES OF NATURAL REMANENT MAGNETISM

Since the NRM in a rock recorded the inclination and declination of the earth's field at the time of its formation, it was possible to measure this information in a rock sample of a known age and compare the NRM inclination with the present day inclination of the magnetic field at the rock's location. If these two inclinations did not coincide, then either the magnetic pole was at a different position at the time of the rock's formation *or* the rock had formed at a different location on the earth's surface and continental drift had occurred.

These two possible interpretations are shown in Figure 3.5. Part A of the figure shows the earth's present magnetic field lines with arrows indicating the inclination in the field at different latitudes on the surface of the earth. The rectangular figure in the right of Part A indicates a situation that might be found at the boxed "NH" point on the earth to the left. Here the inclination in the NRM of a 125-million-year-old rock sample (the lines in the ground below the stick figure) do not correspond to the current magnetic field inclination for the rock's present latitude (the lines above the stick figure). Parts B and C in the figure illustrate the two alternative interpretations of the differences in these two magnetic inclinations. In Part B it is assumed that 125 million years ago the magnetic poles were at a different location. In this "polar wandering"[13] explanation the NRM inclination of the rock is consistent with the 125-million-year-old magnetic field inclinations. In other words, the magnetic pole had moved and the continent containing the rock sample had not moved.

The continental drift interpretation is given in Part C of Figure 3.5. In this explanation the magnetic poles are assumed to have always been near the geographic poles and the 125 million year old rock is assumed to have formed in the southern hemisphere, at point "SH," where its NRM inclination is consistent with the past (and present day) magnetic inclinations. Then during the 125 million years since its formation, the rock's continent moved North to point "NH," while the earth's magnetic field pole position was essentially constant. By the end of the 1950s paleomagnetic researchers doing directional studies agreed upon the continental drift interpretation, but most other geoscientists either doubted the quality of the paleomagnetic evidence or preferred the polar wandering interpretation. The first paleomagnetic group to argue for conti-

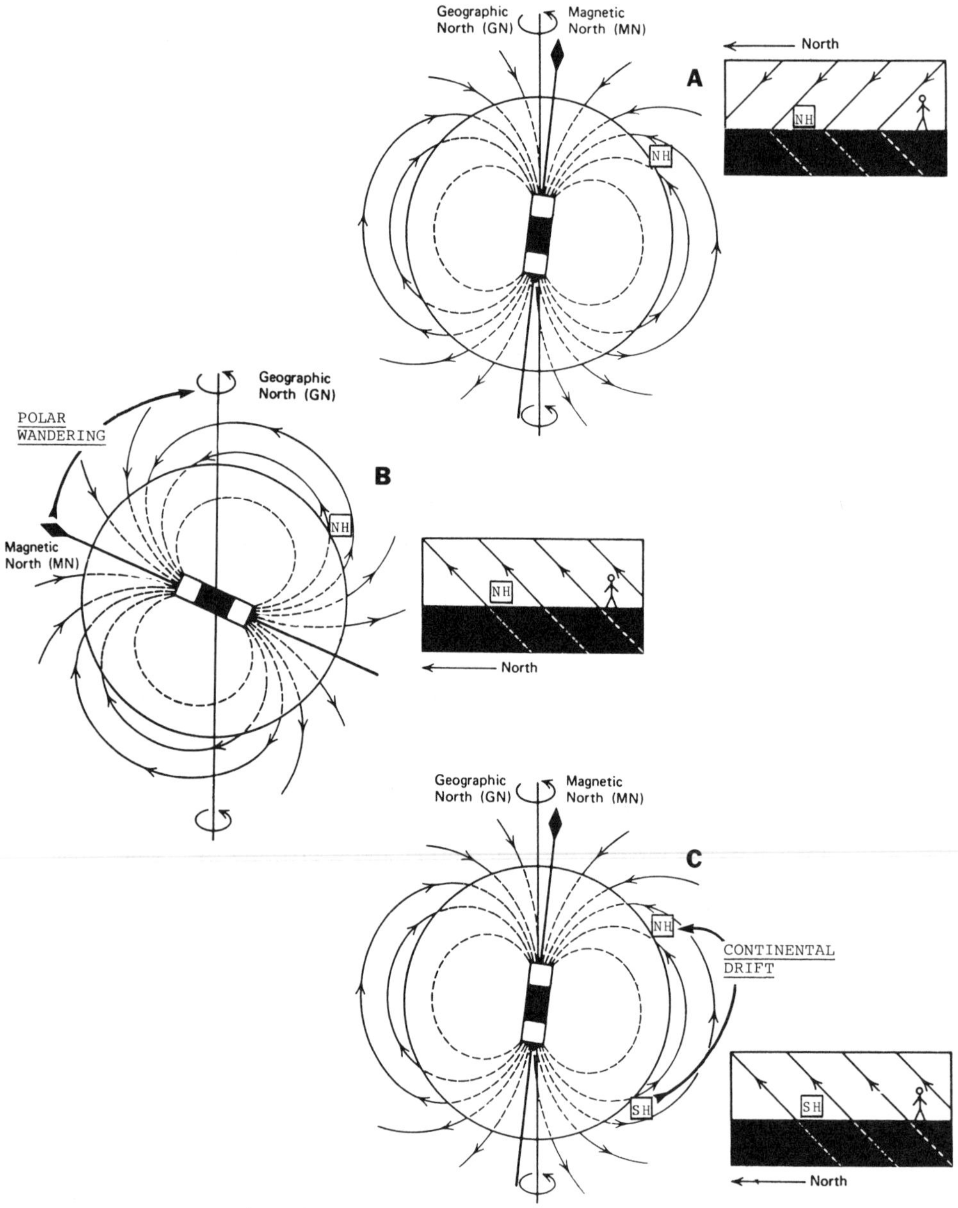

Figure 3.5 The polar wandering and the continental drift interpretations of a discrepancy between the earth's present magnetic field inclination and the inclination recorded in the natural remanent magnetism (NRM) of a rock sample. Part A shows the inclination of the present magnetic field of the earth and how it does not agree with the inclination recorded by the NRM in the ground. Part B shows the "polar wandering" interpretation of this difference, in which the magnetic pole was at a different position at the time of rock formation. Part C shows the continental drift interpretation, in which the magnetic pole has always been close to the geographic pole of rotation and the continent has drifted since the time of the rock's creation. (Adapted from Wyllie, *How the Earth Works*, with permission, copyright © 1976 by John Wiley & Sons.)

nental drift was headed by Patrick Blackett,[14] a British physicist and Nobel Laureate.

Blackett had hypothesized that the earth's magnetic field arose from an unknown law of physics requiring any rotating mass to produce a magnetic field. He developed the very sensitive "astatic magnetometer" to test his hypothesis in the laboratory, but found no support for it. Since his magnetometer was sensitive enough to detect the very weak NRM in sedimentary rocks, he soon had his graduate students collecting such data. This greatly facilitated the study of paleomagnetism because sedimentary rocks were more common and contained fossils, which permitted easier dating.

In 1954 Blackett's group (Clegg, Almond, and Stubbs, 1954) provided evidence from paleomagnetic studies of Triassic sandstones that England had rotated and drifted to the north. Later studies by Blackett's group indicated that India had been in the southern hemisphere in the Jurassic period and had drifted north to its present location in the northern hemisphere, as represented in Part C of Figure 3.5. This involved a movement of over 4,000 miles in about 125 million years and provided support for Wegener's explanations of the distribution of the Permo-Carboniferous glaciers. Blackett's group interpreted these results as the first quantitative evidence for continental drift.

This interpretation was soon challenged by a competing paleomagnetics research group under S. Keith Runcorn[15] at the University of Newcastle. Runcorn, who was a former student and colleague of Jeffreys, pointed out that Blackett's group implicitly assumed the magnetic poles were always located near the geographic poles, whereas polar wandering was a plausible alternative explanation. This interpretation, however, implied that the NRM of the rocks on *all* of the continents should show the identical pattern of magnetic pole positions over the same geological ages. That is, the continents should have the same "polar wandering curves."[16] Runcorn tested this explanation by gathering data from North America and plotting its polar wandering curve with the one for England, see Figure 3.6. Continental drift would imply that the curves would diverge, whereas the polar wandering explanation implies that the lines for separate continents should be identical. At first, Runcorn (1956*a*) considered the systematic difference between these two curves to be measurement error, but then changed his mind (Runcorn, 1956*b*) and interpreted the curve to show the drifting apart of the two continents. This interpretation was again in general agreement with Wegener's predictions.

Despite the fact that these two groups agreed that some form of continental drift had occurred, possibly with some polar wandering, their conclusions were challenged by others. Some geologists noted that both groups were assuming that the earth's field had always been dipolar, whereas it was possible that a multipole field had existed in the past. Furthermore, Runcorn's evidence for the east-west separation of the continents was based on changes in the *declinations* of the NRM, and rotations of the continents in place without drift could produce the separation in his two polar wandering curves. (Only differences in *inclinations* were unambiguous evidence for drift, *provided* one accepted that the magnetic

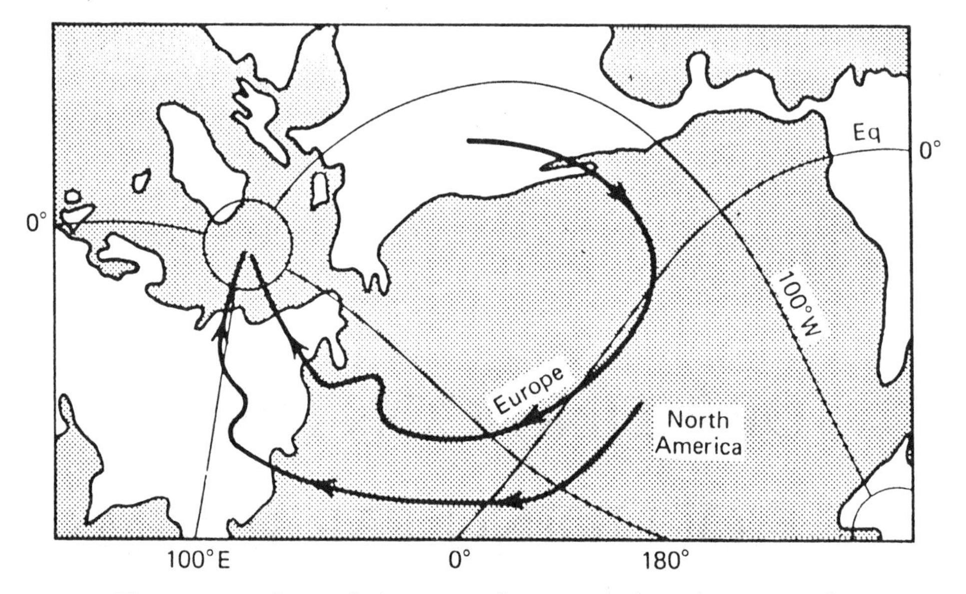

Figure 3.6 "Polar wandering curves" for Europe and North America. The curve for each continent is constructed by connecting together average paleomagnetic pole positions for rocks of different ages. The arrows on the curves indicate locations at more recent geological periods, ending with the paleomagnetic poles of both continents located at the present North Pole. (From Wyllie, *The Dynamic Earth*, with permission, copyright © 1971 by John Wiley & Sons. The figure is a simplification from Runcorn [1962: 23]).

field had always been dipolar with the poles located near the geographic poles.) Others contested the handling of the data because any rocks of the same age and from the same location showed considerable variation in the calculated pole positions. Paleomagneticists used statistical averages of these positions, but skeptics did not trust this technique and wondered if the average positions for different ages were significantly different from each other. Skeptics wondered how the NRM magnetism in a rock was affected by being drilled and hammered during the collection of samples or by the "cleaning" techniques used to remove the "secondary" NRM (Takeuchi, Uyeda, and Kanamori, 1967).

In recognition of these problems, the paleomagneticists developed increasingly careful and standard procedures of data collection and sought paleoclimatic support for their reconstructions of past continental positions. However, hardly any geologists accepted or even understood the evidence and only a few geophysicists accepted continental drift because of it. Even though this new line of evidence for drift did not rally many geoscientists to drift theory, it probably helped its cause for several reasons. Some of those persuaded by the evidence were esteemed scientists, such as Bullard and Blackett. Furthermore, it was quantitative evidence. Finally, paleomagnetism became a specialty within geophysics and this provided an organizational basis from which they recruited new members, gave invited lectures, and provided recognition for the contributions of its members. Blackett and Runcorn gave many invited talks in the United States, and paleomagneticists were important figures in sponsoring continental drift conferences that incorporated other new evidence. Ultimately, however, it was the paleomagnetic studies of the magnetic field reversals that provided the critical support for seafloor spreading in the 1960s.

MAGNETIC REVERSAL STUDIES

In the 1920s a Japanese geoscientist, Matuyama, noted that some volcanic lava flows had a magnetic declination that was the reverse of the present field and of the lava flows beneath and above them. That is, the declination of the NRM was reversed so that the North pole in the rock's NRM pointed toward the earth's South pole. Two possible interpretations were proposed. Either the earth's entire field had reversed its magnetic poles for a period in the past or the reversed rocks were composed of particular minerals that spontaneously reversed their polarity as they cooled. The latter explanation was used to explain reversed NRM until the late 1950s.[17] In the late fifties, a few geophysicists started to study the pattern of reversals in lavas only a few million years old. It became clear that self-reversals were rare and most rocks with reversals must have formed when the earth's field was actually reversed. This interpretation was favored by the growing evidence that throughout the world all lavas of the same age had the same reversal pattern.

Once total reversals of the earth's magnetic field became accepted, the next step was the construction of a chronology of field reversals by studying the NRM in lavas of different ages. By the end of the fifties the most important workers in this area were Richard Doell[18] and Allan Cox[19] at the U.S. Geological Survey in Menlo Park. They had an important advantage over other researchers because they had access to the most accurate method of dating rocks less than four million years old—the improved Potassium-Argon method, which had been developed at Berkeley by one of Cox's professors. Glen (1982) provides a detailed history of the magnetic reversal studies.

Before turning to the attempts at global theories incorporating the specialty studies described in this chapter, it must be emphasized that the study of magnetic reversals proceeded independently of the study of linear magnetic anomalies in the ocean basins. Only in 1963 would these two areas of research be related together as a means of testing the seafloor spreading model.

Global Theories

In the period after World War II few disciplines grew as fast as geophysics and oceanography. Fewer could match the unexpected results from oceanographic research mentioned earlier. To those familiar with these results it was clear that previous global theories were obsolete if they ignored or could not be related to this new evidence from the oceans. However, few global models were proposed and actively considered after the war. Contraction theory remained the implicit model for most geologists, but was only relevant as the force behind the less global geosyncline model. Drift theory was a global model, but it lacked an accepted mechanism and most research related to drift theory was concerned with providing additional evidence for drift (e.g., paleomagnetic evidence), not with developing the theory itself.

Two southern hemisphere geologists, S. Warren Carey[20] at the University of Tasmania in Australia and Lester King[21] in South Africa, strongly defended drift theory. Du Toit had died in 1949, but King was a former student and continued to write on drift and occasionally gave lectures in the United States in the fifties, as did Carey (Marvin, 1973). Carey was perhaps the most outspoken geologist with a new global theory that implied drift of the continents.

In contrast to most geologists of the fifties, Carey returned to the earlier style of geology and looked for the major structural features of the earth's surface, such as the relationships between the major mountain belts and the fit of the continents. He proposed new types of rifts to explain the origins of seas between continental fragments and argued that closing the rifts would fit the continents together and straighten many of the prominent bends in mountain chains. He used a small model of the earth and its continents to produce accurate alignments of continental edges, such as Figure 3.7 showing the alignment of South America and Africa. He argued that the fits between individual continental fragments were adequate, but when he tried to combine them to form Pangaea, he "could never put it all together on a globe, or a rigorous projection. [He] could reconstruct satisfactorily any sector [he] might choose but never the whole" (Carey, 1958*b*: 316). Only when he abandoned the common assumption of a relatively constant size of the earth would his reconstructions fit together. He argued that the continents once formed a continuous sial crust on a smaller earth. Consequently, he held the earth had been expanding, not contracting, through geological time. As the earth expanded to its present size the sial crust was fractured and ocean basins formed as the mantle material welled up to fill the cracks. The mid-ocean ridges marked the present location of crustal tension.

Carey's theory was global and encompassed much of the recent evidence from oceanography and geophysics that had been collected from 1945 to the late 1950s. He related his theory to the mid-ocean ridge system, earthquake location data, and paleomagnetism. The geosyncline theory, which previously was based on contraction theory, was recast in the context of an expanding earth. Geosynclines were now sediment filled rifts with subsequent folding and uplift of the sediments caused by increased flow of magma under the sediments. He was not sure what caused the expansion of the earth, but suggested it might be caused by the radioactive heating of the earth's interior or perhaps a decrease in the gravitational constant over time—a possibility proposed by some prominent physicists. But he was tentative about these possible explanations and argued the evidence was so strong for expansion that it was up to the physicists to figure out why it happened.

There were several reasons for Carey's influence on other geologists. First,

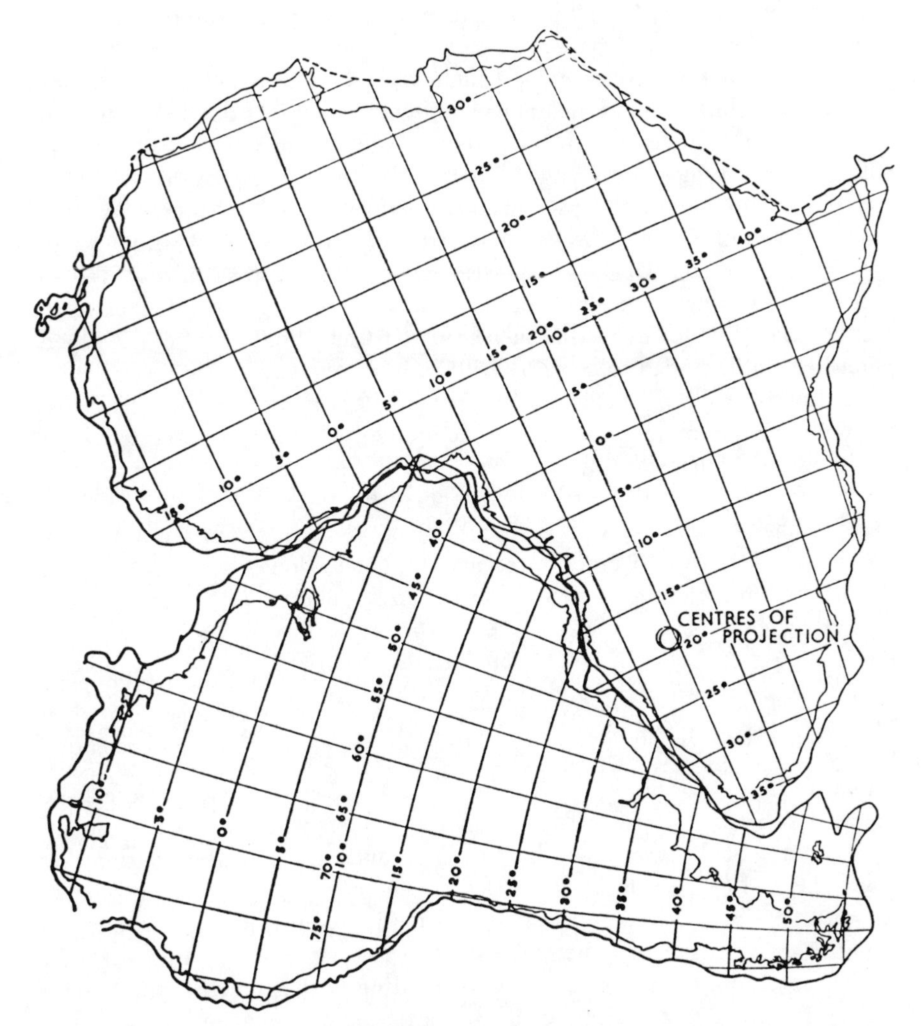

Figure 3.7 Carey's reconstruction of the fit between South America and Africa at the 2000 meter isobath along the continental slopes. (From Carey [1958], published earlier in Carey [1955], with permission, Cambridge University Press.)

he renewed interest in a global perspective (Hallam, 1973). Second, he was a dynamic speaker and his lectures in the United States were influential, especially for students. As one Harvard geologist said, "Carey could sell refrigerators to Eskimos." Finally, he sponsored a symposium on continental drift in 1956, which brought together many of the proponents and their new evidence for drift.

At this symposium Lester King presented two papers and Edward Irving,[22] a recent member of Runcorn's group, presented recent paleomagnetic evidence. The principal guest was Chester Longwell, who had attended the AAPG symposium on drift thirty years earlier. Longwell gave the opening presentation and maintained his skeptical neutralism, but was "convinced that the arguments on continental drift have been a wholesome stimulus in earth science." In an epilogue to the symposium Longwell noted the good fit between the continents and the importance of the paleomagnetic evidence, and even suggested the "radical" hypothesis of an expanding earth may prove a useful stimulus. Altogether about fifteen papers were presented, but Carey's paper constituted about half of the resulting symposium volume (Carey, 1958*a*). It is said (Wilson, 1974) that because of the controversial subject matter Carey could not get the papers published and they had to be distributed in mimeographed form.

Of course, the proponents of continental drift were not the only geologists who attempted to build global models incorporating the new evidence provided in the fifties. Perhaps the most global theory relating the new evidence to a contraction model of the earth was proposed by J. Tuzo Wilson[23] (Wilson, 1959; Jacobs, Russell, and Wilson, 1959). He proposed that contraction was not due to heat loss, but to eruption of mantle material onto the surface of the earth, which reduced the volume below the crust and caused it to contract. The Mohorovicic discontinuity was viewed as the original surface of the earth and the volume of rock above it was erupted mantle material. Since the composition of the continents was quite different from the ocean basins, Wilson suggested that there were two chemically different layers in the mantle. The upper layer was basaltic and the deeper layer was "andesitic"—a type of granite (sial) and similar to the composition of the continents. The Benioff zones were held to be faults penetrating into the deeper andesitic layer, which explained the andesitic nature of the associated island arcs. In contrast, the shallow earthquakes along the mid-ocean ridges marked faults penetrating to the basaltic layer in the mantle. Thus, the ocean basins were basaltic. As mantle material extrudes along either of these fault systems, crustal contraction occurs and this shrinking of the earth's crust is taken up by thrust faulting at the trenches (the Benioff zones). He noted that the surface of the earth could be broken into segments whose edges were marked by the interconnections between these various fault systems, where most geological activity took place.

Although Wilson attempted to incorporate into his theory much of the postwar research results, he considered the paleomagnetic evidence for drift ambiguous and only indicative of polar wandering. However, as we will see in the next chapter, within two years Wilson declared himself a "reformed anti-drifter" and made major contributions to seafloor spreading and plate tectonics. But only after he briefly considered the expansion hypothesis. He would abandon his contraction model, but retain his global view of the fault systems breaking the earth's crust into segments, which eventually became identified as plates.

Discussion and Conclusions

Although the geological research after the war tended to focus on detailed empirical studies with less interest in global theories, these studies still produced a number of dramatic results. One might think that these results were "anomalies" for the accepted fixist or contractionist "paradigm" and should have generated a "crisis" in the geosciences. This did not seem to happen. Only one interviewed person, Bruce Heezen, seemed to have a personal sense of intellectual crisis as a result of these studies. In 1974 he recalled his earlier dilemma.

> We can't explain anything. We can't analyze our data; we can't write our papers because we can't decide whether the basins of the oceans are formed by drift or whether they are permanent, and if you can't decide that, you can't do anything. So you have got to make a decision—just get going. So I said, "Okay, I'm going to decide for drift and go that way, but I've got to be very careful. I can't call it 'drift' because if it was called 'drift' I'd be in trouble." So I called it "continental displacement."

Heezen noted that he used the expansion version of drift because at that time geologists, especially his colleagues at Lamont, had not accepted any form of crustal shortening, say at trenches.

Unlike Heezen most marine geologists did not use drift theory to interpret these results, especially in North America, where geology students were unlikely even to hear about drift theory before graduate school, and even then it was presented as an unlikely hypothesis (Dott, 1981). Advocates of drift were likely to get into the "trouble" mentioned by Heezen, who described one example of a scientist who published a "drift" paper and could not find academic employment as a result. In fact, Heezen's endorsement of expansion apparently created the initial rift between him and Ewing, which later resulted in his loss of access to Lamont data and ship time (Menard, 1986: 106). Later chapters will describe this loss and provide other examples of the hostility toward drift theory.

Even though the results of these studies were dramatic, they did not generate a lot of interest. Table 3.2 summarizes the number of citations to selected articles from the 1950s. The major articles by the paleomagneticists (e.g., Clegg, Almond, and Stubbs, 1954; Runcorn, 1956*b*), which supported continental movements, were never cited very highly. The Tasmanian symposium (Carey, 1958*a*), which presented evidence for both drift theory and the expanding earth theory, created a little interest in the early 1960s, but more in the late 1960s, as did the oceanic studies summarized by Heezen, Tharp, and Ewing (1959). Bullard, Maxwell, and Revelle's (1956) studies of heat flow in the oceans reached its peak in the middle 1960s, as did the studies of the linear magnetic anomalies in the ocean basins (Mason, 1958; Mason and Raff, 1961) and the seismic studies of the Benioff zones associated with trenches (Benioff, 1954). Interest in Wilson's (1959) contraction theory was minimal, perhaps this was because there was little interest in global theories in general and because he quickly changed his mind and accepted continental drift (Allègre, 1988: 62).

TABLE 3.2 Citation Counts to Selected Geoscience Publications from 1950 to 1960

	Science Citation Index citation counts for:															
Publication / Topic	1955	'56	'57	'58	'59	'60	'61	'62	'63	'64	'65	'66	'67	'68	'69	'70
Clegg, Almond, and Stubbs (1954) / Paleomagnetic evidence for drift	5	5	12	5	3	5	2	5	2	1	1	2	4	0	1	1
Runcorn (1956*b*) / Paleomagnetic evidence for drift	—	—	3	0	1	2	0	1	0	1	0	0	0	0	3	1
Carey (1958) / Tasmanian Conference on Drift	—	—	—	—	0	3	12	6	10	9	13	10	13	28	22	31
Heezen, Tharp, and Ewing (1959) / Topography and structure of ocean basins	—	—	—	—	1	6	8	13	12	12	17	22	24	17	22	27
Bullard, Maxwell, and Revelle (1956) / Heat flow in the oceans	—	—	0	2	4	4	2	3	6	6	7	4	9	4	2	5
Mason (1958) / Linear magnetic anomalies	—	—	—	—	3	4	5	3	1	3	3	4	3	6	4	1
Mason and Raff (1961) / Linear magnetic anomalies	—	—	—	—	—	—	1	0	2	5	7	7	5	10	12	10
Benioff (1954) / Trenches and Benioff Zones	0	2	4	0	0	1	3	2	1	1	2	2	6	7	7	9
Wilson (1959) / A contraction model for the Earth	—	—	—	—	1	3	3	0	3	4	3	0	0	0	2	2

NOTE: Citation counts exclude self-citations. The source is Institute for Scientific Information (1984, 1972).

This chapter has covered a very selected portion of the developments in the earth sciences after World War II, but it includes most of the important ones related to continental drift. Other areas have been mentioned because they are necessary for understanding the developments in the 1960s. It is clear that the research conducted after the war produced a number of specific findings that geologists had not expected. Yet these findings did not produce a "crisis." Verhoogen suggests that more important was a "drastic change" toward a "more quantitative, more analytical attitude" that resulted in a quantification of geology and a realization that mantle processes were involved in crustal geology (Verhoogen, 1983: 4–5).

Some geoscientists, such as Ewing, realized that these changes would produce a revolutionary change in our view of the earth. When the new view developed, however, it encountered strong resistance from many geologists, including Ewing, because it implied that some form of continental drift had occurred.

CHAPTER FOUR

Plate Tectonics: Its Origins, Development, and Opponents

The first three sections of this chapter summarize major publications for (a) 1960–1963, which cover the development and modification of the "seafloor spreading" model, (b) 1964–1966, which included persuasive, quantitative evidence for seafloor spreading, and (c) 1967–1970, which cover the development and testing of the simple "plate tectonics" model. The fourth section mentions some later developments in the theory that increased its appeal to other geoscience specialties. The final section describes some of the published reactions of those opposing plate tectonics. This chapter concludes the presentation of the public record "data" on this revolution. Subsequent chapters compare the different perspectives on science, which were outlined in Chapter 1, to this history and the information from personal interviews.

The Seafloor Spreading Hypothesis: 1960–1963

In 1961 the hypothesis of seafloor spreading was introduced by Robert Dietz,[1] but it appears that his article (Dietz, 1961) was influenced by a presentation given by Harry Hess[2] in 1960. Hess's article, "The History of the Ocean Basins," was part of a grant report (Hess, 1960), and was published later in the scientific literature (Hess, 1962). Since Hess's priority is acknowledged, even by Dietz (1968), and the two articles are similar in many respects,[3] the discussion of the 1960–1963 period begins with a description of how Hess attempted to integrate the research results from the previous decades.

Hess suggested that the recent oceanographic research required the complete abandonment of earlier theories of ocean basin development, but cautiously labelled his attempt at a synthesis as "geopoetry." He proposed that after the primitive earth aggregated from the gravitational collapse of cold particles and melted from the heat of radioactive decay, a single convection cycle caused differentiation of the basic layers in the earth. This convective overturn of the earth's materials formed the earth's core and floated most of the lighter sial (or granitic) materials to the surface to form a single large continent, as shown at the top of Part A in Figure 4.1. Subsequent convection affecting the earth's surface was confined to the mantle layer above the core as shown in Part B of Figure 4.1. Hess assumed that the upper surfaces of the present convection cells in the mantle were represented by the ocean basement and that convection cells rose at the mid-ocean ridges and descended at the trenches. As the upper surface

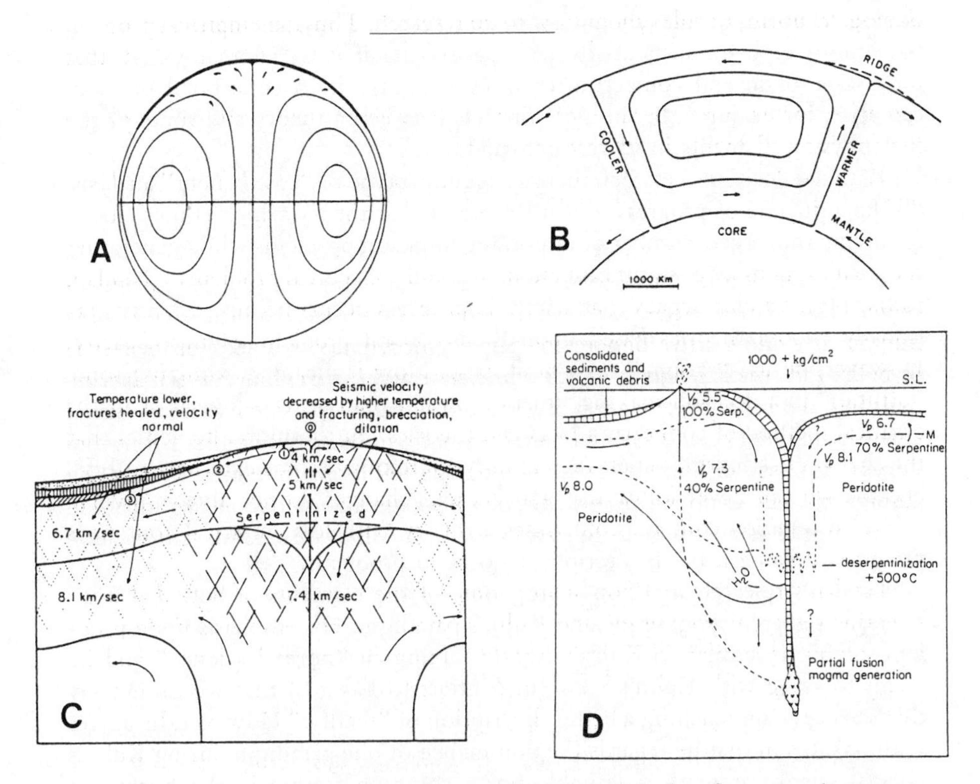

Figure 4.1 Various aspects of Hess's (1962) model of seafloor spreading. Diagram A represents the results from the first convection cycle in the early history of the earth, during which the continental (less dense) material separated from the mantle and core material. Diagram B illustrates the proposed model for present day convection, where only the mantle is convecting and the upwelling convection limbs correspond to mid-ocean ridges and the downwelling limbs to oceanic trenches. Diagram C provides more detail for the processes occurring at ridges. The seismic velocity is slower below the ridge because of the increased temperature and fractured material. Note the increasing accumulation of sediments with distance from the ridge crest. Diagram D is Fisher and Hess's (1963) model for the processes occurring at trenches. The island arc volcanos are formed as the sediments on the mantle convection cells are melted at depth. Note that the mantle material movement is essentially vertical or horizontal without any structures corresponding to the inclined Benioff Zone that extends from the trench to below the volcanos. (Diagrams A, B, and C are from Hess [1962]: 599–620 in Engel, James, and Leonard [eds.] *Petrologic Studies: A Volume in Honor of A.F. Buddington*, with permission, Geological Society of America. Diagram D is from Fisher and Hess in M. Hill [ed.] *The Sea, Vol. 3*, with permission, copyright © 1963 by Wiley Interscience.)

of these cells moved away from the ridges they accumulated more sediments, lost their heat, and carried along any continents embedded in their surface. The continents did not plow their way through the ocean crust, but were passive rafts floating on top of a convecting, denser medium. Since continents were less dense than the mantle material, they would not descend into the mantle at the trenches, but their leading edges would be deformed into mountain ranges parallel to the trenches.

Hess suggested that whole "realms of previous unrelated facts fall into a regular pattern" with this use of the convection current model. His model implied that the ocean basins were not very old since they were being continually formed and destroyed, which explained the thinner sediments at the ridges (Part C of Figure 4.1) and the failure to find any material from the oceans that was older than 180 million years. Furthermore, paleomagnetic studies provided new evidence for drift. The observed increase in oceanic heat flow nearer the ridges was consistent with newer and hotter ocean floor at the ridges. The negative gravity anomaly over trenches was consistent with the idea that the ocean crust was being pulled down into the mantle (Part D of Figure 4.1). Finally, the lower seismic velocity under the ridges suggested the presence of hotter, less dense material rising from the mantle.

Hess described how his model could be consistent with other results and offered testable implications. For example, he suggested that guyots—the flat-topped seamounts—were old volcanos that had formed on mid-ocean ridges and reached sea level. Subsequently, sea waves eroded the volcanos to a flat top and the spreading seafloor eventually carried the flat top island down the flanks of the ridges. He noted that his theory predicted that islands and sea mounts further away from the ridges should be older. Although he noted that Carey's expanding earth hypothesis (see Chapter 3) was consistent with much of this evidence, he argued that it lacked a plausible mechanism and was inconsistent with the geological evidence for a relatively constant sea level over the last few hundred million years.

In the 1962 paper Hess did not propose a model for the processes occurring at the trenches, but did so in a 1963 paper (Fisher and Hess, 1963); see Part D of Figure 4.1. In this model the trenches formed where the convecting mantle material descended back into the mantle, but now it carried a veneer of sediments. As it descended into higher temperatures and pressures the old, hydrated mantle material reverted to its denser form and the sediments melted to form a less dense andesitic magma—melted rock similar to the sial or granite of the continents. This less dense magma rose to form the andesitic volcanos in the island arcs that were associated with trenches.[4] It is interesting to note that Fisher and Hess (1963) described the association between trenches and the inclined Benioff zones, but did not give a representation of these zones in their trench diagram. Chapter 5 considers possible reasons for this omission.

Although few geoscientists considered seafloor spreading a plausible model for the origin of ocean basins, J. Tuzo Wilson was one of the first to modify it and test some of its implications. In 1959 Wilson had proposed the contraction model described at the end of Chapter 3, but shortly after Dietz (1961) proposed seafloor spreading Wilson (1961) defended it and even proposed that a modest expansion of the earth could have produced the mid-ocean ridges and their flanks, which was much less than the one-third expansion proposed by Carey. In 1963 Wilson published four articles that not only elaborated the seafloor spreading model, but tested its implications and applied it to new phenomena.

First, Wilson (1963*a*) followed Hess's suggestion that islands further away from the ridges should be older than those closer to the ridges. Figure 4.2 shows that the ages of islands increase with the distance from the ridges, with an average spreading rate of two centimeters (less than one inch) per year. In two other articles Wilson (1963*b*, 1963*c*) modified the seafloor spreading model by proposing that there was a rigid layer over the convection currents, see Figure 4.3. The lower boundary of this layer was not defined by the Moho discontinuity, but by the recently discovered "asthenosphere"—the low-velocity zone deeper in the mantle that seismologists had identified in the late 1950s. The region above the asthenosphere was called the "lithosphere" and was about sixty miles thick. It included the crust defined by the Moho and part of the

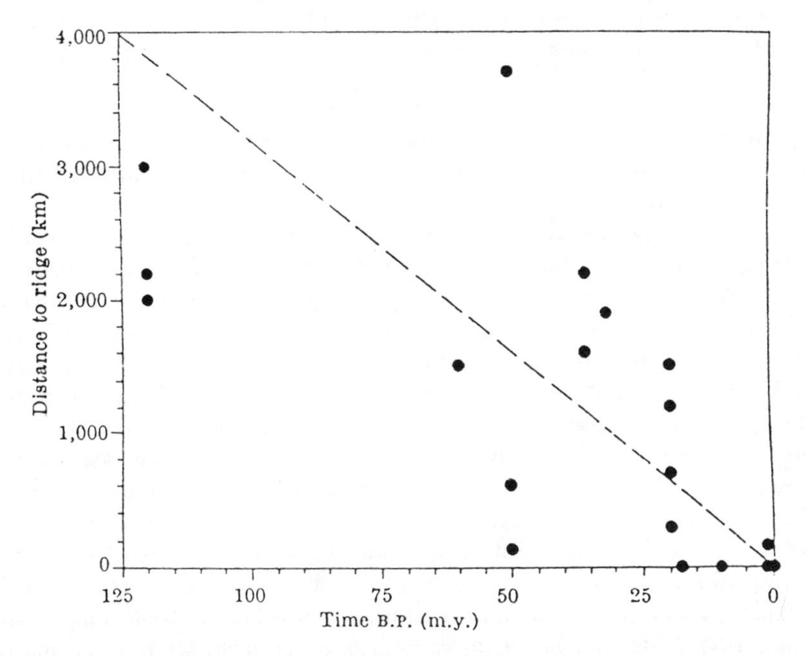

Figure 4.2 Wilson's data illustrating the general increase in island ages with distance from the ridges in the Atlantic and Indian Oceans. The dashed line is the distance predicted if both oceans were spreading at 3.2 cm/year. (From Wilson, "Evidence from Islands of the Spreading of the Sea Floor," *Nature* 197 [1963]: 536–538, reprinted by permission of Nature, copyright © 1963 by Macmillan Magazines Limited.)

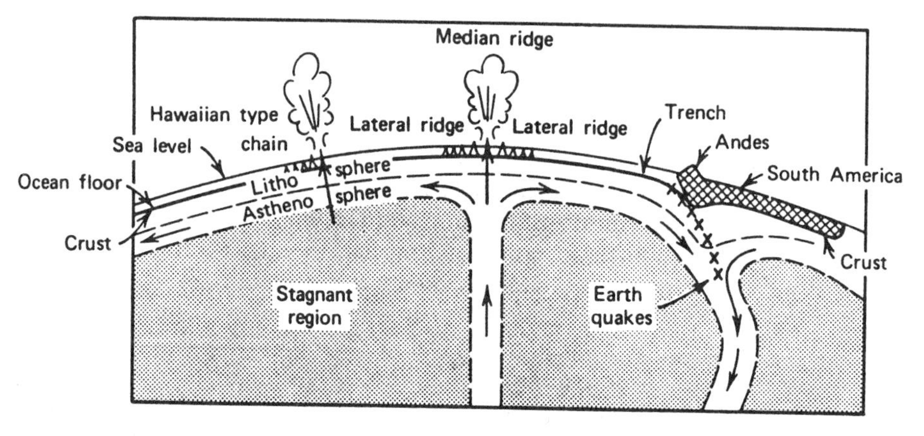

Figure 4.3 Wilson's model for seafloor spreading with a lithosphere that would produce a Benioff Zone when returning to the mantle at trenches. This figure also illustrates how the linear chain of the Hawaiian Islands could be produced as the lithosphere moves over a fixed hot spot in the mantle below. (From Wyllie, *The Dynamic Earth*, with permission, copyright © 1971 by John Wiley & Sons, as adapted from Wilson, "Hypothesis of the Earth's Behavior," *Nature* 198 [1963]: 925–929, reprinted by permission of Nature, copyright © 1963 by Macmillan Magazines Limited.)

upper mantle, and might have continents embedded in it. The Benioff zone marked the descent of the lithosphere as it was carried back into the mantle by the convection currents.

Figure 4.3 also illustrates how Wilson used seafloor spreading to explain linear chains of islands. He suggested that highly localized regions of hot material originating in the mantle formed "hot spots" that were fixed with respect to the lithosphere moving over them. If these hot spots existed, they might melt a linear chain of holes through the lithosphere as it moved over them, which would yield a chain of volcanic islands. The Hawaiian Islands form such a chain and show increasing ages as one moves away from the presently active islands. This explanation for the origin of linear island chains was virtually ignored until the early 1970s, so it will be discussed in the fourth section of this chapter.

Although Wilson proposed that the asthenosphere separated the convection currents from the earth's surface, he believed the surface features still expressed the underlying pattern of currents. This is apparent in his explanation (Wilson, 1963*d*) of the fault known to run through New Zealand, which he explained as the zone between two convection cells moving in different directions; see Figure 4.4. Within two years, however, Wilson would give up attempts to relate surface features to underlying convection cells and use his new concept of "transform faults" to explain many of these fault zones on the earth's surface.

Wilson did not mention the paleomagnetic evidence for continental drift,

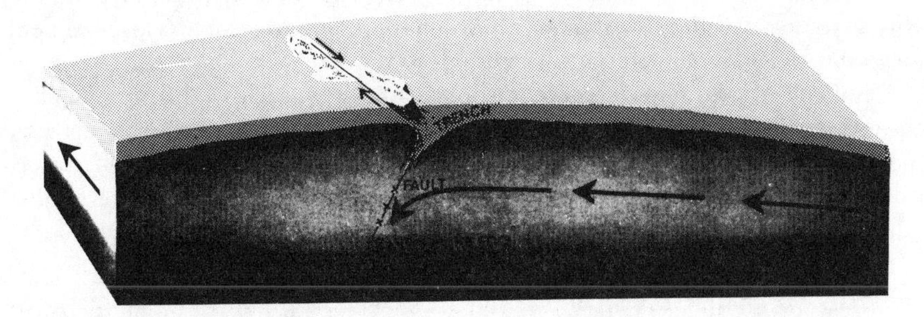

Figure 4.4 Wilson's explanation for the fault offsetting parts of New Zealand as due to the meeting of two mantle convection cells that are moving in different directions. (Adapted from "Continental Drift" by Wilson. Copyright © April 1963 by SCIENTIFIC AMERICAN, Inc. All rights reserved.)

but research in this area continued despite continuing controversy over the reliability of the data and their interpretation. In 1962 Runcorn sponsored a symposium on drift, which included some of the recent evidence for drift. Runcorn gave the paleomagnetic evidence; Benioff described the seismic evidence for thrust faults at trenches; and Vacquier described the linear magnetic anomaly evidence for large displacements of the ocean floors. The convection current hypothesis was reviewed by Vening-Meinesz. Heezen reported on recent oceanographic research, including Hess's model, but he endorsed the expansion hypothesis as the best explanation. The symposium volume (Runcorn, 1962) included a reprint of Dietz's seafloor spreading model.

The study of magnetic reversals in lavas was proceeding independently of both the paleomagnetic research aimed at proving continental drift and the mapping of the linear magnetic anomalies in the oceans. Although most geologists did not believe that the earth's magnetic field had periods of reversed polarity (Glen, 1982), those studying reversals were convinced and were developing a time scale for past field reversals (Cox and Doell, 1960). The major researchers on reversal studies were Doell and Cox at the USGS Menlo Park branch. In the early 1960s they recruited Brent Dalrymple[5] to do the potassium-argon dating. At the same time Ian McDougall,[6] a postdoctoral fellow at Berkeley, took the dating method to Australia, and eventually started work on magnetic reversals. This resulted in a "friendly spirit of competition" between the USGS group and the Australian group as each sought to be the first to identify new reversal periods in the earth's magnetic history (Cox, 1973: 139–146).[7]

In 1963 Fred Vine[8] and Drummond Matthews[9] of Cambridge University proposed that the linear magnetic anomalies in the ocean basins could be explained by combining seafloor spreading with magnetic reversals of the earth's field—neither of which were accepted widely among geoscientists.[10] They suggested that if material is added to the seafloor at the mid-ocean ridges and cools through the Curie point, it would acquire a natural remanent magnetism (NRM) in line with the earth's magnetic field. When the earth's field reversed, the new material added at the ridge would have a NRM of a reverse polarity compared to adjacent material formed before the reversal. The result would be that present oceanographic surveys should find a pattern of linear magnetic anomalies parallel to the mid-ocean ridges. This explained the linear magnetic anomalies observed in the Pacific, although at that time Vine and Matthews did not know that the area surveyed contained a segment of the mid-ocean ridge system.

Vine and Matthews provided some supporting data from a recent survey of the Carlsberg Ridge in the Indian Ocean. They first estimated the magnetic susceptibility of the seafloor material by examining the magnetic field over two seamounts that had opposite magnetic polarities. They then provided Figure 4.5, which shows the magnetic profiles recorded on three traverses of the Carlsberg Ridge. The solid lines are the observed magnetic profiles across the ridge crest. The dashed lines are predicted magnetic profiles after adjusting for seafloor topography and assuming an uniformly magnetized seafloor. In profile B, how-

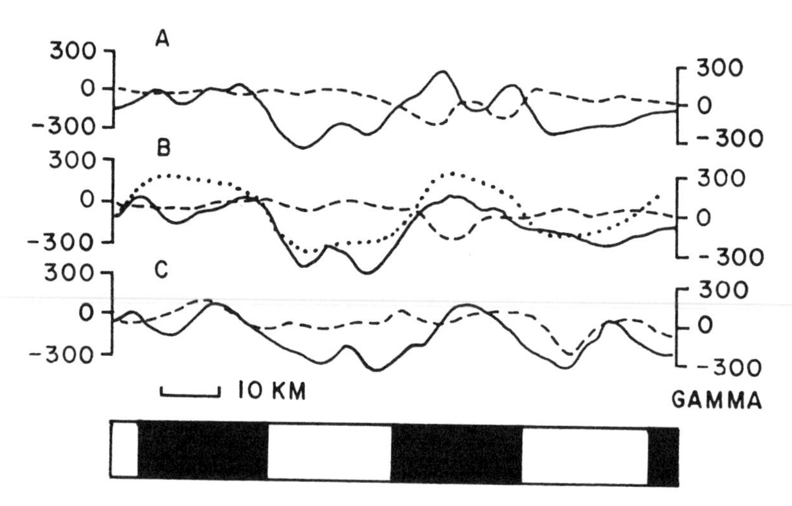

Figure 4.5 Vine and Matthews's illustration of how uniform blocks of alternately magnetized seafloor could produce a linear magnetic anomaly pattern under the assumptions that seafloor spreading occurs and that the earth's magnetic field reverses at regular intervals. Three observed magnetic anomaly profiles across the Carlsberg Ridge are shown as solid lines. The dashed lines are the predicted profiles assuming a uniformly magnetized ocean floor with adjustments for the observed topography of the floor. In profile B the dotted line is the predicted magnetic profile obtained when the ocean floor is assumed to consist of 20 km wide blocks of alternately normal and reversed magnetization, as shown at the bottom of the figure. (Adapted from Vine and Matthews, "Magnetic Anomalies over Ocean Ridges," *Nature* 199 [1963]: 947–949, reprinted by permission of Nature, copyright © 1963 by Macmillan Magazines Limited.)

ever, the dotted line shows the profile predicted after assuming the seafloor consisted of twelve-mile-wide blocks of alternating normal and reversed magnetization, as shown at the bottom of Figure 4.5. This is the closest they came to a quantitative analysis of their hypothesis. They did not base the widths of the bands on what was known about the chronology for field reversals and to generate the predicted profile they had to make assumptions about the thickness of the magnetized layer in the ocean floor and its magnetic susceptibility.

The response to the Vine-Matthew hypothesis was minimal for the first few years. As Matthews recalled, "the paper dropped into a sort of vacuum, as we expected it to. . . . Teddy Bullard used to proselytize for it a bit, but American labs wouldn't hear anything of it—thought it was all nonsense" (Glen, 1982: 303). As Table 4.1 indicates, this initial reaction changed quickly; Vine and Matthews (1963) article had fewer than ten citations per year until 1965, but more than forty per year after 1967. The seafloor spreading articles by Dietz (1961) and Hess (1962) also had marked influence after 1967, as did the articles by Wilson and by Cox and Doell. Even the publications promoting continental drift on the basis of other evidence, such as King (1962) and Runcorn (1962), began to receive many citations by the late 1960s. The next section summarizes the evidence for seafloor spreading that led to its widespread acceptance and the theory of plate tectonics.

Growing Evidence for Seafloor Spreading: 1964–1966

Probably the most significant publication in 1964 was Menard's *Marine Geology of the Pacific*, in which he proposed a convection current model to explain some of the characteristics of the ocean basins. However, Menard did not accept seafloor spreading. He proposed that convection currents rose beneath the ridges and descended at trenches along continental margins, but these currents did not actually break and create new overlying crust as in seafloor spreading. Instead, they thinned and faulted the crust at the ridges and compressed and depressed the crust at the trenches. Fracture zones marked the locations where differences in the thinning and compression of the fractured ocean floor caused relative movements along the faults. Thus oceanic crust was not created or destroyed, so it was old. Menard had published a similar model as early as 1960 and later in 1965, but at a special conference in 1966 he accepted seafloor spreading and was one of the first members of Scripps to adopt the model and contribute to its further development. This occurred as he became familiar with the evidence summarized in this section.

The major conceptual advance in the period from 1964 to 1965 was Wilson's (1965*a*) "transform fault" explanation for the mystifying offsets of the mid-ocean ridges and the associated fracture zones. These fracture zones were marked by differences in the heights of the seafloor and displacements of the linear magnetic anomalies along the fault zone; shown earlier in Figures 3.2 and 3.3. Most oceanographers held that the ocean floor had moved along these faults, but within the traditional concept of a "transcurrent fault" these displacements

TABLE 4.1 Citation Counts to Selected Geoscience Publications from 1960 to 1963

	Science Citation Index citation counts for:											
Publication / Topic	1961	'62	'63	'64	'65	'66	'67	'68	'69	'70	'71	'72
Dietz (1961) / first published proposal of seafloor spreading model	—	4	9	8	5	2	16	20	27	19	14	14
Hess (1962) / proposal of the seafloor spreading model	—	—	2	7	7	9	23	23	47	42	28	29
Wilson (1963*a*) / using seafloor spreading to predict island ages	—	—	2	2	6	6	15	8	4	5	5	4
Wilson (1963*b*) / adding a lithosphere to the seafloor spreading model to explain Benioff Zones	—	—	1	5	9	3	11	9	3	7	4	5
Wilson (1963*c*) / explained the Hawaiian Islands with mantle plumes	—	—	0	0	1	1	1	1	0	0	5	9
Vine and Matthews (1963) / seafloor spreading used to explain linear magnetic anomalies in the oceans	—	—	0	3	6	12	25	45	45	42	41	35
Cox and Doell (1960) / review of reversal studies in paleomagnetism	7	11	17	10	12	7	17	10	6	7	5	5
King (1962) / *The Morphology of the Earth* reviewed the evidence for drift	—	—	5	1	5	8	12	9	13	12	15	8
Runcorn (1962) / Symposium proceedings on drift	—	—	7	19	22	12	32	25	17	7	1	10

NOTE: Citation counts exclude self-citations. Wilson (1963*c*) reached 23 citations in 1973 as the subject of mantle plumes became a major research topic in the early 1970s. This development is described later in this chapter. The source is Institute for Scientific Information (1984, 1972, 1975).

should continue until the faulted medium terminated at a place that could absorb the relative movements. Menard's model implied that the medium itself (the seafloor) absorbed the displacements by becoming compressed, whereas others sought extensions of the ocean fracture zones onto the continents. Wilson, however, noted that the transcurrent fault concept assumed that the faulted medium was conserved, whereas the seafloor spreading model implied that faulted medium was being created at ridges and consumed at trenches.

Wilson suggested that seafloor spreading implied a new type of fault mechanism with implications that could be tested and contrasted to the traditional transcurrent fault mechanism. Figure 4.6 shows his model for how a continent can be rifted apart with the formation of an ocean basin by seafloor spreading. The continent splits apart along a north-south line and the two halves drift apart with the formation of an ocean basin with a ridge system that is offset along old lines of weakness in the continent. The most important implications of his model concerned the relative movements of the crust along the offsets or faults connecting segments of the ridge—one of these "transform faults" is the DD′ line in Figure 4.6. No spreading or creation of seafloor occurs along these offsets; the adjacent sides only slide past each other. These offsets extend beyond the ridge crests as fracture zones, but here there is no relative movement between the two sides. Thus oceanic earthquakes associated with ridges should occur only along the "transform faults" connecting ridge crests, not along the "fracture zone" extensions. This implication already had empirical support, but Wilson showed that the relative motion along the transform faults should be just the opposite of that expected from transcurrent faulting. This implication is grasped most easily in the same manner that Wilson discovered it: by working with a paper model (Glen, 1982: 374).

Figure 4.7 is an adaptation of a paper model used by Wilson to communicate his idea to others. When it is cut along the horizontal line and folded as indicated, one has a working model of crustal creation at a ridge crest offset by a transform fault. Pulling the ends of the paper creates new crust with corresponding linear magnetic anomalies, and causes relative movement of the crust just along the transform fault between the ridge crests. This relative movement is just the opposite of that predicted by transcurrent faulting, which is represented by cutting the paper model completely in half along the horizontal cut, lining up the ridge crests, and then moving one half until the original figure is reproduced. The reader is encouraged strongly to make this model because it illustrates how working models provide tacit knowledge not communicated easily by words.

Wilson provided a classification of transform faults based upon what they connected. For example, the faults connected two ridges in Figures 4.6 and 4.7, but they could also connect two trenches or a ridge and a trench. He suggested that the San Andreas Fault in California was a transform fault connecting a segment of ridge in the Gulf of California to a short section of ridge off the northeast coast, the Juan de Fuca Ridge. This implied that the region to the West of the San Andreas fault was moving North relative to the continental U.S., see Figure 4.8. Extending this same approach to the globe, Wilson concluded that the world's system of ridges and trenches were connected together by transform faults to form a "continuous network of mobile belts about the earth which divide the surface into several large rigid plates" (Wilson, 1965*a*: 343).

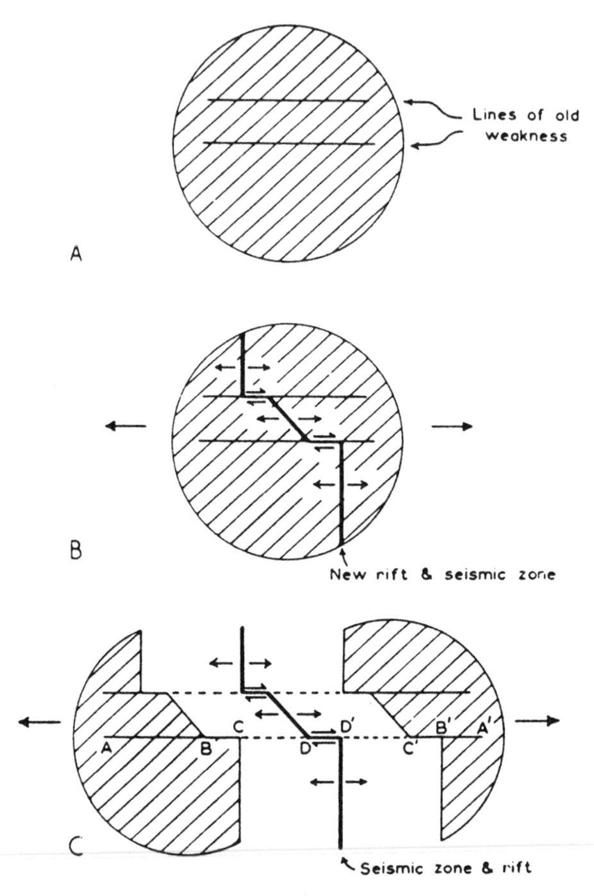

Figure 4.6 Wilson's model for the development of a ridge-ridge transform fault. The dashed lines represent fracture zone extensions of the transform faults offsetting the ridge crest. Seismic activity only occurs along the heavy lines in the figure. (From Wilson, "A New Class of Faults and Their Bearing on Continental Drift," *Nature* 207 [1965]: 343–347, reprinted by permission of Nature, copyright © 1965 by Macmillan Magazines Limited.)

Although Menard (1986) notes that he had described the Juan de Fuca Ridge earlier, Vine and Matthews were unaware of this so they did not realize that they could have tested their prediction that linear magnetic anomalies would 3e parallel to ridge crests. Wilson's transform fault model not only indicated that a ridge (the Juan de Fuca) was in the magnetic survey area studied earlier by Mason, but Wilson also noted that their hypothesis implied that the pattern of linear magnetic anomalies should be symmetrical across a ridge. Thus a more quantitative test was possible, but before describing it we must review the developments in the study of magnetic field reversals.

The development of the magnetic field reversal chronology made rapid

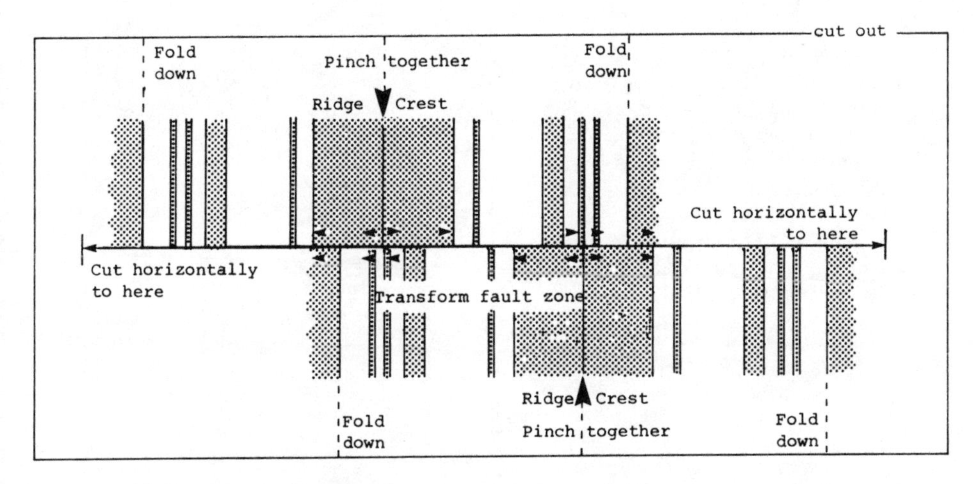

Figure 4.7 Paper model illustrating the creation of linear magnetic anomalies at an ocean ridge offset by a transform fault. When the ends are pulled after making the indicated cuts and folds, the model also illustrates how the seismic zone along a ridge is confined to the transform fault and has a direction of movement opposite from transcurrent faults. The model works best when the pinched-together ridge crests are inserted in narrow slots cut in a 3×5 card.

advances between 1964 and 1966, partly because of the competition between the Australian and USGS groups. Figure 4.9 depicts the evolution of this chronology over a six year period. By 1964 it was becoming recognized that the major periods, "epochs," of a stable direction in the earth's field were interrupted by shorter "events" of the opposite polarity. There was strong competition to identify these events and their identification played an important role in efforts to match them to the pattern of linear magnetic anomalies across ridges. The discoveries of the Olduvia and the Jaramillo events—see Figure 4.9—in 1964 and 1966 were particularly important.

It should be noted that an accurate land-based, reversal chronology was limited to the previous four million years because the error in the potassium-argon dating method increased with the age of the rock. After about four million years the error was large enough that short events could not be dated accurately. Yet within a few years the reversal chronology was extended twentyfold by using a different "yardstick"—the distance of linear magnetic anomalies from the mid-ocean ridges. This new yardstick first had to be calibrated and the Juan de Fuca Ridge data were used in the first attempt.

In 1965 Vine and Wilson gave the first systematic model relating the magnetic reversal chronology to magnetic anomaly profiles across an ocean ridge—the Juan de Fuca. In Figure 4.10 they illustrated the effects on a magnetic profile of different assumptions and conditions. Part (a) of Figure 4.10 shows the profile predicted from the pattern of remanent magnetization in the underlying sea-

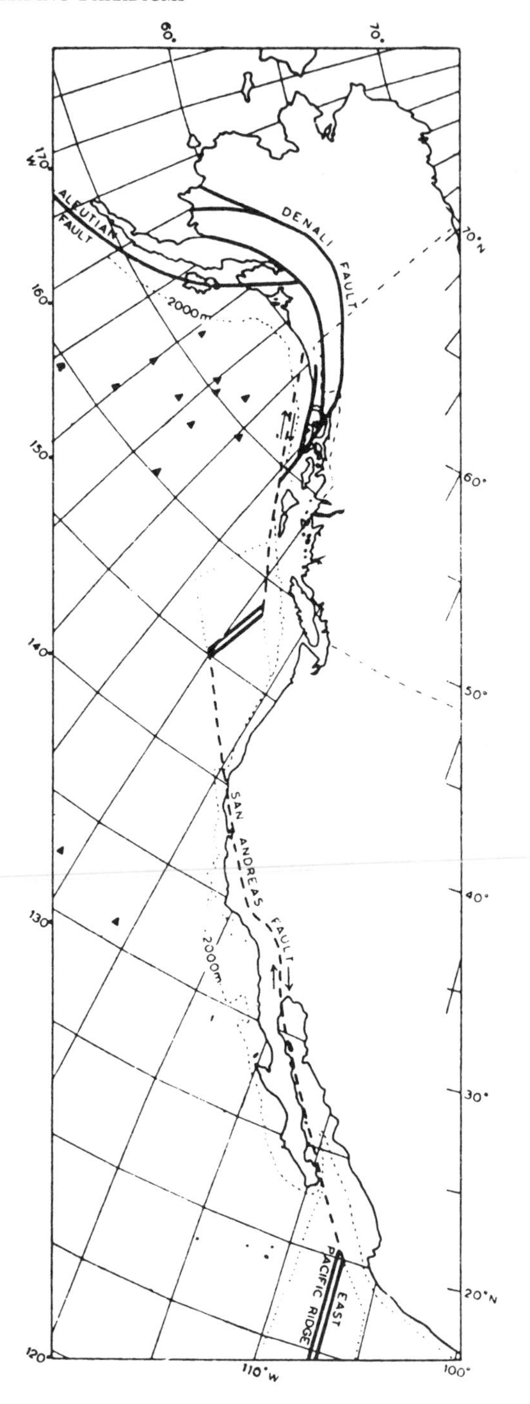

Figure 4.8 Wilson's suggestion that the San Andreas fault is a transform fault connecting the ocean ridge in the Gulf of California and the Juan de Fuca Ridge. (From Wilson, "A New Class of Faults and Their Bearing on Continental Drift," *Nature* 207 [1965]: 343–347, reprinted by permission of Nature, copyright © 1965 by Macmillan Magazines Limited.)

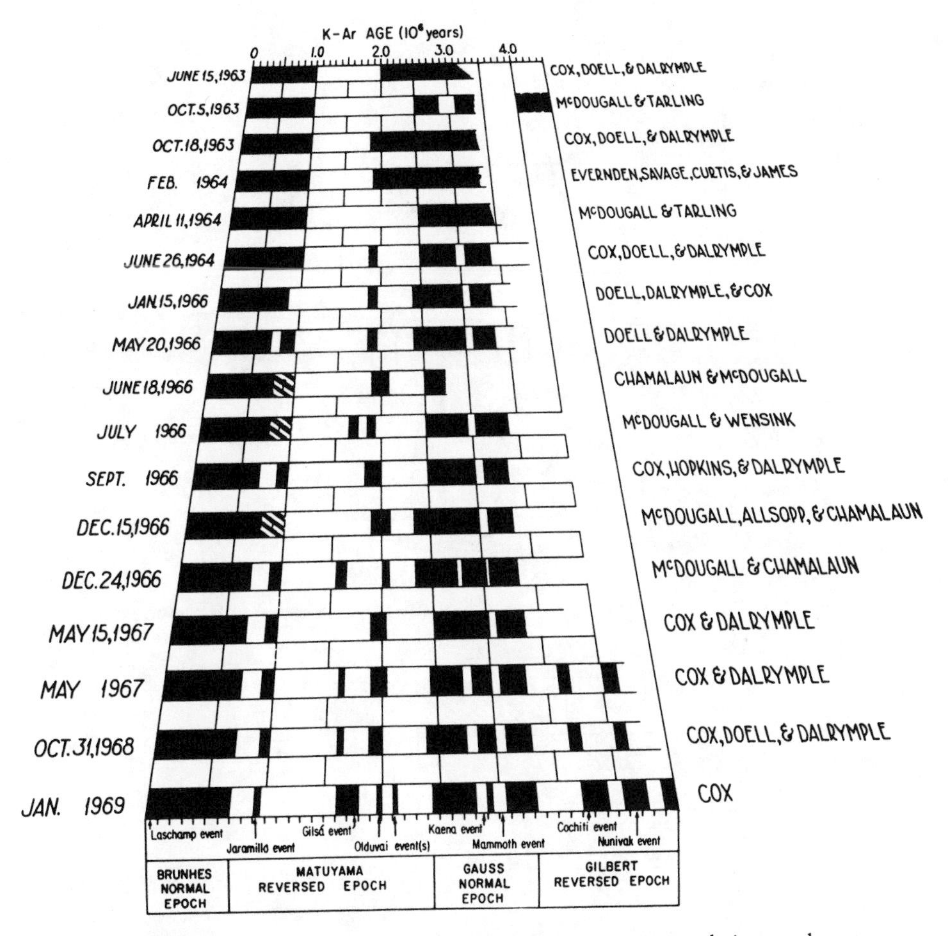

Figure 4.9 Successive versions of the geomagnetic-reversal time scale as determined from the potassium-argon dating of volcanic rocks from the continents. Shaded rectangles represent periods of normal polarity. (From Dalrymple [1972] in Bishop and Miller [eds.], *Calibration of Hominid Evolution*, with permission, Scottish Academic Press.)

floor, where the shaded blocks have normal magnetization and unshaded blocks are reversely magnetized. This particular sequence of seafloor magnetization was produced by combining the accepted magnetic reversal chronology with a *constant* spreading rate of 1.0 cm/year for each side of the ridge. Part (b) is the same, except the spreading rate is a constant 2 cm/year and shows that a higher spreading rate would be able to detect a possible short reversal at three million years—the dashed sections in the profiles and the seafloor model. Part (c) is their first proposed model for the Juan de Fuca Ridge, but they selectively varied the spreading rate at different times in the past to give an *average* rate of 1.5 cm/year for each side.

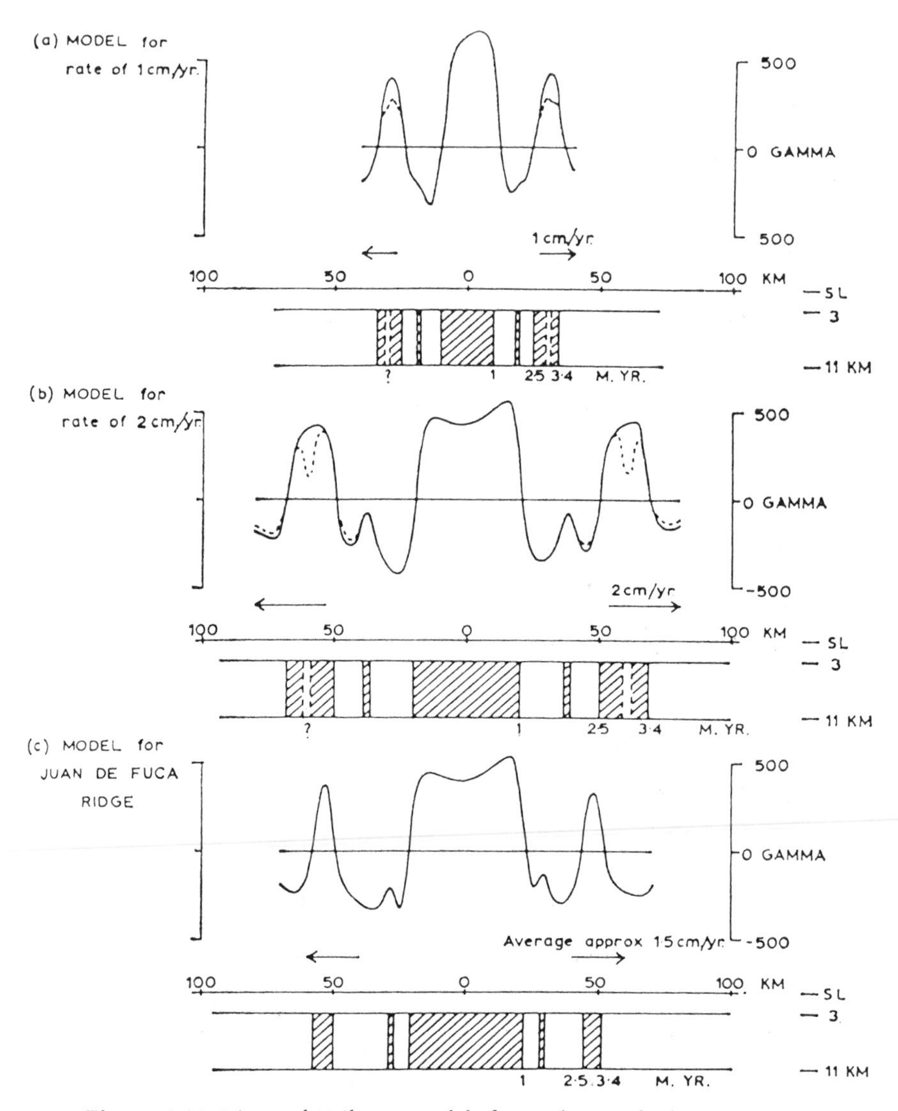

Figure 4.10 Vine and Wilson's models for predicting the linear magnetic anomaly patterns produced by seafloor spreading and the chronology of magnetic field reversals. Models A and B assume *constant* spreading rates of 1.0 and 2.0 cm/yr, respectively. Model C has a variable spreading rate that *averages* 1.5 cm/yr. The dashed lines in models A and B represent the effects of a possible reversal 3 million years ago. (From Vine and Wilson, *Science* 150 [1965]: 485–489, copyright © 1965 by the AAAS.)

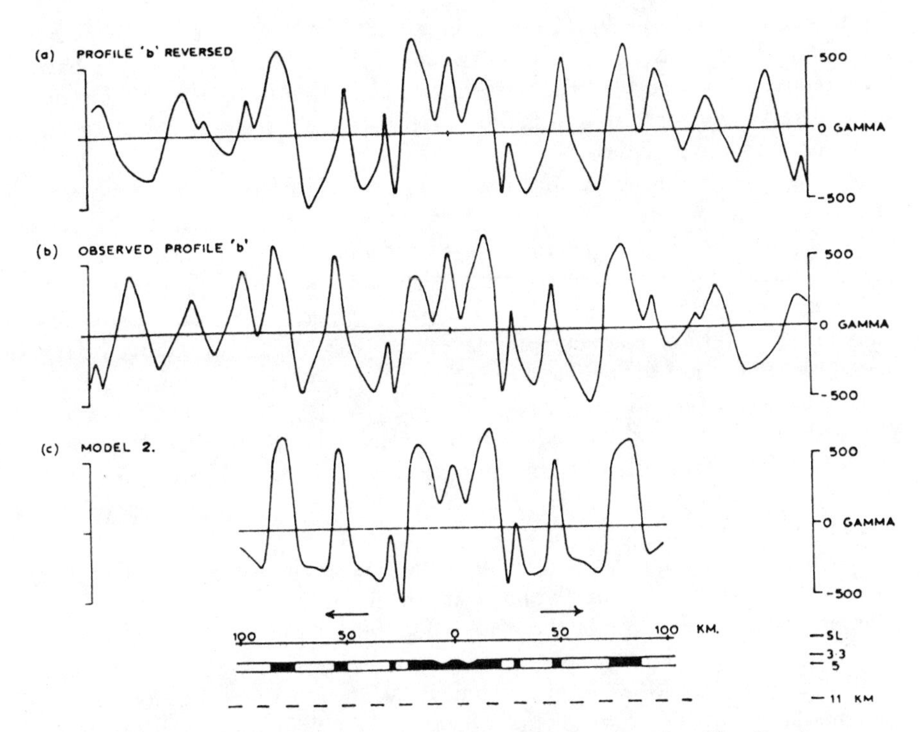

Figure 4.11 Vine and Wilson's model for the Juan de Fuca Ridge. Portions A and B show the observed magnetic field and its mirror image to demonstrate its symmetry. Portion C is the calculated magnetic field assuming an *average* spreading rate of 1.5 cm/yr and a more confined layer of basaltic rock that records the magnetic field reversals than that in Figure 4.10. (From Vine and Wilson, *Science* 150 [1965]: 485–489, copyright © 1965 by the AAAS.)

Figure 4.11 shows the best "fit" obtained between the observed and predicted magnetic profiles for this ridge. Parts (a) and (b) give the observed profile and its mirror image to show the general symmetry of the profile across the ridge crest. Part (c) is the predicted profile using the previous *average* rate of 1.5 cm/year for each side, but now several changes have been made in the seafloor model. They decreased the thickness of the layer in the seafloor preserving the natural remanent magnetization (NRM) and increased its ability to record NRM, which together increased the predicted profile's capacity to reflect short reversals. Finally, they added a pair of "arbitrary" depressions in the seafloor near the ridge crest.[11]

Although the fit between the observed and predicted profiles is quite good for 100 kilometers (62 miles) on each side of the ridge, Vine and Wilson admitted that using different assumptions about the seafloor and a changing spreading rate were so "flexible" that any reversal chronology could be fitted to a symmetrical anomaly profile. Despite the good fit, this article seemed to have little immediate impact on other geoscientists. Part of the reason might have been the arbitrary aspects of the model, the publication in 1966 by Vine using better data, or the opposition of Lamont scientists to the seafloor spreading and Vine-Matthews hypotheses.

During 1965 and 1966 Lamont scientists published a series of articles on the crustal structure of the mid-ocean ridges. These articles examined the geophysical characteristics of the ridges gathered during seismic, magnetic, and gravity studies and concluded that these data were inconsistent with seafloor spreading. For example, James Heirtzler[12] and Xavier LePichon[13] (1965) examined the pattern of magnetic anomalies and noted that they changed character—becoming more irregular with greater fluctuations in intensity—on the distant flanks of the ridges. They concluded that this "completely different character argues against the Vine-Matthews hypothesis" (Heirtzler and LePichon, 1965: 4013). Although the opposition was not strongly worded, it was uniform throughout this series of articles and indicated the general opposition of Lamont researchers (Glen, 1982: 315 322). Since Lamont was one of the most prestigious oceanographic institutes in the world, their opposition may have reduced the acceptance of the revolutionary theory of seafloor spreading. A later chapter will consider some of the reasons for their opposition.

This opposition did not last long. In 1966 Walter Pitman[14] and Heirtzler published an article in *Science* that concluded: "We feel that these results strongly support the essential features of the Vine-Matthews hypothesis and of ocean-floor spreading as postulated by Dietz and Hess . . ." (Pitman and Heirtzler, 1966: 1171). This conversion of opinion was based on Lamont surveys over the Reykjanes Ridge below Iceland and over a section of the Pacific-Antarctic Ridge. For the latter ridge system, one profile—the Eltanin-19 profile—was particularly symmetrical and could be fitted to the most recent magnetic reversal chronology by assuming a *constant* spreading rate of 4.5 cm/year. Furthermore, they showed that when this profile was contracted by a factor of 4.5, it matched the anomaly pattern across the Reykjanes Ridge, which suggested that the latter ridge had a spreading rate of 1.0 cm/year. Finally, they showed how the assumption of a steady spreading rate allowed the extension of the magnetic reversal chronology back to ten million years. It should be noted that some of this evidence had been used previously to argue *against* the Vine-Matthews hypothesis.

This article was the first published report indicating acceptance of seafloor spreading by Lamont scientists and it provided persuasive new evidence from several locations on the earth's surface. However, its impact was overwhelmed by an article by Vine (1966), which appeared two weeks later in *Science*. This article was longer, provided additional evidence, and included the recent Eltanin-19 profile. (A later chapter will describe how Vine obtained this profile and the resulting dispute over priority in publishing the results.) Vine (1966) used the Lamont data on the Reykjanes Ridge below Iceland and the Eltanin-19 profile to show the symmetry in the magnetic anomaly patterns. These and other magnetic profiles were compared to hypothetical profiles based upon seafloor spreading and the magnetic reversal chronology, which now included the just-discovered Jaramillo Event. In the new predicted profiles the first magnetic peak on each side of the central ridge was associated with the one-million-year-old Jaramillo Event instead of the two-million-year-old Olduvai Event used by Vine and Wilson (1965). The latter assumption had forced Vine and Wilson to use a variable spreading rate, but with the new event Vine showed that each ridge seemed to have a unique, *constant* spreading rate. Figure 4.12 shows the very good fit with the Eltanin-19 profile, which later was referred to as the "magic profile" because it helped to persuade many geoscientists to accept seafloor spreading.

Vine then examined the assumption that the spreading at ridges occurred

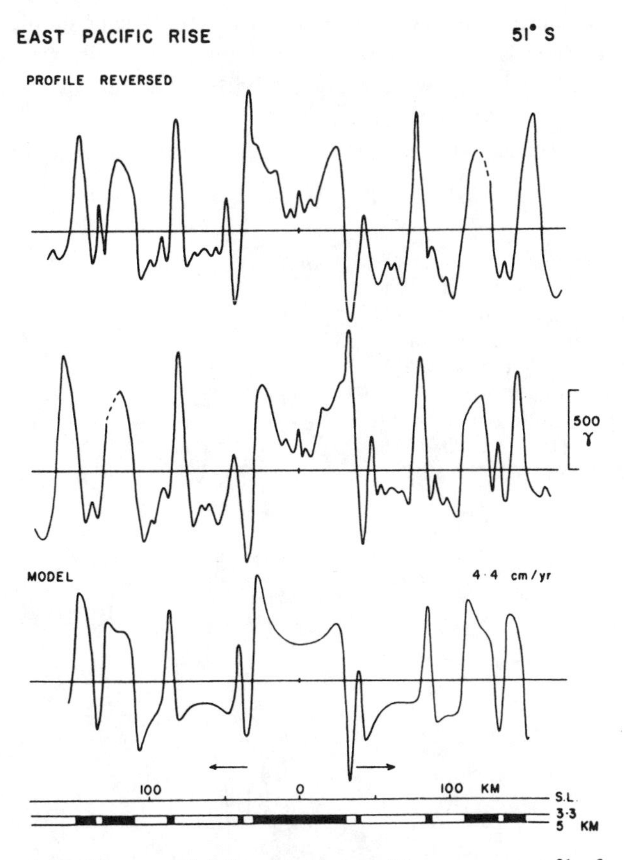

Figure 4.12 Vine's model for the Eltanin-19 magnetic profile from the East Pacific Rise. The top two profiles are the observed profile and its mirror image and the lower profile is calculated assuming a *constant* rate of seafloor spreading of 4.4 cm/yr and the magnetic reversal chronology indicated at the bottom of the figure. (From Vine, *Science* 154 [1966]: 1405–1415, copyright © 1966 by the AAAS.)

at a uniform rate and found it supported for the most recent four million years. By assuming that the *same* spreading rate existed for earlier anomalies—those further from the ridge—Vine was able to reconstruct the magnetic reversal chronology for the last eleven million years. This was a considerable extension of the lava based chronology, which was restricted to the four million years given by the potassium-argon dating method. Figure 4.13 shows Vine's revision of the predicted profile for the Juan De Fuca Ridge, which he and Wilson had studied in 1965, but now using a constant spreading rate and no arbitrary depressions in the seafloor.

The identification of the Jaramillo Event played a key role in the above studies. Not only was it found in reversal studies on land, but it was also identified by studying the reversals recorded in cores of ocean sediments. Since sedimentary rocks on land contain a weak natural remanent magnetism (NRM), it would seem plausible that cores of sediments in the oceans should show some NRM. However, most early attempts to find a pattern of reversals in sediment cores had failed until they were rediscovered[15] at Lamont in 1966 by one of Neil Opdyke's[16] students.

Except for Heezen, who accepted a form of earth expansion, Opdyke was

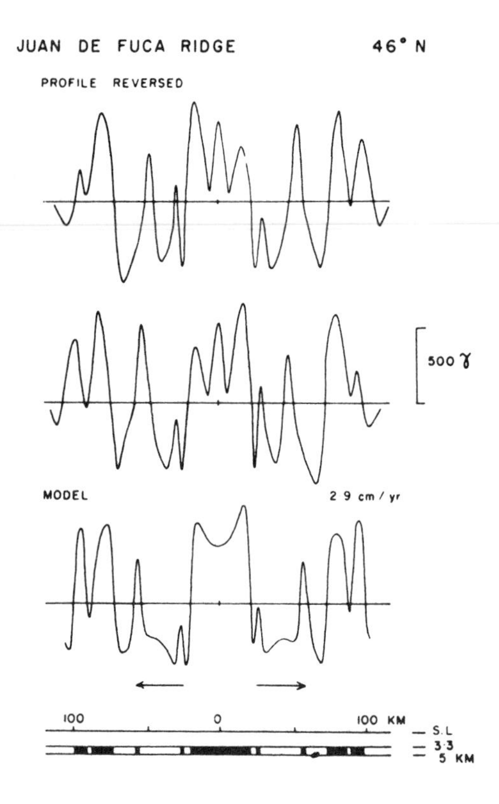

Figure 4.13 Vine's model for the Juan de Fuca Ridge with a *constant* spreading rate of 2.9 cm/yr and a magnetic reversal chronology that differs from that in Figure 4.11 in that it includes the Jaramillo Event. (From Vine, *Science* 154 [1966]: 1405–1415, copyright © 1966 by the AAAS.)

the only geologist at Lamont who accepted continental drift theory at that time. His acceptance derived from his work as an assistant in Runcorn's study of the polar wandering curves of Europe and North America and his later examination of paleoclimatic evidence, which forced him to deal with the Permo-Carboniferous glaciers. Opdyke's group found that the sediments in some Antarctic cores replicated the reversal chronology quite well, if one assumed a constant rate of sedimentation for each core. However, they (Opdyke et al., 1966) found that a very short event appeared in several cores, but not in the known reversal chronology. At the same time this event, the Jaramillo, was identified and named by the USGS group.

The above studies indicated that the magnetic reversal chronology could be established through three separate means—by reversal studies on land, by reversal studies of sediment cores, and by the pattern of linear magnetic anomalies. The chronologies obtained from these three sources were in close agreement and provided persuasive new evidence for the seafloor spreading and Vine-Matthew hypotheses. As a result of informal communication and conference presentations, these articles had a considerable impact before formal publication (Glen, 1982; Menard, 1986). In the 1964–1966 period several conferences were major events in the acceptance of seafloor spreading and continental drift.

Most of these conferences were sponsored by paleomagnetists who were using the inclinations in the NRM of old rocks to show that drift had occurred. Irving (1964) summarized much of this evidence, but it had less impact than the later studies supporting seafloor spreading. Runcorn had organized the 1962 conference on drift, and in 1964 he joined Blackett and Bullard to organize a conference in conjunction with the Royal Society of London (Blackett, Bullard, and Runcorn, 1965). The 1964 conference had reports similar to those in the earlier Runcorn (1962) conference. For example, there were reports on the paleomagnetic evidence for drift, on the characteristics of the ocean basins, and on mantle convection currents. Although the Vine-Matthew hypothesis was virtually ignored, Hess's seafloor spreading model was used by several of the participants. Wilson gave his evidence for increasing ages of islands with increasing distance from ridges and examined evidence for matching structures on opposite sides of the Atlantic. Both Wilson and G. Nicholls examined the different chemical compositions of the lavas forming ridges and islands, thereby providing some of the earliest attempts to relate petrology to seafloor spreading. Ron Girdler used the seafloor spreading model to interpret data on the Red Sea as the initial opening of a new ocean basin.[17]

The most influential paper was by Bullard, Everett, and Smith (1965). They provided a computer generated fit between the continents around the Atlantic Ocean. The computer program minimized the disagreement between the continental edges defined at various depths along the slopes; the best fit was obtained by using the 900 meter depth to define the edges of the continents. To quantify the problem they used a theorem of Euler that described the relative motion between two areas on the surface of a sphere as a single rotation about an unique axis. Figure 4.14 shows that a good fit was obtained between the southern

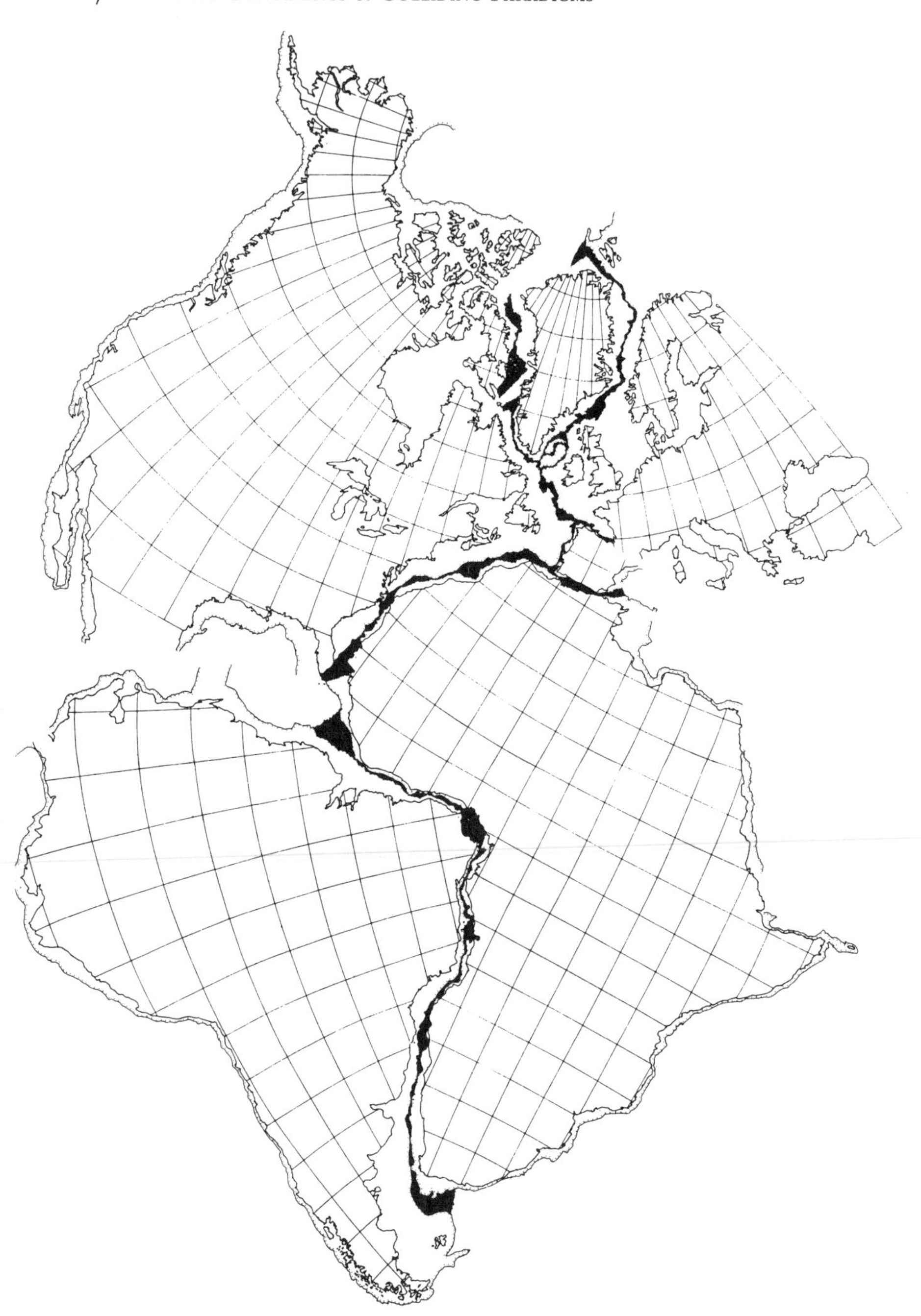

Figure 4.14 A computer generated fit of the continents around the Atlantic Ocean. The shaded areas represent gaps or overlaps of the edges of the continents. (From Bullard, Everett, and Smith [1965], with permission, Royal Society, London.)

continents, but the fit between the northern continents was poorer and required some assumptions, such as rotating Spain and ignoring most of Mexico. Although the results were similar to the qualitative fits obtained by du Toit and Carey, this article became one of the classics in the history of plate tectonics because of its quantitative nature and its use of spherical geometry, which was used later in the quantitative applications of plate tectonics.

The 1966 Goddard conference is probably the most important conference in the history of plate tectonics. It occurred at the Goddard Space Center in November 1966: a time of momentous change in the geosciences. As the editor of the resulting volume noted:

> In July 1966, the results of Vine and Pitman and Heirtzler were known to only a few colleagues. Word got around during the fall, by preprint, and the Goddard Conference, slightly preceding formal publication in *Science*. By January 1967 the impact was such that nearly 70 papers on sea-floor spreading had been submitted for the April meeting of the American Geophysical Union (Phinney, 1968*a*: 9).

The Goddard papers presented much of the recent evidence and tested some of the implications of the seafloor spreading hypothesis. Cox, Doell, and Dalrymple presented the latest magnetic reversal chronology, and the paleomagnetic evidence for drift was presented and related to past ice ages by Irving and W. Robertson. Both Vine and Heirtzler presented their soon to be published results and Opdyke described the reversal sequence recorded in cores of ocean sediments. Marshall Kay[18] and John Dewey[19] compared the geological structures in the Appalachian Mountains with those in the Caledonian Mountains of Britain, and concluded they were part of the same mountain range split apart by drift. Finally, Menard, who now accepted seafloor spreading, described some of the problems with this hypothesis. Most of these concerned the difficulties in matching the pattern of ridges and trenches to simple models of convection cells in the mantle.

Two of these papers were important tests of the seafloor spreading hypothesis. Lynn Sykes[20] described the results of first-motion studies of the earthquakes along the transform faults connecting segments of the Mid-Atlantic Ridge, and found these motions to be in the direction predicted by Wilson (1965*a*); see Figures 4.6 and 4.7. Patrick Hurley[21] and J. Rand tested the computer-generated fit between Africa and South America by comparing the radiometric ages of the rocks on each side of the Atlantic and found similar ages on adjacent sections of the two continents.[22]

The most negative articles used paleontological and geophysical arguments. Boucot, Berry, and Johnson examined fossil evidence from the Paleozoic Era and concluded that it was ambiguous regarding drift, but tended to emphasize the inconsistencies with drift theory. McConnell argued the mantle did not have properties compatible with convection currents. However, Dan McKenzie[23] provided arguments that mantle convection was possible.

This conference persuaded many of the attending geoscientists to accept seafloor spreading, and others were persuaded as these reports (Phinney, 1968*b*) became available in the scientific literature. The articles appearing in the 1964–1966 period were very influential, see Table 4.2. For example, Vine's 1966 article had over sixty citations in 1968. This is about thirty times the number of citations received by the average geoscience article in the second year after publication (Nairn, 1976). Yet even more important articles would appear within two years as seafloor spreading evolved into plate tectonics and a majority of geoscientists accepted this new view of the earth.

TABLE 4.2 Citation Counts to Selected Geoscience Publications from 1964 to 1966

	Science Citation Index citation counts for:									
Publication / Topic	1965	'66	'67	'68	'69	'70	'71	'72	'73	'74
Menard (1964) / *Marine Geology of the Pacific*	21	28	43	48	56	48	35	23	25	23
Irving (1964) / *Paleomagnetism* . . .	11	26	37	43	32	42	47	35	32	28
Cox, Doell, and Dalrymple (1964) / magnetic field reversals of the Earth	7	10	16	12	9	5	6	7	0	2
Wilson (1965*a*) / the transform fault concept	2	8	11	30	33	30	38	41	23	24
Vine and Wilson (1965) / testing Vine & Matthews on the Juan de Fuca Ridge	1	5	11	13	10	7	6	5	2	3
Bullard, Everett, and Smith (1965) / computer generated fit of the continents	—	3	7	24	14	43	54	47	49	35
Opdyke et al. (1966) / new evidence for magnetic reversals in deepsea cores	—	3	15	26	10	10	17	10	9	8
Pitman and Heirtzler (1966) / support for the Vine-Matthews hypothesis with a constant rate of seafloor spreading	—	2	20	30	19	18	6	14	6	6
Vine (1966) / additional evidence for a constant rate of seafloor spreading	—	—	30	65	60	69	50	49	17	22

NOTE: Citation counts exclude self-citations. The source is Institute for Scientific Information (1972, 1975).

The Rise and Partial Acceptance of Plate Tectonics: 1967–1970

Figure 3.1, which indicated the growth in the number of articles related to continental drift from 1910 to 1970, showed that seafloor spreading and plate tectonics became major topics from 1967 to 1970. In these four years over 600 articles were published on these or closely related topics. Of course, many of these were "minor" articles, but even a count of those cited five or more times yields over 150 articles. Consequently, this section will focus only on the articles reporting major developments or on related topics that facilitate understanding the material in later chapters.

Both Vine (1966) and Pitman and Heirtzler (1966) supported the seafloor spreading hypothesis and the quantitative implications of the Vine-Matthews hypothesis. Although Lamont scientists may have resisted the idea of seafloor spreading for a short time, they quickly applied it to the interpretation of their extensive collection of data on marine magnetics. During 1967 and 1968 Lamont scientists published numerous papers that analyzed magnetics data from the major ocean basins of the world. These studies culminated in the Heirtzler et al. (1968) paper that provided evidence for the following conclusions. First, they gave additional evidence that the linear magnetic anomalies in different oceans could all be modeled by the same reversal chronology by simply assuming that each ocean basin had a different rate of spreading. Second, the spreading rates have been fairly uniform at the major ocean ridges. If the spreading had been intermittent, it had stopped and started in all oceans simultaneously. Finally, they proposed a tentative reversal chronology for the last *80 million years* by assuming a constant rate of spreading in the South Atlantic. This was a twentyfold increase over the reversal chronology based on the potassium-argon dating of lavas, and it allowed them to predict the ages of the seafloor at considerable distances from the ridges. These predictions were supported by the deep sea drilling project, which is described later in this chapter.

Other important publications by Lamont scientists included Sykes's (1967) verification of the direction of movement along the transform faults connecting segments of the mid-ocean ridges, which had been presented at the 1966 Goddard Conference. John Ewing and Maurice Ewing (1967) related the amount of sediments in the ocean basins to seafloor spreading, but interpreted the sudden increase of sediment thickness on the flanks of the ridges to mean that spreading rates varied over time for the Atlantic Ocean. Later studies indicated that the spreading rate was constant and other factors explained the distribution of the sediments.

Finally, a key article by Jack Oliver[24] and Bryan Isacks[25] (1967) provided evidence supporting Wilson's earlier model that the lithosphere returned to the mantle at the trenches. They examined the deep earthquakes associated with the trench and island arc system near Fiji. Although the study was started as part of the international Upper Mantle Project without considering any possible relevance to the seafloor spreading hypothesis, they found that the Benioff zone itself conducted seismic waves faster than the surrounding mantle material. The faster time suggested that the Benioff zone was composed of a denser medium than the surrounding mantle, so they suggested that the Benioff zone was composed of a cooler (hence denser) slab of the ocean's crust that penetrated into the mantle below the trench.

During 1967 and 1968 the plate tectonics model was developed and applied in four major articles (McKenzie and Parker, 1967; Morgan, 1968; LePichon, 1968; Isacks, Oliver, and Sykes, 1968). Each of these articles assumed the earth's surface was composed of a set of rigid plates that were separated from each other by one or more types of boundaries where most geological activity occurred. These boundaries were (a) mid-ocean ridges where the lithosphere was created by seafloor spreading, (b) trenches where the lithosphere was consumed by being thrust down into the mantle, and (c) transform faults connecting together ridges and/or trenches. In contrast to Hess's (1962) model of seafloor spreading, where the seafloor below the sediments is the upper surface of a mantle convection cell, the plate tectonics model placed these currents below a rigid and thicker lithosphere. These articles were not concerned with the causes of plate movements, nor with relating surface features to hypothetical patterns of mantle convection. Their only concern was the *kinematics* of plate motion, that is, calculating and comparing the relative directions and rates of plate movements and examining some of the implications.

The emphasis on kinematics was possible because the plates were assumed to be rigid. This rigidity implied that creation of the lithosphere at the ridges must be balanced by destruction at the trenches. The key to quantifying these processes was Euler's theorem on spherical geometry, which had been used by Bullard, Everett, and Smith (1965) to fit together the continental edges. The theorem stated that the relative motion of two objects on the surface of a sphere could be described as a rotation about a hypothetical pole of spreading. Figure 4.15 illustrates the spreading pole perspective and indicates how three methods could locate the spreading pole. The first method examines the rate of spreading at various points along a mid-ocean ridge axis separating two plates. As indicated in Figure 4.15, the rate of spreading will be highest at the "equator" of the pole of spreading and diminish to zero as one approaches either end of the pole. Given spreading rates at different points along a ridge system, the mathematical relationship between spreading rate and latitude from the spreading pole allows calculation of the pole's location. This relationship can be expressed as $W_A = W_O \text{ Sin } A$, where A is the arc distance from the spreading pole, W_A is the velocity of spreading at this arc distance, and W_O is the spreading rate at the spreading pole's "equator." The second method uses the orientation of transform faults on the boundaries between the plates. Figure 4.15 shows that these faults should lie on concentric circles centered at the spreading pole. Thus circles perpendicular to these transform faults should intersect at the spreading pole. Finally, first motion studies of earthquakes along the transform fault boundaries between two plates should indicate the relative motion of the two plates. Per-

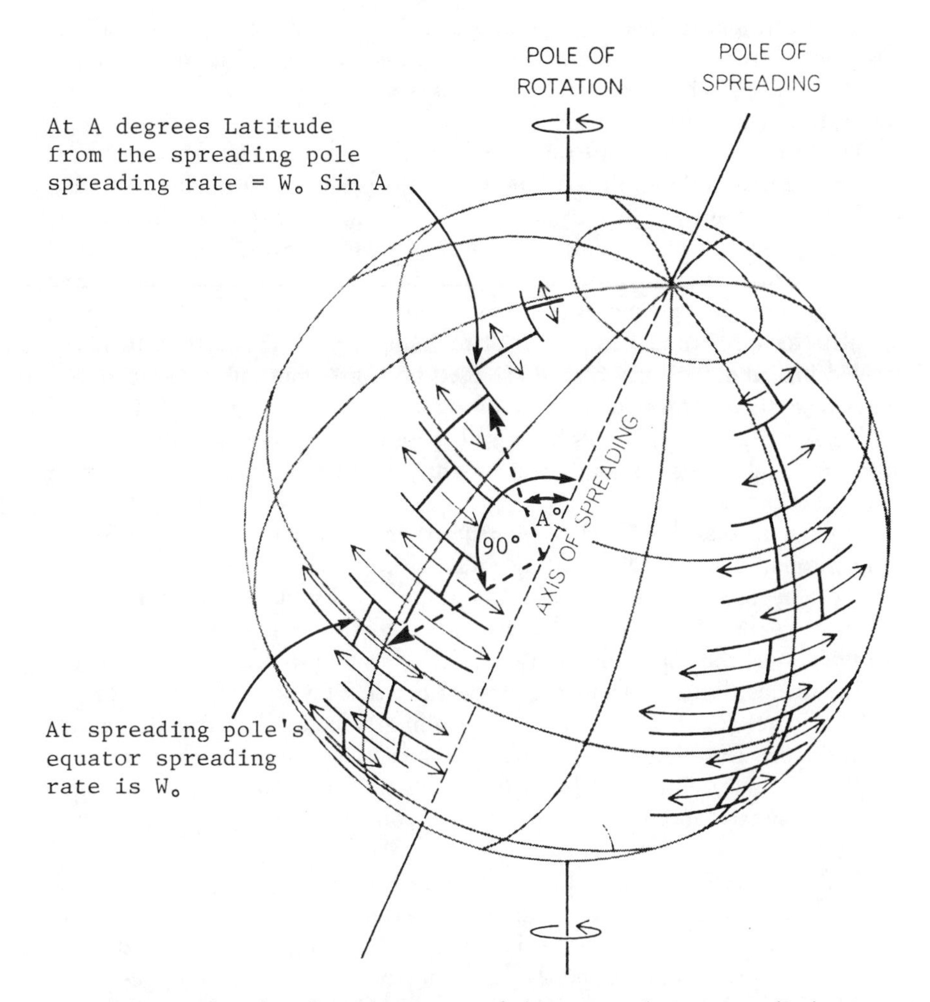

Figure 4.15 The spherical geometry of plates on a sphere. Perpendiculars from transform faults between two plates will intersect at the pole of spreading. The width of new seafloor at A degrees from this pole is given by W_0 Sin A, where W_0 is the width generated at the spreading pole's equator. (Adapted from "The Origin of Oceans" by Bullard. Copyright © September 1969 by SCIENTIFIC AMERICAN, Inc. All rights reserved.)

pendiculars to these directions of relative motion should also intersect at the spreading pole.

Note that these methods used different information and applied to specific types of boundaries. The spreading rate method relied on the Vine-Matthews hypothesis and only worked at ridges. The second method only worked with maps of transform faults, whereas the last method used seismological techniques. The identification of plate boundaries was facilitated by recent data showing the major earthquake zones on the surface of the earth (Barazangi and Dorman, 1969), which replicated earlier data on earthquake locations. Figure 4.16 shows this distribution, where the bands of intense seismic activity were taken as the boundaries of the plates.

The first publication to use these methods was by Dan McKenzie and Robert Parker (1967) at Scripps. They examined eighty published first-motion studies of earthquakes between the Pacific and North American plates. They found that 80 percent of these studies were consistent with a spreading pole centered in the north-central part of North America. Very soon after Parker and McKenzie's results were published,[26] Jason Morgan[27] (1968) at Princeton applied the plate tectonics model to several plate boundaries on the earth, which was divided into about twenty plates. He used both first-motion studies and perpendiculars to transform faults to locate the same pole studied by McKenzie and Parker. Figure 4.17 shows how well these two methods agreed with each other, although Morgan's pole was at a slightly different location than McKenzie and Parker's.

Morgan discussed a number of issues, but particularly striking were his comparisons of predicted spreading rates with observed spreading rates. For example, he used the transform faults along the Pacific-Antarctic Ridge to calculate its spreading pole. After he estimated this pole's location and the spreading rate at its equation, he used Euler's theorem to predict the spreading rate at various distances from the spreading pole. These predicted spreading rates for increasing distance from the spreading pole are plotted as the dashed line in Figure 4.18, which shows that the observed spreading rates are close to the predicted rates. This provided evidence that transform fault data were related to linear magnetic anomaly data in a manner consistent with the plate tectonics model.

Later in 1968 LePichon at Lamont assumed six major plates and used Morgan's approach with data from all the major oceans of the world. Pole positions were determined by using a statistical procedure to combine data on the orientation of transform faults and the spreading rates along oceanic ridges. Once the different spreading poles were determined—one pole for each pair of plates—the predicted rates and observed rates were compared and found to agree reasonably well. LePichon also followed the earlier suggestion of Morgan and noted that the relative motions between the plates should sum to zero over the surface of the earth because the plates were assumed to be rigid. With these assumptions LePichon was able to use the calculated spreading poles for the movements between the six plates to estimate rates of spreading or convergence at plate boundaries lacking relevant data. In general, his results were consistent with known seafloor topography; he predicted spreading at known ridge crests and consumption at known trenches. He concluded that "none of the ridges can be understood as an isolated feature but that each is part of an intricate pattern that transfers new earth's surfaces from sources . . . to sinks" (LePichon, 1968: 3679).

LePichon also gave a tentative reconstruction of the pattern of continental

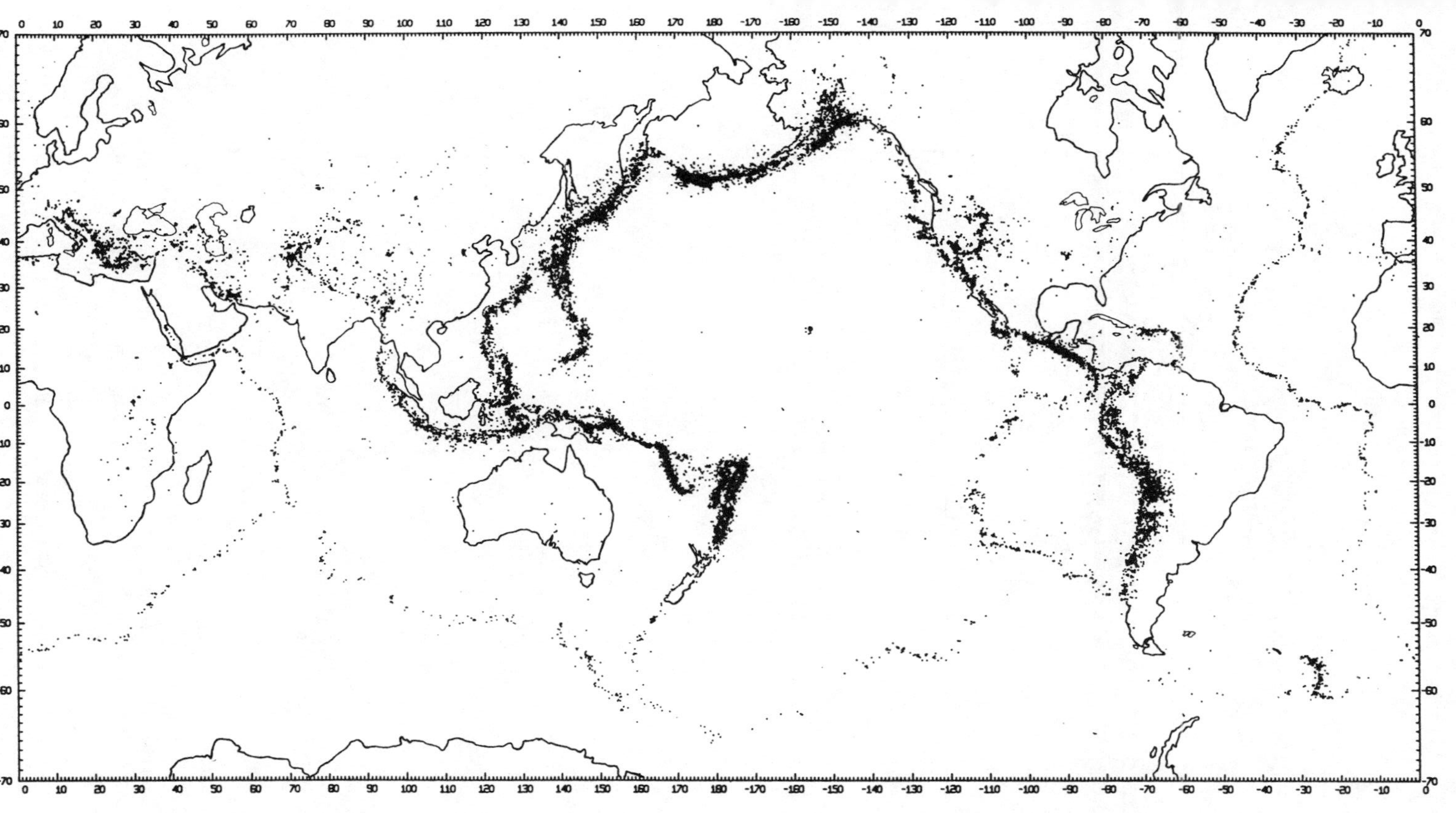

Figure 4.16 Worldwide distribution of all earthquake epicenters for the period 1961 through 1967. The earthquake zones are assumed to mark the edges of plates. (From Isacks, Oliver, and Sykes, "Seismology and the New Global Tectonics," *Journal of Geophysical Research* 73 [1968]: 5855–5899. Copyright © 1968 by the American Geophysical Union. As adapted from Barazangi and Dorman, "World Seismicity Map Compiled from ESSA Coast and Geodetic Survey Epicenter Data, 1961–1967," *Seismological Society of America Bulletin* 59 [1969]: 369–380. Reprinted with permission from Seismological Society of America.)

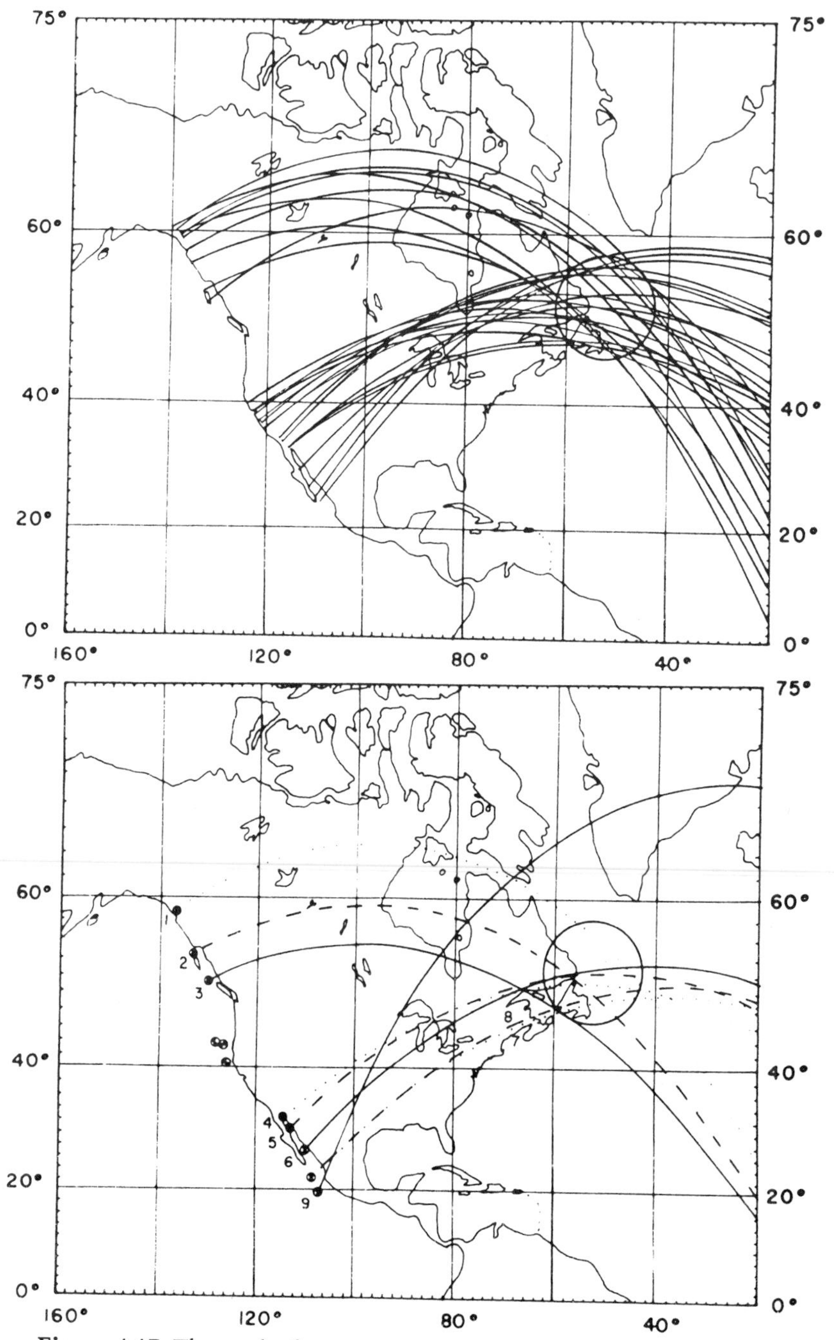

Figure 4.17 The results from two methods of calculating the pole of rotation between the Pacific and North American plates. The upper figure shows the area of intersection for circles perpendicular to transform faults between the plates. The lower figure shows the circles perpendicular to fault strikes determined from earthquake mechanism solutions. (From Morgan, "Rises, Trenches, Great Faults, and Crustal Blocks," *Journal of Geophysical Research* 73 [1968]: 1959–1982. Copyright © 1968 by the American Geophysical Union.)

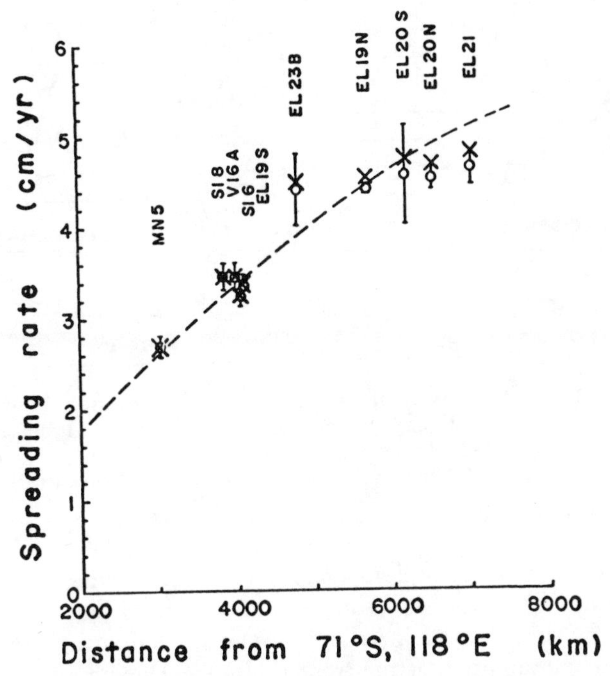

Figure 4.18 Observed and predicted spreading rates on the Pacific-Antarctic Ridge. The dashed line is the predicted rate based upon a spreading pole located by circles perpendicular to transform faults along the ridge. The circles are the spreading rates measured perpendicular to the strike of the ridge; the crosses are these rates projected parallel to the axis of spreading. The vertical bars represent estimated errors in calculating spreading rates for different profiles. (From Morgan, "Rises, Trenches, Great Faults, and Crustal Blocks," *Journal of Geophysical Research* 73 [1968]: 1959–1982. Copyright © 1968 by the American Geophysical Union.)

drift over the last sixty million years, which accepted Ewing and Ewing's (1967) suggestion that there had been episodic spreading. An additional comment was made with respect to the expanding earth hypothesis. He noted that spreading on north-south ridges generally was faster than spreading on east-west ridges, which was inconsistent with an uniformly expanding earth hypothesis.

The last of the four major articles in 1967 and 1968 was by Isacks, Oliver, and Sykes (1968) at Lamont. They not only outlined the essential features of the new model, but examined some of its implications for seismology. They noted that its impact on seismology would be "one of revolutionary proportions" and presented the first pictorial model of plate tectonics (Figure 4.19).

This model summarized several key components in the plate tectonics model. In the center of the model new lithosphere is created at a mid-ocean ridge, which is offset by two ridge-ridge transform faults. At the right the created plate descends into the mantle at a trench to create the Benioff zone.

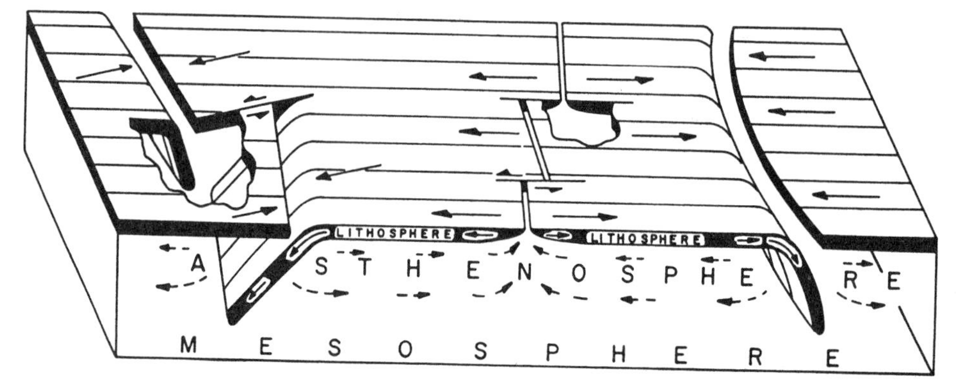

Figure 4.19 Isacks, Oliver, and Sykes's (1968) block diagram illustrating several key components and processes in plate tectonics. (From Isacks, Oliver, and Sykes, "Seismology and the New Global Tectonics," *Journal of Geophysical Research* 73 [1968]: 5855–5899. Copyright © 1968 by the American Geophysical Union.)

On the left side is a more complicated system of trenches connected by a trench-trench transform fault. Ridges, trenches, and transform faults constitute the three types of plate boundaries. As indicated in the figure the creation at the ridges and the consumption at the trenches imply that some form of mantle convection must occur.

Their paper began by outlining the development of continental drift theory and the dramatic results and syntheses obtained in the 1960s. They suggested:

> Certainly the most important factor is that the new global tectonics seem capable of drawing together the observations of seismology and observations of a host of other fields, such as geomagnetism, marine geology, geochemistry, gravity, and various branches of land geology, under a single unifying concept. Such a step is of utmost importance to the earth sciences and will surely mark the beginning of a new era. (Isacks, Oliver, and Sykes, 1968: 5860)

They adopt the perspective of plate tectonics, noting that it requires that they abandon the "traditional divisions of seismology," and organize their paper on the "principal effects predicted by the new global tectonics."

Accordingly, they discuss seismic evidence for the three types of plate boundaries: ridges, transform faults, and trenches. Seismic data from ridges are consistent with spreading. For example, ridge earthquakes occur only at shallow levels—deeper quakes could not occur in the hot, ductile environment of a spreading center. Furthermore, ridge earthquakes only occur along the ridge crests or the transform faults offsetting the ridge crests. The first-motion studies of the earthquakes on the ridge crests indicated that the ridges are under tension—splitting apart—and similar studies of earthquakes along the transform faults indicated that the sides of the transform faults are sliding past each other in the manner predicted by Wilson (1965*a*) and supported by Sykes (1967).

The seismic data from the third type of plate boundary—the trenches—was more extensive and in some cases seemed inconsistent with plate tectonics, but they provided a plate tectonics interpretation of these data. For example, most geophysicists had interpreted the first-motion studies of earthquakes in the Benioff zone as strike-slip faulting (horizontal movement) instead of a thrust of one side below the other, although both movements were possible interpretations of the first motion studies (Cox, 1973: 286). Isacks, Oliver, and Sykes noted that these previous studies had less reliable data and had neglected the thrust faulting interpretation of the same data. Some previous studies indicated tensional stresses along the front edge of the trench, which meant that the surface was splitting apart, instead of being compressed as one might expect from the plate tectonics model. They suggested, however, that the evidence for tension at trenches was associated only with the outermost surface of the lithosphere where it was bent and stretched before descending into the mantle. Other first-motion studies of earthquakes along the Benioff zone were consistent with compression along the deeper areas of the zone.

Isacks, Oliver, and Sykes speculated about the processes occurring in the Benioff zone: plates might slowly melt, break up, or bend, which could explain the distinct bends in some Benioff zones. They suggested that if plates required a certain amount of time before melting completely, then trenches with faster consumption should have deeper Benioff zones. Using data from LePichon (1968) on the estimated the rate of consumption at different trenches, they found that trenches with deeper Benioff zones had higher predicted consumption rates.

Other seismological data were found compatible with the plate tectonics model. The only anomalies seemed to be the presence of within-plate earthquakes and a few extremely deep earthquakes. However, these were not viewed as reasons for doubting plate tectonics, but as problems that needed to be reconciled with the new global tectonics. Some of the potential applications of plate tectonics to seismology were mentioned, such as earthquake prediction, revision of the simple layered model of the earth's mantle, and relating Benioff zones to the types of volcanism in the island arcs above them. They concluded:

> Surely, the most striking and perhaps the most significant effect of the new global tectonics on seismology will be an accentuated interplay between seismology and the many other disciplines of geology. The various disciplines which have tended to go their separate ways will find the attraction of the unifying concepts irresistible, and large numbers of refreshing and revealing interdisciplinary studies may be anticipated. . . . Even if it is destined for discard at some time in the future, the new global tectonics is certain to have a healthy, stimulating, and unifying effect on all the earth sciences. (Isacks, Oliver, and Sykes, 1968: 5845).

Each of these four articles noted the implausibility of some of their assumptions and the problems of data quality or interpretation, such as completely rigid plates, a limited number of plates, indeterminacy in first-motion studies, and the distorting effects of fracture zones on the symmetry of the magnetic profiles. Despite these problems the authors concluded that the simple model of plate tectonics provided a new approach for the study of the earth. Others seemed to agree, since the citation counts in Table 4.3 indicate that the articles from the 1967–1968 period were very influential. For example, the Isacks, Oliver, and Sykes (1968) article had over 100 citations in 1970, as did the LePichon (1968) article. These articles laid out the evidence for plate tectonics and the research that followed in 1969–1970 provided further tests and applications of this new theory.

A good example of an "anomaly" or "puzzle" for plate tectonics is the "magnetic bight" in the northeast Pacific, where there is a definite *bend* in the magnetic anomalies, which is represented in Part D of Figure 4.20. The simple seafloor spreading model predicts that magnetic anomalies should be *linear* strips parallel to a ridge, but the strips in Part D of Figure 4.20 are not parallel and there is no ridge associated with them. Pitman and Hayes (1968) proposed a solution that assumed the magnetic anomalies were formed at an earlier "triple junction" of three ridges, whose evolution is shown by the dark lines in Parts A, B, and C in the figure. Subsequent plate motions forced the ridge down the trench (the hatched areas in the figure). That is, the North American plate overrode the spreading center.

Other studies by Menard and Tanya Atwater[28] (1968) noted that discontinuities in the pattern of magnetic anomalies could be produced by changes in the direction of plate motions. Other irregularities were explained by the McKenzie and Morgan (1968) analysis of possible triple junctions and their development. They noted that sixteen possible types of triple junctions could exist between ridges, transform faults, and trenches. They used these types and simple geometrical reasoning to account for different types of discontinuities in magnetic anomalies.

Heat flow and ocean basin topography were related to plate tectonics by John Sclater[29] and Jean Francheteau[30] at Scripps. They proposed that ocean ridges were elevated features because they were composed of recently formed crust, so they were hotter and less dense, which caused them to "float" higher on the underlying asthenosphere. This explained the higher heat flow of ridges and the height differences on the adjacent sides of fracture zones because the two sides were of different ages, temperatures, and densities. Sclater and Francheteau (1970) were able to use the estimated spreading rates to predict the ocean depth and the rate of heat flow for different distances from the ridges by making assumptions about the thickness, composition, and heat conductivity of the lithosphere and about the rate of heating from below and the rate of cooling by the waters above. Their quantitative results indicated that oceanic variations in heat flow were not due to underlying convection currents, but were simply the result of a cooling lithosphere. This simple model explained why the Pacific Ocean was shallower than the Atlantic: It was spreading faster, so its lithosphere was hotter and less dense.

The early results of the Deep-Sea Drilling Project (DSDP) provided an important test of the implications of plate tectonics. This project was an out-

TABLE 4.3 Citation Counts to Selected Geoscience Publications from 1967 to 1968

Publication / Topic	*Science Citation Index* citation counts for: 1967	'68	'69	'70	'71	'72	'73	'74	'75
Sykes (1967) / seismological tests of Wilson's (1965*a*) transform fault concept	6	32	34	28	24	27	21	17	18
Oliver and Isacks (1967) / seismological evidence that the Benioff Zone is composed of a slab of the lithosphere	1	9	22	25	24	21	17	13	19
Ewing and Ewing (1967) / seafloor spreading and the distribution of oceanic sediments	5	24	28	28	9	12	3	2	6
McKenzie and Parker (1967) / application of the plate tectonics to the North Pacific	0	9	23	31	39	33	35	10	15
Heirtzler et al. (1968) / using the Vine-Matthews hypothesis and magnetic reversal dates to map the ages of the oceans' floors	—	13	48	66	81	66	57	36	56
Morgan (1968) / application of plate tectonics to the major ocean basins	—	9	48	65	83	77	49	27	27
LePichon (1968) / application of plate tectonics to the major ocean basins, especially to the opening of the Atlantic Ocean	—	5	58	106	98	97	76	49	44
Isacks, Oliver, and Sykes (1968) / seismology and plate tectonics	—	1	50	103	89	85	66	35	56

NOTE: Citation counts exclude self-citations. The source is Institute for Scientific Information (1972, 1975).

growth of the earlier "Mohole" Project, which had sought to drill though the ocean's crust to get samples of the mantle below the Moho discontinuity. Initial work on the project developed techniques for deep-sea drilling, but the Mohole Project was terminated because of social and political pressures described by

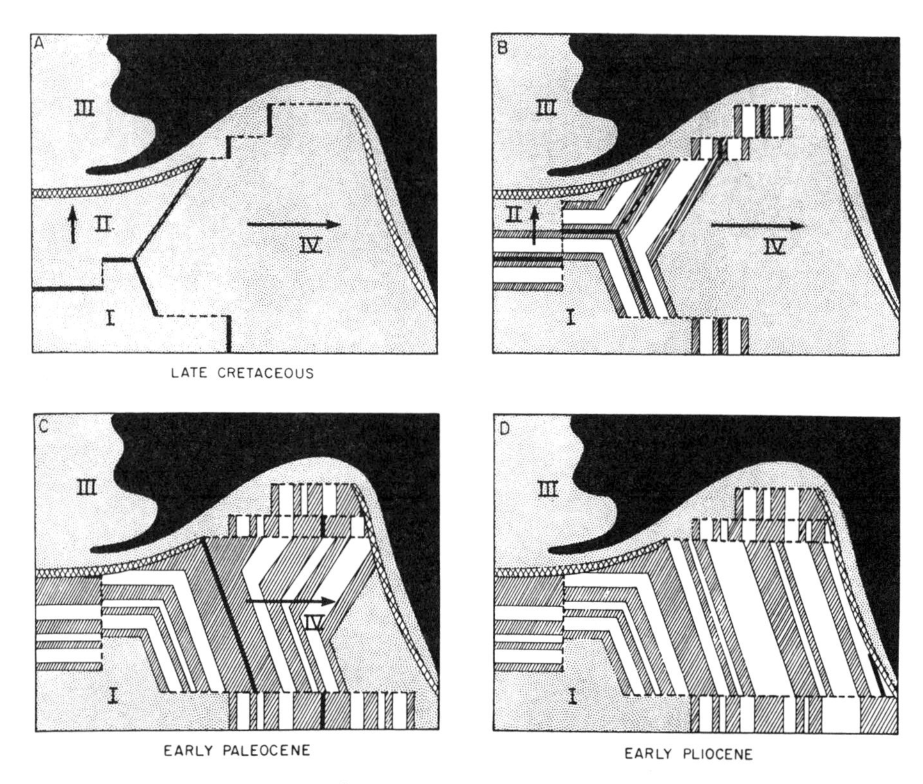

Figure 4.20 Pitman and Hayes's model for how a triple junction between three ridge crests could produce the magnetic bend or "bight" observed in the Northeast Pacific. Part D represents the linear magnetic pattern presently observed in the NE Pacific. Parts A through C represent stages of seafloor spreading from a junction of three ridge crests (dark lines) with offsets (dashed lines), most of which are consumed at the trenches (hatched areas) over time. (From Pitman and Hayes, "Sea-floor Spreading in the Gulf of Alaska," *Journal of Geophysical Research* 73 [1968]: 6571–6580. Copyright © 1968 by the American Geophysical Union.)

Sullivan (1974) and Greenberg (1967). In 1966 the National Science Foundation funded the DSDP to obtain sediment cores down to the basaltic ocean floor.

By 1970 initial DSDP results provided striking support for seafloor spreading. Maxwell et al. (1970) reported the results of drilling a series of holes through the sediments in the South Atlantic. These holes were drilled at various distances from the Mid-Atlantic Ridge and sediments immediately above the basaltic basement were dated by the microfauna fossils in these sediments. Figure 4.21 shows the ages of the fossils plotted against the distance from the ridge. The straight line is the predicted plot using the linear magnetic anomalies to date the age of the ocean floor at these distances. The close agreement between these two means of dating the seafloor provided support for seafloor spreading and indicated that spreading had been continuous, not episodic, for the last ninety million years.

These studies illustrate some of the ways that plate tectonics was used to explain oceanographic data, but during the late 1960s other studies applied plate tectonics to land geology. For example, Hurley (1968) compared the ages of rocks on different sides of the South Atlantic to test the computer fit of Bullard, Everett, and Smith (1965). Other studies sought explanations for how plate tectonic processes could produce the rocks and geological structures observed on land. William Dickinson[31] of Stanford and Trevor Hatherton[32] of Australian National University noted that the potassium content of the lavas in island arcs was correlated positively with the depth to the underlying Benioff zone. They suggested this correlation arose from the increased heat and melting of the

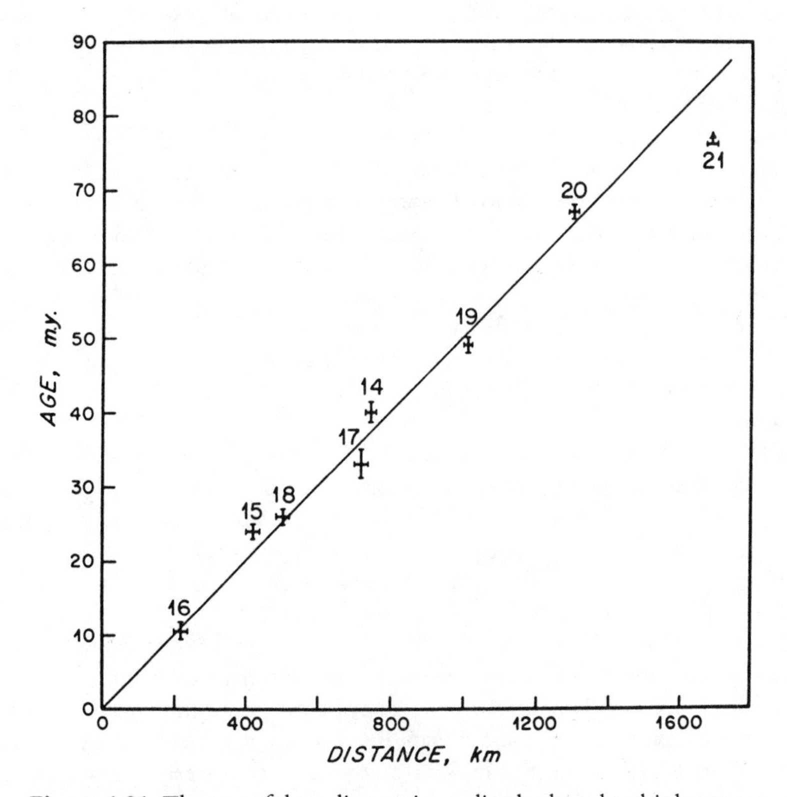

Figure 4.21 The ages of the sediments immediately above basaltic basement are plotted against the distances from the Mid-Atlantic Ridge. The ages were determined by identification of micro-fossils in the sediments. The line is the predicted ages assuming a constant spreading rate of about 2 cm/yr, which was similar to the rate of spreading estimated by magnetic anomalies. (From Maxwell et al., *Science* 168 [1970]: 1047–1059, copyright © 1970 by the AAAS.)

sediments carried down as oceanic plates were consumed at trenches (Hatherton and Dickinson, 1969).

In the middle 1960s Dietz had related the seafloor spreading model to traditional geosyncline concepts, but he was almost alone in doing so (Menard, 1986). The same was done with the plate tectonics model in the late 1960s, but now by land geologists. In particular, the 1969 Dickinson-Penrose Conference on "The Meaning of the New Global Tectonics for Magmatism, Sedimentation, and Metamorphism in Orogenic Belts" sought to reinterpret key aspects of geosyncline models in terms of plate tectonics (Dickinson, 1970). One of the first published efforts to relate plate tectonics to mountain building and the geosyncline model was by John Dewey and John Bird[33] (1970).

They summarized the major features of mountain belts that must be explained by any theory of mountain formation and noted that the ". . . understanding of mountain belts can only come from a full integration of their features with observed sedimentary, volcanic, and tectonic processes of modern oceans and continental margins" (Dewey and Bird, 1970: 2626). They considered the likely structure and composition of both the seafloor created at ridges and the sediments accumulated on continental margins, and then proposed hypothetical models of the events occurring in the two plate tectonic processes that form mountains: when an oceanic plate descends under a continental margin or when two continental margins collide. Figure 4.22 shows their schematic for the sequence of events occurring when ocean lithosphere descends under a continental margin. Allègre suggests that these integrating schemes were "quite badly received by geologists," but it was clear to all that Dewey and Bird had "invited geologists to use their methods and observations to build a new geology. The proposed method consists of linking a defined geologic entity to the site of its geodynamic origin within the framework of plate tectonics" (Allègre, 1988: 175–176).

Table 4.4 shows the citation counts to this and the other articles applying plate tectonics to oceanographic and continental data in the 1968–1970 period. Although these articles are not as highly cited as the articles laying out the foundations of plate tectonics (see Table 4.3), they had considerable influence on subsequent researchers. For example, the Dewey and Bird (1970) article was cited over fifty times per year shortly after publication.

The basic plate tectonics model did not gain instant success, especially among continental geologists. In general, geophysicists and marine geologists adopted it quickly. In North America, where geophysicists were numerous and influential, most continental geologists rapidly accepted the new theory within a couple of years. Allègre (1988: 120–121) suggests that European geologists were much slower, in part because geophysicists were a smaller fraction of their geoscience community. He notes that as late as 1978 half of the French geologists at a national conference were hostile to plate tectonics. The theory was not taught in Tokyo University's geology department until 1975, even though the geophysicists had accepted it much earlier. However, the longest resistance was by the geologic community in the Soviet Union, most of whom were hostile

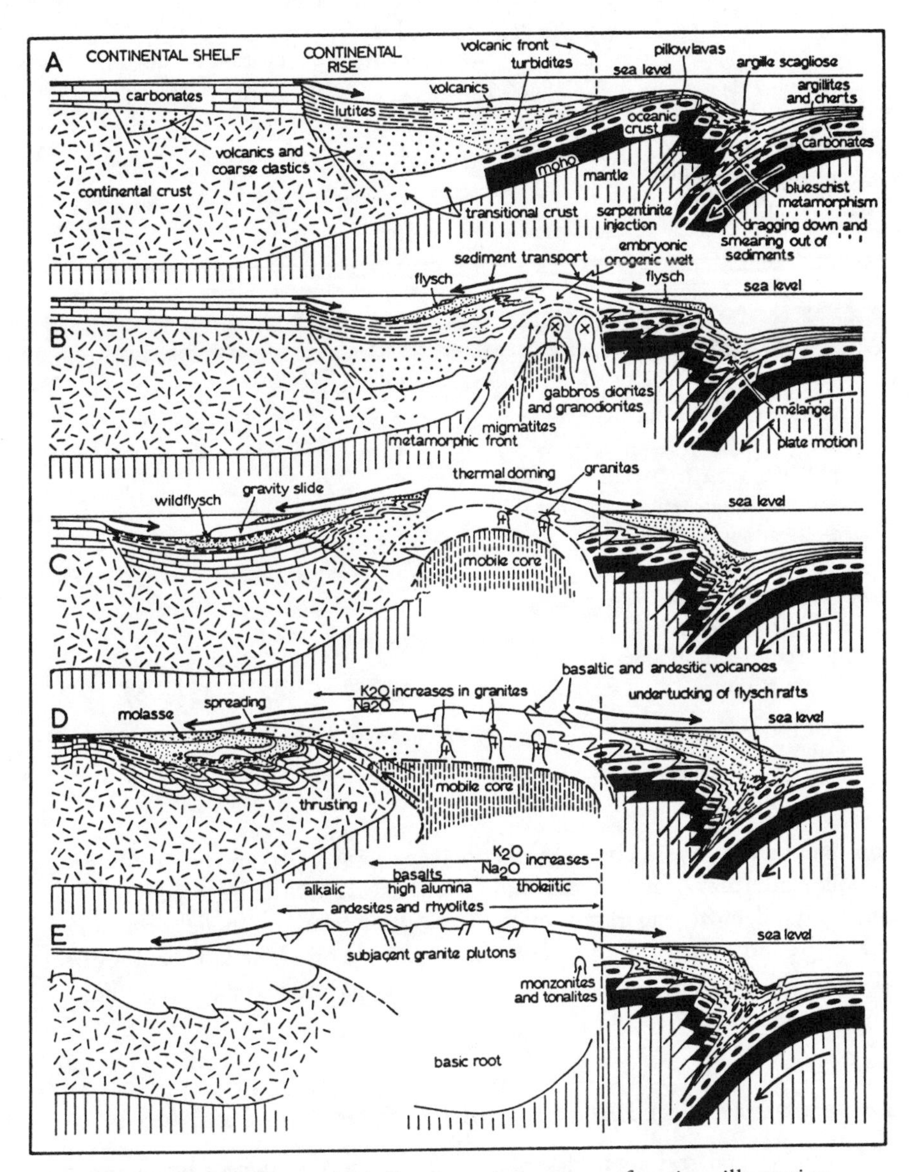

Figure 4.22 Dewey and Bird's schematic sequence of sections illustrating how a mountain belt might form from the underthrusting of a continent by an oceanic plate. (From Dewey and Bird, "Mountain Belts and the New Global Tectonics," *Journal of Geophysical Research* 75 [1970]: 2625–2647. Copyright © 1970 by the American Geophysical Union.)

TABLE 4.4 Citation Counts to Selected Geoscience Publications from 1968 to 1970

Publication / Topic	*Science Citation Index* citation counts for:							
	1968	1969	1970	1971	1972	1973	1974	1975
ˆPitmanand Hayes (1968) / triple junction explanation of major bend in magnetic anomalies	15	9	13	6	6	7	1	2
McKenzie and Morgan (1969) / geometry of triple junctions	—	0	17	29	31	21	11	13
Sclater and Francheteau (1970) / relating oceanic depth and heat flow to the age of the seafloor	—	—	1	5	23	25	36	35
Maxwell et al. (1970) / deepsea drilling supports age of seafloor predicted by magnetic anomalies	—	—	5	18	22	7	6	12
Dewey and Bird (1970) / schematic models relating plate tectonics to mountain range formation	—	—	2	37	52	42	57	49

NOTE: Citation counts exclude self-citations. The source is Institute for Scientific Information (1972, 1975).

throughout the 1970s. Undoubtedly these patterns of acceptance are due to a complex interplay of intellectual and social factors, but the next section focuses on the intellectual elaboration of the basic plate tectonics model occurring in the 1970s and some of the problems facing the theory. The last section of this chapter discusses the initial (and continuing) opposition to plate tectonics.

Developments in the 1970s and Continuing Problems

There are several reasons to describe briefly a few more recent developments. First, some of them are quite exciting and illustrate how a scientific theory evolves as it is tested and applied to more diverse data. Second, the interview excerpts used in later chapters mention several of the later developments. Third, some problems still remain unsolved. These deserve mention because they are areas of intense research and provide some of the reasons others continue to oppose this theory.

The literature on plate tectonics grew very rapidly in the 1970s. The 200 or so articles in 1970 were scattered under numerous subject headings in the *Bibliography and Index of Geology*. In 1976 the *Index* finally included a general "plate tectonics" category, which had over 750 entries distributed under various subcategories. In 1977 the general category had over 1100 entries. After 1977, the "plate tectonics" category simply referred one to over 200 different topic or area categories that included articles on plate tectonics. These numbers and changes not only show the rapid growth of the literature, but also suggest the diffusion of the plate tectonics model throughout the geosciences. Consequently, the following discussion only touches on a small portion of the literature and is based mostly on review articles and brief summaries given in general science journals. Allègre (1988) provides an accessible summary of the many ways that plate tectonics has integrated the different geoscience specialties.

Allègre (1988) suggests that some of the following results increased *geologists'* interest in plate tectonics. First, there was a change in the assumption of plate rigidity. The simple plate tectonic model implied that most phenomena of interest to continental geologists should be associated with the boundaries of the plates, especially the boundaries at trenches, where the ocean lithosphere was "subducted" into the mantle, and the boundaries caught between colliding continents. But the rigidity assumption used in kinematics calculations did not allow much internal deformation of the plates. During the 1970s some researchers argued that the plates may not behave in the rigid manner assumed in studies of plate movements. The plates may deform internally near ridges (Silver, 1974) and in some contact areas, such as in the western United States (Atwater, 1970) and in the Himalayas (Molnar and Tapponier, 1975, 1978). Two competing models for this process of deformation range from assuming that continental blocks consist of material that "flows" in plate boundary areas to assuming that along boundaries the continents break into smaller, *rigid* fragments that move as complete units (Molnar, 1988).

Another example is the use of plate tectonics theory to explain the origin of ophiolites, which were of special interest to structural geologists. Ophiolites are assemblages of rocks—deep-sea sediments above basaltic rocks, which are above "ultramafic" (high in manganese and iron) rocks—that occur in mountain ranges as contrasting units from the more typical sedimentary and granitic rocks. Previous theories had trouble explaining how these unusual assemblages of rocks were formed or placed among the more typical rocks. Dewey and Bird (1971) and Coleman (1971) argued that ophiolites represented slivers of the igneous ocean floor and its overlying sediments that were emplaced in mountains formed by collisions between continents and/or island arcs.

Perhaps the most exciting application of plate tectonics to continental geology has been the recent recognition of "exotic terranes," which builds upon a suggestion by Wilson (1968*c*) that plate tectonics implies that continental margins might contain fragments of other continents that were either left behind during the separation of super-continents or added to the margins by subducting plates carrying island arcs or continental fragments. One recent reviewer of this area suggests that exotic terranes imply that "continental geology stands on the threshold of a change that is likely to be as fundamental as plate-tectonic theory was for marine geology" (Saleeby, 1983: 45).

Exotic terranes are continental regions that differ from adjacent regions in several respects. Early evidence for exotic terranes was based upon different fossils in adjacent regions. For example, parts of the Pacific Northwest had fossils similar to those in China but not present elsewhere in North America. Before plate tectonics was accepted widely, this evidence was either doubted, ignored, or explained away by assuming that various shallow seaways had existed in the past and permitted limited, one-way migration of the organisms from China. Subsequent research identified more exotic terranes on the basis of fossils, structural incompatibility, rocks of different types and ages, and discrepant paleomagnetic poles in the NRM of the rocks. The results have major implications for continental geology and even the basic theory of plate tectonics. By the end of the 1970s the study of exotic terranes suggested that perhaps one-fourth of North America was composed of exotic terranes that had been accreted to the North American craton by subduction of oceanic plates carrying "micro-continents" that were added to the continental margins. Figure 4.23 portrays some of the exotic terranes that have been proposed for the Pacific Northwest and Alaska.

Coupled with the study of exotic terranes came changes in the concept of how plate tectonics can form mountains. The early version of plate tectonics held that mountains could be formed by two processes: (a) continental collisions and (b) subduction of oceanic plates under continents, such as the Andes in South America. However, some geoscientists have argued that subduction alone cannot create mountains like the Andes, but they only form when the subducting plate also carries micro-continents or ocean plateaus that are added to the continental margins (Ben Avraham, 1981; Ben Avraham et al., 1981). In other words, only collisions between continents and micro-continents or island arcs create mountains. Reviews of the miniature revolution created by exotic terranes are given by Saleeby (1983), Kerr (1980), West (1982), Jones et al. (1982), and Schermer et al. (1984).

A final example of the evolution of plate tectonics theory is the changing views on the "subduction" process, where oceanic lithosphere descends into the mantle at a trench. The simple subduction model of the early 1970s implied two processes. First, most of the sediments on the descending oceanic plate should be scraped off and welded to the edge of the other plate. Second, the trench area should be in a state of compression as the seafloor was shoved into the mantle. The first assumption was challenged by evidence that some trenches did not seem to have deformed sediments in them or stacked on the edge of the adjacent plate. The second assumption seemed inconsistent with the discoveries of Karig (1970, 1971*a*, 1971*b*) that sometimes a small "back-arc" ocean basin existed behind an island arc. The descent of the seafloor at the trench on the other side of the island arc implied compression of the region, but the back-arc basin seemed to be formed by seafloor spreading, which implied extension. One of the proposed explanations argued that the subduction process evolved

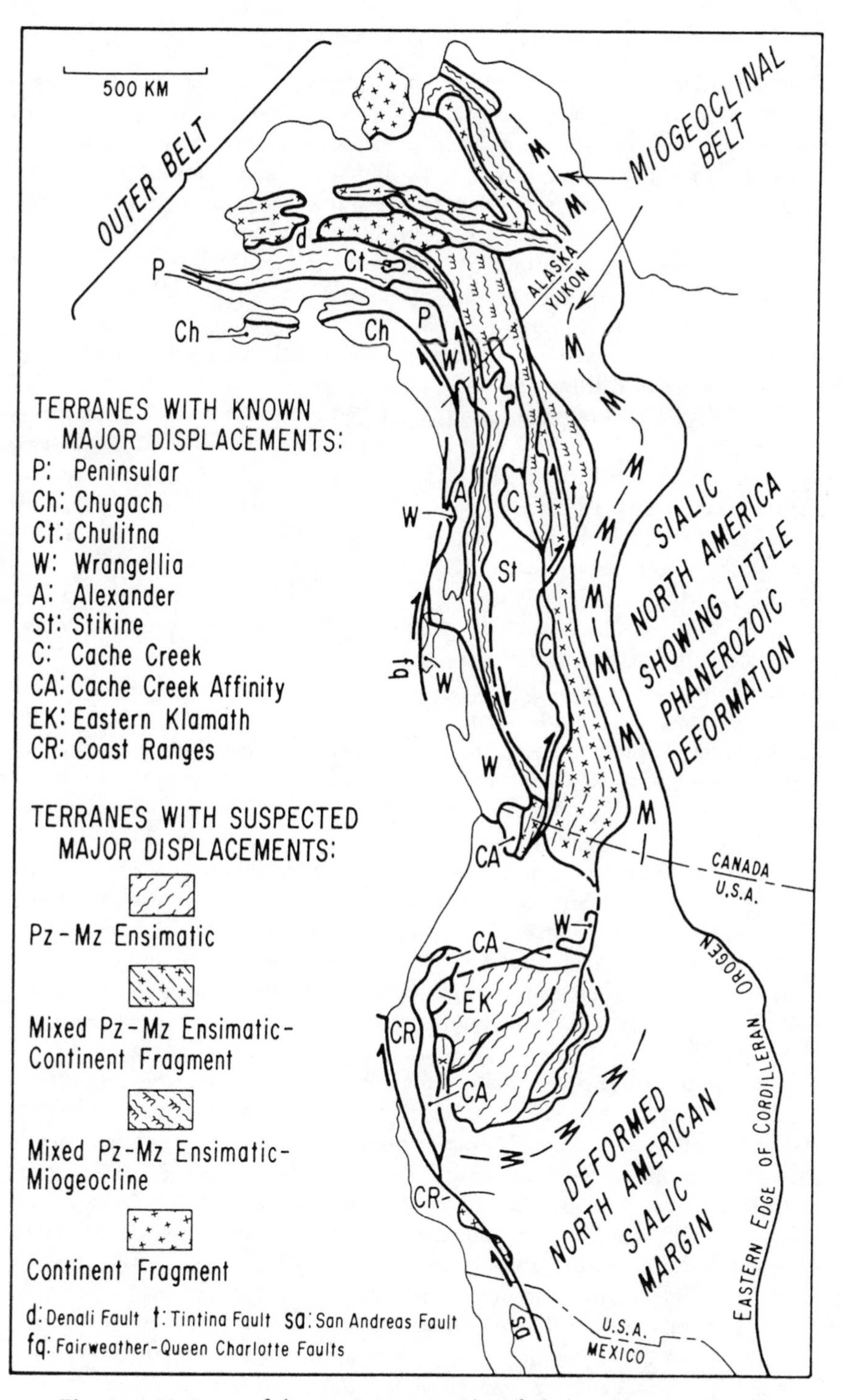

Figure 4.23 Some of the exotic terranes identified along the West Coast of North America. (From Saleeby [1983], reproduced, with permission, from the *Annual Review of Earth and Planetary Sciences*, Vol. 15, 1983 by Annual Reviews Inc.)

through various stages. It started with compressive forces and the scrapping of sediments, as assumed in the basic plate tectonics model. But then subduction evolved toward a steeper descent of the seafloor with the trench moving back from its old location. This allowed the sediments to descend with the plate and caused seafloor spreading in the back-arc basin to fill the extension of the earth's surface created by this backward shift in the trench location.

These are just a few examples of some developments in plate tectonics theory after 1970 and each could be elaborated in more detail. They all illustrate how the basic model was elaborated as researchers extended its ability to cover a wider range of geoscience phenomena or tried to explain known or new "anomalies." They seem to illustrate Kuhn's idea of "normal science" activities, where the basic plate tectonics model itself is not doubted, but relatively minor modifications are made to it in the course of research. Some geoscientists, however, never did accept the basic model or any of its elaborations. Before discussing some of their views, we need to consider one additional "add-on" to the basic model and then some of the continuing anomalies for plate tectonics. Both of these provide some of the basis for continued opposition.

In the early 1970s Jason Morgan revived and elaborated an earlier suggestion by Tuzo Wilson (1963*c*, 1965*b*) that "hot spots" existed in the mantle. Wilson had argued that linear island chains, such as the Hawaiian Islands, were formed as mantle hot spots melted a series of holes through the crust as it drifted over the stationary spots. Jason Morgan developed this mechanism in several articles that considered different island chains and suggested that hot plumes of material from the lower mantle provided both the heat for the hot spots and the driving force for plate motions. Two other possible advantages of plumes made them a major research topic in the 1970s. First, if they were relatively fixed in the mantle, they could provide an *absolute* framework for mapping plate movements, whereas the previous studies of plate dynamics were restricted to the *relative* movements between the plates. Second, mantle plumes might explain the vertical movements observed on some continents because a hotter, less dense plume would uplift an area in the overlying continent. This idea generated considerable research in the 1970s as researchers tested some of its implications, suggested alternative causes of "hot spots," or challenged the basic concept.

One of the reasons that mantle plumes appealed to some geoscientists is that plumes might solve some of the continuing problems or "anomalies" for the basic plate tectonic model. One of these problems was explaining some of the broad uplifts on the surface of the earth. Plate tectonics provided mostly horizontal forces, but areas like the Colorado Plateau seemed to be formed by a *vertical* uplift of a broad region. An underlying mantle plume might cause such uplifted areas, but plumes might explain an even more critical problem that also plagued earlier continental drift theory: what are the major forces that cause plate movements? There is little doubt that some form of mantle convection must occur to counter-balance the addition of mantle material to plates at mid-ocean ridges and the subduction of plates at trenches. The basic question concerns whether this convection drives the plates or is driven by the plates.

If the plates drive the convection currents in the mantle, then something else must drive the plates. Forces that might drive the plates have included the pull exerted by a dense plate as it returns to the mantle, the "push" exerted at mid-ocean ridges as new material is injected at the ridges, and gravitational sliding of plates down from the topographically higher mid-ocean ridges. Alternatively, mantle plumes or other forms of mantle convection could cause plate movements. Kerr (1978*a*) provides a review of these different theories and how they are being tested, but notes that the basic problem is the lack of data constraining these different models.

Some new data of possible relevance to this issue are now developing from the new technique of "seismic tomography," which is helping researchers develop three dimensional models of convection patterns in the earth's mantle by combining the information from many earthquakes recorded at many seismic stations. The seismic waves recorded at different stations pass through different parts of the mantle and travel more slowly in the hotter parts. With a sufficient number of paths through the same parts of the mantle it is possible to use computer analysis to model temperature differences, which presumably reflect convection patterns. Evidence presented by Dziewonski and Anderson (1984) indicates that mantle convection occurs throughout, not just in the upper half of, the mantle and that some mantle plumes seem to have origins in the lower mantle. More recent studies are even mapping "structures" on the boundary of the earth's core.

Another key debate is the degree that plate tectonics applies to the early history of the earth, especially Precambrian geology. Oceanic data, such as the linear magnetic anomalies, provides the strongest evidence for plate tectonics, but this evidence is limited to the last couple hundred million years—less than one-twentieth of the estimated age of the earth. The most relevant data for earlier periods are provided by paleomagnetic studies of the apparent polar wandering curves for continental cratons, but the errors in the estimated ages increase with age so that results are open to alternative interpretations. Kerr (1978*b*) reports that some geoscientists believe that Precambrian "plates" were too hot and weak to act in the coherent manner of modern plates so that subduction did not occur. Instead areas of deformation were limited to jostling and faulting between more stable cores without major movements between the different "plates."

In addition to these major problems, numerous "details" still remain unsolved. For example, the existence of old limestone at the Mid-Atlantic Ridge (Bonatti and Honnorez, 1971; Honnorez et al., 1975) does not fit into the simple plate tectonics model, which requires only young rocks to be present at ridges. Other "details" include linear magnetic anomaly patterns that do not fit the simple seafloor spreading model. To most proponents of plate tectonics these problems are either bypassed as less important or tackled within the framework provided by plate tectonics. However, opponents see the neglected "details" as more important and even question the relevance of the basic geophysical techniques and data providing the basic foundation of this revolution.

Opposition by a Minority

The history presented in this and the previous two chapters has emphasized the evidence that became incorporated into plate tectonics. As a result this account can be criticized as a biased reconstruction of the development of plate tectonics. To partly counter such criticism it is important to present the alternative theories and evidence given by the few geoscientists who continued to resist plate tectonics after the early 1970s.

After World War II proponents of continental drift were a small minority and more often ignored than criticized. This pattern basically continued up through the quick acceptance of seafloor spreading and plate tectonics in the late 1960s and early 1970s. After plate tectonics became the new basis of research only a few geoscientists publicly criticized it. The most outspoken critics included Harold Jeffreys, who still maintained that drift was not possible, Howard[34] and Arthur Meyerhoff,[35] a father and son pair of geologists, who have criticized much of the evidence for plate tectonics, Vladimir Beloussov,[36] a prominent Soviet geologist, and, surprisingly, S. Warren Carey, who continued to advocate an expanding earth hypothesis in order to explain the phenomenon of continental drift.[37]

Plate Tectonics: Assessments and Reassessments (Kahle, 1974) is a collection of mostly skeptical articles and indicates that some other geoscientists had reservations about selected aspects of the theory. However, this section deals primarily with Jeffreys, the Meyerhoffs, and Beloussov because they consistently denied the overall validity of plate tectonics and offered alternative theories. After describing these different theories, the remainder of the section examines the published debates between Beloussov and proponents of plate tectonics.

Harold Jeffreys opposed continental drift theory in every edition of *The Earth*, which was published in 1924 with revisions through 1970. He continued to advocate a contracting earth and when discussing continental drift he gave progressively less attention to the evidence for drift and focused instead on the physical properties of the mantle that seem to deny the possibility of continental movements. For example, Jeffreys (1974) argued that the "Lomnitz Law" for the viscosity of rocks implied that convection currents cannot exist in the mantle and that continents cannot move a significant distance over a rigid mantle material. These arguments seem to carry little weight among proponents, who had developed alternative mathematical models of the mantle that allowed convection (e.g., McKenzie, 1969). Furthermore, proponents (e.g., Bullard, 1975*a*) have noted that the Lomnitz Law is based on studies at ordinary temperatures and pressures with short time intervals, so its applicability to the mantle is dubious.

The Meyerhoffs (in collaboration with several other geoscientists) have published nearly a dozen articles against plate tectonics. These articles critique the published evidence for plate tectonics or cite evidence inconsistent with this theory. For example, Meyerhoff and Meyerhoff (1974) emphasize inconsistencies between the published polar wandering paths for the same continents.

They note that proponents have ignored linear magnetic anomaly data that were not symmetrical about ridges or that gave inconsistent spreading rates for the same ocean basins. To them this is evidence against the seafloor spreading interpretation, but to the proponents this means that better mapping of areas would show the faults that break up the simple pattern predicted by plate tectonics. Both sides have been able to cite cases where better mapping has or has not reduced the complexity.

They cite other examples of selective presentation of data, including the Deep-Sea Drilling Project (DSDP) results in Figure 4.21, which shows the ages of fossils at different distances from the Mid-Atlantic Ridge. The Meyerhoffs note that the original DSDP records indicated that some of the sediments above the basaltic basement were "baked," which they took to mean that the drilling had stopped at the level of a more recent basaltic intrusion between the sedimentary layers. Thus the true basement had not been reached and more sediments could be below the level of drilling, which implies that the seafloor was older than some of the ages plotted in Figure 4.21.

Aside from disputing the evidence used to defend plate tectonics, they also emphasize evidence that appears inconsistent with the theory. For example, they present paleoclimatic evidence consistent with the assumption of stable continents. They argued that the distribution of rocks indicating past climatic conditions, such as coal, evaporite deposits, and ancient deserts, is only compatible with present continental positions and ocean currents. The Permo-Carboniferous glacial evidence used by proponents is dismissed as due to mountain glaciers because, they suggest, continental glaciers could not exist in the centers of the reconstructed continents where precipitation would be insufficient. Finally, they emphasize that along the Mid-Atlantic Ridge a segment of ocean floor has been dated as over 100 million years old, whereas seafloor spreading implies only new crust should be at ridges. Proponents recognize this anomaly and consider the segment to be a crustal block that was somehow trapped between transform faults and sections of the ridge. The Meyerhoffs regard this as only one of many "ad-hoc" explanations used by proponents to protect plate tectonics from similar problems.

The Meyerhoffs and Briggs (1972) proposed an alternative global theory to account for both the evidence cited in favor of plate tectonics and the anomalies overlooked by proponents of plate tectonics. This theory is essentially a version of Jeffreys's contraction theory. They note that the asthenosphere is thicker under oceans, thinner under the continents, and appears to be absent under the most ancient parts of the continents—the continental cratons. They argue that this means the cratons and hence the continents are anchored in the mantle and cannot slide on the asthenosphere as assumed by proponents. When this coupling is combined with a rather small contraction of the earth due to thermal cooling, it suggests that the crust between the cratons will be buckled, which produces the doming and fractures along the mid-ocean ridges. The presence of both old and new rocks at the ridges is consistent with this model: old rocks have always been there, but may be covered by newer rocks as mantle material is squeezed out by the forces of contraction. The fracture zones in the oceans are attributed to uneven stresses created in the oceanic crust by the uneven shapes of the continental cratons. They argue that the linear magnetic anomalies are more parallel to the edges of the cratons than to the ridges, and that the anomalies probably are related to pressure stresses in the crust, which explains why some anomalies seem to extend onto continental crust. The major mountain chains of the world are explained as the result of contraction.

Beloussov's (1970, 1974) theory of crustal development is quite different from the Meyerhoffs', but like them he emphasizes geological data from the continents and the traditional models used to interpret these data, such as the geosyncline model. He had helped develop the Soviet model of crustal evolution with an emphasis on vertical, not horizontal, movements of the earth's crust. His model had the earth's crust broken into blocks separated by vertical faults that may extend into the mantle below. These blocks move independently of each other and whether they rise or subside depends on cycles of heating and differentiation in the deeper mantle. The buildup of radioactive heat under a block causes an increase in the melting and thickness of the asthenosphere, which causes a doming of the overlying crustal blocks. Continued melting of higher regions in the mantle, however, eventually causes the overlying crustal blocks, which are cooler and denser, to sink into the hotter and less dense mantle material, where the crustal blocks are absorbed and transformed into basaltic types of rocks. This transformation of granitic (sial) rocks into basaltic (sima) rocks was called "oceanization" because this process is used to account for the formation of the ocean basins.

Beloussov's evidence for his model included the observed uplift of wide areas within continental regions, such as the Colorado Plateau, and the existence of fault block mountains—those created when a large segment of the earth's crust is broken and tilted. The model also explains the equality of heat flow from the oceans and continents because the oceans have mantle material containing the "digested" continental rocks, where the radioactive elements are concentrated. Since he held that ocean basins formed from the oceanization of previous continental blocks whose sides are defined by vertical faults, his model explained some of the oceanographic data. For example, he suggested that the volcanic extrusions at island arcs and the mid-ocean ridges defined some of these vertical faults. Even the linear magnetic anomalies are explained as successive lava flows from the central rift valley with a distinct pattern where the older flows extend further than more recent flows. This produces a "shingle" effect, an elevated ridge crest, which, when combined with the magnetic reversals in the earth's field, produces linear magnetic anomalies somewhat parallel to ocean ridges. Beloussov argued that the initial stages of oceanization are observable today where massive basaltic flows have erupted through continental areas and spread over large areas, such as those that exist in the American Northwest.

More than other opponents of plate tectonics theory, Beloussov (1968, 1979) has engaged in published exchanges with proponents. A brief summary of some of their features will provide a fitting conclusion to this history and serve to raise some of the issues discussed in subsequent chapters. The first debate was in response to an article by Tuzo Wilson (1968*a*), who outlined the development of plate tectonics and suggested some of the implications for the earth sciences, university curricula, and industrial applications. Beloussov (1968) gave an invited critique, which was followed by comments from Wilson (1968*b*). Beloussov's main critique was that plate tectonics "schematized natural phenomena" even more than the contraction theory which it replaced, because plate tectonics emphasized oceanic data and ignored more abundant continental data. He gave some specific objections against plate tectonics, such as some of those mentioned earlier, to which Wilson (1968*b*) responded. Beloussov argued that the best methodology for geology was the "method of multiple working hypotheses," which should include his concept of oceanization.

Ten years later Beloussov provided some more objections to plate tectonics, and his article (Beloussov, 1979) was given a paragraph-by-paragraph response by A. Sengör and Kevin Burke (1979), which was followed by Beloussov's replies. Plate tectonics was the established viewpoint at this time and the responses of Sengör and Burke had a more philosophical emphasis with less hope for agreement. They began by noting:

> Our emphasis has not been on citing detailed evidence that shows the weakness of Beloussov's position because we believe that such evidence is both ample and familiar. Rather, we draw attention to what we see as his philosophical attitude and the historical perspective [which,] we conclude, derives from those of the German masters of contraction tectonics. Long before the advent of plate tectonics the inadequacies of this approach had been recognized and its place taken by hypotheses involving large-scale continental displacements in the writings of such tectonicians as Suess, Argand, and du Toit.
>
> Our general feeling is that dialogue with Beloussov is unlikely to be very fruitful, since the premises on which his interpretation stands are so different from those on which both plate tectonics and the mainstream of tectonic studies rest. (Sengör and Burke, 1979: 207)

It is interesting to note how Sengör and Burke have reinterpreted the history of geological thought. Eduard Suess argued for a contraction theory of mountain belts (Hallam, 1973; Greene, 1982). Emile Argand did question the contraction theory and advocated some continental movements, but these aspects of his and du Toit's writings were not adopted by later geologists until the sixties (Marvin, 1973: 86–87). Perhaps what we have here is something similar to Kuhn's suggestions that proponents of a new theory will reinterpret the historical developments of the theory. Some other aspects of this debate have a Kuhnian "flavor." For example, these geoscientists tend to emphasize different philosophies, models, data sources, and interpretations of research results. A few examples from this debate illustrate these problems.

The participants in this debate could not agree on which data sources were most relevant to theories of crustal evolution. For example, Beloussov emphasized the importance of continental data and argued that the evidence for the average equality of heat flow from the continents and oceans must be explained by any theory. Oceanization explains this equality, but Sengör and Burke saw it as an "accident of history" since it appeared that heat flows were quite different in the past. They favored the explanation of ocean basins and mountains in terms of the "Wilson cycle" for the opening and closing of ocean basins, where mountain belts are "suture zones" created by the closing of an ocean basin. They gave the Ural Mountains in the Soviet Union as an example of a suture zone, but Beloussov claimed this interpretation is contrary to all the studies that he and other Soviet geologists had conducted in the Urals.

The principle of uniformitarianism was brought into the debate. Sengör and Burke suggested that uniformitarianism means geological theories must be based upon processes seen in operation today, which in turn means plate tectonics is a uniformitarian theory because it is based on presently observable processes occurring in the oceans and at continental margins. In contrast they argue that Beloussov's use of the geosyncline model, which is based upon fossil evidence, is not uniformitarian. Beloussov responded that to deny the fundamental importance of the fossil record is to deny geology as a "natural historical science."

The reader may notice a curious twist in justifications for plate tectonics. In the debate over drift theory in the 1920s, drift theory was cited as not uniformitarian, the fossil record was used as evidence for drift, and drift proponents argued for the method of multiple working hypotheses. In 1979 it was just the opposite.

The participants raised philosophical issues, which included statements about the proper philosophy of science. Furthermore, they could not agree on how to apply these philosophies to geoscience research. For example, Beloussov argued that defining plate boundaries by the location of earthquake belts and then turning around and using plate tectonics to explain the location of these belts was a "vicious circle." "Consequently, the plates explain nothing. They only repeat in other terms what has been well known long ago" (Beloussov, 1979: 209). He argued that the knowledge about the formation of ocean crust was still insufficient, and we must know the "real (not just supposed) origins of the magnetic anomalies."[38] In response Sengör and Burke suggested that geologists tacitly or consciously use a Popperian method and that the best explanation of the anomalies was the Vine-Matthews hypothesis, which had not yet proven to be inadequate. However, when Beloussov noted that the seafloor spreading hypothesis could not explain the older blocks of sediments found near the ridges, Sengör and Burke argued that the rocks "need to be studied further, but we are not prepared to abandon a theory that explains so many aspects of Atlantic geology for one curious observation" (Sengör and Burke, 1979: 208). Note how this assertion contradicts Popper's argument that theories should be abandoned when such a *striking and accepted* anomaly occurs.

Later in the article Sengör and Burke accuse Beloussov of following a "Baconian or inductive approach" to science and suggest that Popper had established the inadequacy of this method: "what scientists do . . . is to advance hypotheses and then try as hard as possible to prove them wrong" (Sengör and Burke,

1979: 209). Beloussov agrees with this statement, and asks: "But why then do plate tectonicians stubbornly disregard things that are wrong in plate tectonics theory, and why are they so aggressive toward those who try to point out the wrong sides of their theory? Is this attitude evidence of the strength of the theory?" (Beloussov, 1979: 209). Finally, both debaters argue that the other's theory contains more ad-hoc modifications than their own.

It is unusual for such debates to appear in the scientific literature, but their presence and the nature of these debates indicates that there are fundamental differences between the majority who accept plate tectonics and the minority who oppose it. These differences can cause considerable hostility between the two groups, especially when the debaters tend to make rather categorical judgments about their opponents' views, such as "there simply is no foundation to [plate tectonics]" (Beloussov, 1968: 17) and "the basic assumptions of the new global tectonics range from unproven speculations to errors in fact" (Meyerhoff and Meyerhoff, 1974: 47). In a letter to the editor written in response to Arthur Meyerhoff's (1972) review of a book on the history of plate tectonics, Warren Hamilton (1972) replied: "Meyerhoff's zeal in this review—as in his other recent articles and talks—leads him to denounce the intelligence and honesty of proponents of continental motions. He is a master of one-line assertions that are irrelevant where technically correct but which too often are incorrect" (Hamilton, 1972: 10).

Any perspective hoping to describe how scientific knowledge develops should help us understand the nature of these debates, including those occurring when only a minority accepted continental drift. Both debates share many similar features. Before concluding this chapter we should note that most geo[chscientists have accepted plate tectonics and tend to ignore the alternative theories and criticisms of the opponents. This conclusion is supported by contrasting the citation counts in earlier tables with those in Table 4.5, which shows that most of the articles by the opponents were rarely cited after publication.

Summary and Discussion

In 1960 Harry Hess presented a piece of "geopoetry"—the seafloor spreading model—as a way to synthesize the surprising results of postwar oceanographic research. It was virtually ignored even though Vine and Matthews indicated how it could explain the linear magnetic anomalies. By 1966 Vine and others presented new, more quantitative evidence that persuaded many geoscientists that Hess's geopoetry was "geofact." Within two years Hess's theory was incorporated into plate tectonics as Jason Morgan and others showed how the newer theory quantitatively interrelated even more of the geophysical characteristics of the ocean basins.

For most geophysicists and marine geologists the "revolution" was over by 1970. They immediately pursued several directions of research. Some produced further "verifications" of the basic model, such as those obtained by the Deep-

TABLE 4.5 Citation Counts to Selected Publications Opposed to Plate Tectonics

	Science Citation Index citation counts for:						
Publication / Topic	1971	1972	1973	1974	1975	1976	1977
Beloussov (1970) / critique of evidence for plate tectonics and defense of "oceanization"	1	10	2	7	3	1	1
Meyerhoff, Meyerhoff, and Briggs (1972) / a contracting Earth alternative to plate tectonics	—	0	3	3	2	0	0
Meyerhoff and Meyerhoff (1974) / critique of evidence for plate tectonics	—	—	—	—	0	3	1
Jeffreys (1974) / argues against possible mantle convection	—	—	—	—	0	0	0
Beloussov (1974) / critique of evidence for plate tectonics and defense of "oceanization"	—	—	—	—	0	0	0

NOTE: Citation counts exclude self-citations. The source is Institute for Scientific Information (1975, 1976, 1977, 1978).

Sea Drilling Project. Others used it to explain the magnetic anomalies in other ocean basins. Yet others focused their efforts on the geological and geophysical processes occurring at the key areas defined by the new theory: the plate boundaries. Most importantly, some geologists adopted the basic framework and elaborated it so that it was seen as relevant to continental geology. The latter efforts were so successful that almost all geologists eventually accepted the basic framework by the end of the 1970s. The revolution was over and the many specialties in the geosciences were integrated together for the first time in their history. A pattern of "normal science" research was established and continues today.

Kuhn's "normal science" is not quite established because there are still some opponents of the theory. However, they are small in number, disagree on their alternative theories, and have almost no effect on the ideas of the majority. Furthermore, they are diminishing in numbers as they literally die off. Yet it would be a mistake for us to ignore them. To do so would make it all too easy to see this revolution as the "obvious" product of "rational" scientists collecting data and simply developing increasingly better theories to account for the empirical results. Even with an awareness of the opponents' views, it is difficult to avoid a tendency to label them as "irrational" holdouts. This tendency is accentuated because the history presented here has emphasized the major evidence incorporated into plate tectonics. The relationship between data and theory now seems so "logical" that it is difficult to see the new theory as a socially constructed belief system that is created by the interactions, negotiations, and decisions of a cultural group.

This chapter concludes the presentation of the historical "data" on the plate tectonics revolution. The next two chapters will use this historical (and interview) information to compare the different perspectives on science outlined in Chapter 1. By the end of Chapter 6, it is hoped we will have accepted a broader view of what occurred in this revolution—not one that minimizes the tremendous intellectual accomplishments of geoscientists, but one that sees these accomplishments as the product of normal social beings using typical methods of reasoning and arguing within a unique type of social organization.

CHAPTER FIVE

Philosophical and Historical Perspectives

This chapter compares the philosophical and historical perspectives outlined in Chapter 1 to some "data" on this revolution. These data include the preceding history, which not only mentioned some of the major intellectual events and individuals in the development of plate tectonics theory, but also gave us some sense of how geoscientists think about the earth, use empirical data, and argue among themselves. This essential information will help us understand three other sources of data: (a) the comments obtained from personal interviews, (b) some excerpts from the geoscience literature, and (c) previous efforts to describe this revolution in terms of particular historical and philosophical perspectives.

The first four sections consider the perspectives of logical empiricism, Popper, Lakatos, and Laudan. All of these perspectives emphasize the rationality and progressiveness of scientific knowledge. The discussion of each perspective emphasizes its particular contributions and problems for our understanding of this history. The fifth section considers some of their common problems. The remaining two sections describe Kuhn's perspective and examine the "models and analogies" perspective as one contribution to solving some of the problems in Kuhn's perspective.

The following discussion cannot "prove" that one of these perspectives is the best one of those considered. This occurs for several reasons. First, the data sources are too incomplete to meet the standards of many of the readers, especially historians of science. Second, the different perspectives vary in their degree of specificity, their assumptions, their models for how scientists reason, and, hence, the data most relevant to their perspective. Finally, the readers of this book do not share the same set of presuppositions that would allow a "proof," even if the other two problems could be removed.

This basic description of our present situation should sound familiar: It has many parallels in the 1920s' debate about continental drift theory. In that debate, the above problems did not stop geoscientists from arguing for or against continental drift theory, and I also will adopt a particular perspective. Although each of these perspectives helps us understand some aspects of this history, none seem completely adequate, especially as we consider more of the specific reactions of individual scientists. Yet I shall adopt Kuhn's basic perspective and suggest how some modifications increase its applicability to the history of continental drift and plate tectonics. But there are still some problems of the "causal mechanism" of scientific choice in Kuhn's perspective. The problem is not one of an inadequate mechanism, but an incomplete specification of the mechanism.

Thus, the last section of this chapter and the next chapter try to elaborate a mechanism fairly consistent with Kuhn's perspective.

Logical Empiricism

Given the emphasis on deductive or mathematical reasoning as the key "rational" process in science, it is not surprising that logical empiricists have focused on the highly formalized sciences, such as physics, with fewer applications in less formalized sciences, such as geology. As is apparent from our brief history, the rise of plate tectonics gives a new mathematical and quantitative rigor to the geosciences. We can use this aspect to illustrate some of the essential points and problems of the logical empiricist tradition.

Cox (1973: 40–43) formalizes some of the basic assumptions of plate tectonics. He suggests there are two key "postulates" in plate tectonics: "the plates are internally rigid but uncoupled from each other" (Cox, 1973: 41) and "the pole of relative motion between a pair of plates remains fixed relative to the two plates for long periods of time" (Cox, 1973: 42). After defining such terms as "[spreading] pole of relative motion" and "plate boundaries," Cox provides several "theorems" that can be derived from the above postulates with the mathematical rules of spherical geometry. These theorems include the following hypotheses: (a) transform faults between plates must lie on circles centered at the pole of rotation, (b) this pole may be located by the intersection of lines perpendicular to local segments of transform faults, and (c) the width, W_A, of new lithosphere formed at ridges is given by $W_A = W_O \text{ Sin A}$, where A is the arc distance from the spreading pole to the point of observation and W_O is the maximum rate of spreading, which occurs at the spreading pole's equator (see Figure 4.15).

The development and subsequent support for this formal model clearly makes logical empiricism more relevant to plate tectonics than to the earlier continental drift theory. However, there are still several problems. For example, some of the basic "postulates" of plate tectonics theory are known to be false: (a) it is impossible to have stable poles of spreading on a sphere with more than two plates (Cox, 1973: 408–409), (b) the plates are not rigid, and (c) the earth is not a perfect sphere. Thus the basic model cannot be a "true" description of reality. Another problem is the assumption of theory-free data. For example, the magnetics data used to test the formalized plate tectonics model depend on the theories of electromagnetic radiation and atomic structure as expressed in the design of the proton-precession magnetometer. More importantly, it is necessary to assume certain properties of the seafloor in order to translate the accepted reversal chronology into a predicted magnetic anomaly pattern that can be compared to the measured magnetic profile. These are just some of the secondary assumptions needed to test the formal model's predictions.

A third problem concerns the logical "circularities" in the explanations within plate tectonics theory. For example, the theory *predicts* that most earthquake activity occurs along the edges of plates, but plate boundaries are *defined* by the pattern of earthquakes on the earth's surface. Additionally, the theory predicts that transform faults will lie along concentric circles centered at the spreading pole, but instead of using the transform faults to test the theory, they might be used to *define* the spreading pole. A fourth problem is the difference between what scientists regard as "confirmations" and what would count as *logical* "confirmations." For example, plate tectonics implies that volcanos will always be associated with ocean trenches. Scientists would regard the observation of a trench with the Aleutian volcano chain as a confirmation, but would not consider the *absence* of a trench/volcano pairing in Ohio as a confirmation, even though symbolic logic implies that this joint absence is logically consistent with the implied pairing. Finally, there are other problems dealing with how scientists select which facts are important, decide when the "fit" between predictions and observations is "adequate," or invent elaborations of the basic model to handle geological data. These and other problems pertain to several of these perspectives, so they will be discussed later.

Critical Rationalism

The critical rationalism or "falsificationism" of Karl Popper (1959) shares many of the above problems. Since Popper also assumed that the "rational" thought of scientists was equivalent to the use of deductive logic, the problems with logical reasoning identified above will apply here as well. In addition, his emphasis on trying to falsify theories and to reject a theory that was "falsified" does not fit the behavior of geoscientists, even though we have seen that some geoscientists explicitly state they are following a "Popperian" philosophy.

Plate tectonics was born and still exists in a "sea" of anomalies, but these do not cause its rejection. Instead, they provide "problems" that need to be integrated into the theory or are simply ignored for the time being. As noted in the previous chapter, plate tectonics predicts that only very young rocks will be found at the mid-ocean ridges, where new crust is being formed, but few geoscientists doubt that a limestone layer over 100 million years old has been found at a segment of the Mid-Atlantic Ridge. For the opponents this "disproves" plate tectonics, but proponents simply try to figure out how this happened within the basic plate tectonics framework.

Perhaps the most important strength of Popper's perspective was his recognition that empirical data were not theory-free. That is, the test of a theory required the assumption of other secondary theories and results, so a negative result could be due to using a false theory or using false secondary assumptions. For example, Peter Molnar at MIT was studying whether mantle "hot spots" had stable locations with respect to each other. His work combined data on magnetic anomalies, paleomagnetic determinations of past continental positions, and linear island chains that were assumed to be produced by oceanic plates moving over hot spots in the mantle. In a 1974 interview he noted that some problems had developed.

> We just found that we couldn't fit the paleomagnetic data [from the continents surrounding the Atlantic and Indian Oceans with similar data from the Pacific] right now and this means that either plate tectonics is wrong, which is not really a viable alternative, or the reconstructions that we got are wrong, which I don't think is the case, or it means the paleomagnetic data are wrong, which may be right, or it means the dates for the paleomagnetic rocks are wrong, which I think is the most likely. . . . It's kind of neat. If we can turn to the geochronologists and say that we believe plate tectonics and we believe paleomagnetics and, therefore, your data are wrong, they aren't going to like it.

It is clear that empirical tests of important theoretical propositions assume the use of other theories, and scientists believe some of these assumed theories are more viable than others. Furthermore, there may be disagreements among scientists and possibly even conflict over what assumptions may be invalid. Although Popper did not deal adequately with how such disagreements were resolved, he did recognize the importance of these secondary assumptions in the test of a particular theory.

For Popper the failure of geoscientists to follow his methodology only means that they illustrate poor scientific practice. However the remaining philosophical and historical perspectives are more interested in simply describing how scientists actually behave. Since Lakatos and Laudan strive to maintain some of the features of logical empiricism, such as its presumption that science is "rational" and "progressive," we will examine them before considering the fundamentally different approach of Thomas Kuhn.

Sophisticated Falsificationism

In contrast to Popper, Lakatos (1970) accepted that confirmations played a role in theory acceptance, which is certainly true for the acceptance of plate tectonics. He agreed with Popper that "rational thought" was equivalent to the use of deductive logic and that theories could not be empirically tested without accepting some secondary theories, assumptions, boundary conditions and agreed upon facts. Thus his approach has the same problems with the use of deductive logic that were noted above, but his suggestion that a tested theory and the necessary secondary assumptions represented an evolving, larger "scientific research programme" avoids some of Popper's problems. Now a "falsifying" fact does not cause rejection of the tested theory residing in the "hard core" of the research programme, but it does require an adjustment in its "auxiliary belt" of secondary assumptions.

This addition certainly characterizes much of current plate tectonics research. For example, the old limestone near the mid-ocean ridge problem mentioned above did not cause the rejection of plate tectonics, but some modifications in other assumptions. Bonatti and Honnorez (1971) reported this anomaly and suggested that the limestone formed on a block of the earlier "proto-Atlantic" seafloor and that this block was trapped and held in a transform fault zone between two segments of the Mid-Atlantic Ridge. This explanation protected the "hard core" of the plate tectonics model.

However, Lakatos suggested that these "ad-hoc" modifications are not "rational" unless they also imply "novel facts" that can be empirically supported. That is, modifications in the auxiliary belt of a research programme must not only protect the hard core, but also increase its domain of explained facts beyond the addition of the explained anomaly. If the novel facts are later verified, then the research programme is making "progressive problem shifts." If they are not verified, then the programme becomes littered with ad-hoc modifications and is in a "degenerative phase." Thus the above explanation for the anomalous data must imply, for example, the existence of previously unobserved faults that allow this block of seafloor to remain trapped at the Mid-Atlantic Ridge for millions of years.

One can find numerous examples of these ad-hoc adjustments in modern plate tectonics research, so this aspect of Lakatos's perspective seems supported. Yet there seems less concern with Lakatos's emphasis on stating and empirically supporting the implied "novel facts" (Saull, 1986). This provides some of the reasons that opponents of plate tectonics think it is being protected by a proliferation of ad-hoc adjustments. Finally, the current dominance of plate tectonics among today's geoscientists conflicts with Lakatos's suggestion that science will be characterized by competition between research programmes. Although we discussed some alternative theories in the previous chapter, plate tectonics is the dominant perspective today and none of the alternatives to it have received even the limited attention bestowed on drift theory in the earlier debates (Hallam, 1983). Lakatos's perspective only describes selected aspects of current research in the geosciences, but perhaps it describes better the earlier debate on continental drift theory.

Frankel (1979*a*) has used Lakatos's perspective to analyze the earlier debate about continental drift and the eventual acceptance of seafloor spreading. He identified three research programmes in his examination of the history of drift theory: contraction, permanence, and drift. The oldest programme was the contraction programme, which held that the earth had been contracting since its formation and that seafloor and continents had exchanged positions periodically. The permanence programme's core belief was that, after an initial contraction, the surface of the earth was composed of two different materials, which formed the ocean basins and the continents, and these surface features were permanent features that did not interchange or move with respect to each other. In the early part of this century Europeans generally accepted a contraction model and North Americans accepted the permanence programme. Drift theory was the newest programme and directly challenged both of these programmes' assumption of no or little horizontal movement of continents.

Frankel suggested that each of these programmes had to protect their hard cores from falsifying evidence. The contraction proponents had to adjust their theory to the discovery of radioactive heating and the principle of isostasy, so Thomas Chamberlin and Harold Jeffreys each proposed new models for a contracting earth that tried to account for these features. Proponents of permanence had to account for the paleontological evidence of past land connections within the hard core of permanently fixed continents, so Bailey Willis suggested that isthmian connections and sunken land bridges created by changing sea levels could account for the distribution of fossils. Those sympathetic with the drift programme, such as du Toit and Holmes, proposed mantle convection currents as a way to bypass the problems created by Wegener's weak forces for drift.

Here at least we seem to find competing research programmes with the protective modifications of each programme's auxiliary belt. However, Frankel's major emphasis is the inability of the modifications of drift theory to make confirmed, novel predictions. For example, he argues that the convection models added to drift theory only saved it from "extinction" and did not make testable predictions. The development of paleomagnetic evidence in the 1950s would seem to add additional support for drift, but Frankel suggests that geoscientists were rational to continue to resist drift because polar wandering provided an alternative, non-drift interpretation of the same data. Similarly, geoscientists were rational to resist Hess's seafloor spreading model because the new evidence explained by his theory was precisely the same evidence he considered in developing his version of drift theory. It didn't predict any novel facts. According to Frankel, it was only rational to accept drift theory after Lamont's Eltanin-19 profile supported the Vine-Matthews hypothesis and Sykes (1967) supported Wilson's (1965*a*) concept of a transform fault. Both of these predictions were novel facts not mentioned by Hess, but shown by later researchers to be implied by seafloor spreading.

There are a number of problems in Frankel's account of the development of drift theory. For example, Frankel ignores Hess's (1962) prediction of a novel fact: that islands further away from the mid-ocean ridges would be older. Within a year Wilson (1963*a*) offered support for this novel fact (see Figure 4.2) and, furthermore, suggested that seafloor spreading could explain linear island chains and Benioff zones. Earlier modifications of drift theory also implied novel facts, but without much effect on the beliefs of others (Oreskes, 1988). Frankel himself had suggested in an earlier article (Frankel, 1978) that Holmes not only provided a better mechanism for drift, but also increased drift's "explanatory power" by relating it to new geological phenomena, such as the distribution of volcanos and earthquakes. In this article Frankel argued that Holmes's contribution did not help drift theory as much as it should have because the "greater popularity of the contractionist programme," especially because it was promoted by Jeffreys, made most geoscientists unwilling to consider such an unpopular theory (Frankel, 1978: 146–147).

Finally, Frankel's (1979*a*) application of Lakatos's perspective emphasizes the generation of novel facts and ignores that one of the key criteria in evaluating programmes is the total number of facts explained by the programme. Again, in an earlier work Frankel (1976) suggests that the greater explanatory power of drift theory was not appreciated by geoscientists because of their narrow specialization. Within any given specialty, drift theory was no better than other theories, so the specialists rejected it without recognizing (a) drift's ability to solve problems in other specialties and (b) that their own ad-hoc modifications of their preferred theories conflicted with beliefs in other specialties. These comments would seem to imply that drift theory did explain more facts, but the specialists were unable to recognize its virtues.

The "Problem-Solving" Perspective

In another article, Frankel (1979*b*) argues that when Laudan's (1977) science as "problem-solving" perspective is supplemented with Lakatos's emphasis on "novel facts," one will find ". . . that the history of continental drift theory can be explained as completely rational so long as close attention is paid to the historical developments involved, and narrow and mistaken analyses of rationality are eschewed" (Frankel, 1979*b*: 51). Following Laudan, scientific "rationality" is choosing those theories or "research traditions" that solve the most "problems." The judgement of overall problem-solving ability requires considering both empirical *and* conceptual problems and recognizing that these two types of general problems have different components with different "weights." Laudan's research traditions are similar to "scientific research programmes," except that Laudan argues that their "hard cores" are more likely to change in response to attacks. Laudan also differs from Lakatos in emphasizing the importance of "conceptual" problems.

These aspects are supported by Frankel's analysis of the continental drift debate and the acceptance of seafloor spreading. For example, Wegener's mechanisms for drift are not so much empirical problems as ones arising from conceptual conflict with the widely accepted idea of a rigid mantle. Furthermore, critics argued that drift theory was conceptually inconsistent because it held that continents could plow through a resisting seafloor, but still retain their basic shape so that matching continental margins are observed. In the realm of empirical problems, Frankel simply notes that "opponents argued that their own theories provided equally good or better solutions to some of the empirical problems" (Frankel, 1979*b*: 64–65).

The decreased interest in continental drift theory in the 1940s is attributed to its inability to increase the number of solved problems, but this changed with Hess's seafloor spreading model, which "turned much puzzling data into solved problems" (Frankel, 1979*b*: 71), which then became "anomalous" empirical problems for the other research traditions that could not solve them. Furthermore, seafloor spreading removed the conceptual problem created by continents plowing through the seafloor: now they were passive rafts riding on top of a convection current. But Frankel notes that seafloor spreading was not accepted until support was gained for the Vine-Matthews hypothesis and Wilson's transform fault hypothesis. Thus, he suggests that Laudan's weights assigned to empirical problems should include components of Lakatos's "novel facts." If a theory is shown to imply a new fact that was not used as a basis of the theory's development, then if the new fact is validated, this solved empirical problem has a much "inflated" value in the assessment of the problem-solving ability of a research tradition. Thus Frankel argues that geoscientists were "rational" to resist the idea of continental drift until about 1966.

Frankel's analysis supports Laudan's emphasis on the importance of "conceptual" problems and his argument that the "hard core" of research traditions can change over time. The existence of several competing research traditions for the same sets of problems also fits his suggestions, as well as those of Lakatos. Yet Frankel's analysis does not pay the "close attention" to historical developments that is needed to properly apply Laudan's perspective. This would require a careful analysis of the different conceptual problems avoided and empirical problems solved by each tradition, the assignment of "weights" to each of these problems, and then an overall comparison of each tradition's problem-solving capacity: all within the context of what was known and accepted at that time. This would be a difficult and probably impossible task, but it is necessary to fully test Laudan's perspective.

Subsequent analyses by Frankel (1981*a*, 1984, 1987) more thoroughly document the way that "drifters" and "fixists" presented arguments and counter-arguments for the ability of their respective traditions to explain the various data used to argue for drift. Now, however, Laudan's rules for the evaluation of the problem-solving ability of different traditions are presented more as descriptions of the "rhetorical" logic used by the defenders of each position. For example, he argues that this history shows that:

1. Proponents of competing theories attempt to provide solutions to a common nest of problems that constitute the subject matter of the controversy.
2. Proponents of competing theories attempt to bring up difficulties with their opponents' solutions.
3. Such attacked solutions are defended against the difficulties raised by altering the solutions so as to avoid the difficulties or showing that the difficulties are ill founded. . . .
4. Closure of the controversy comes about when one side enjoys a recognized advantage in its ability to answer the relevant questions. In light of points 2 and 3, this comes about when one side develops a solution that cannot be destroyed by its opponents. (Frankel, 1987: 203–205)

Clearly, the first three strategies for the defense of one's beliefs derive from Laudan's perspective and actually describe the nature of the arguments in the debate about continental drift. The fourth strategy differs from Laudan in that he would require a thorough comparison of each tradition's problem-solving ability with respect to the known problems, where each problem is assigned specific weights. Moreover, this must be done repeatedly over the course of the debate as new facts are discovered and the traditions change. This is probably an impossible task, both for us as outside observers *and* for the scientists involved.

Thus I think that Frankel has adopted Laudan's view in its most appropriate form. It is a description of the general way that scientists try to persuade others to adopt their viewpoint. Others have noted some of the "rhetorical" aspects in this debate. For example, Le Grand (1986*b*: 105) notes that a typical feature of these debates is asymmetrical criticism: opponents emphasize critical problems with proponents' evidence and cite other evidence supporting opposition to drift, without subjecting the latter evidence to the same critical evaluation. Thus the statements of opponents (and proponents) cannot be accepted at face value. Instead they are more like rhetorical arguments aimed at persuasion than the careful comparison of the problem-solving ability of the opposed traditions as Laudan predicts will occur in science. Le Grand also notes how proponents and opponents managed to support their positions by giving different interpretations of such principles as uniformitarianism and the "method of multiple working hypotheses."

We might note some other rhetorical aspects of the earlier debate. Critics of drift emphasized that drift theory was inconsistent with geophysical data for the rigidity of the mantle. Yet isostasy implies some degree of flow in the mantle. Furthermore, Schuchert believed in land bridges, but accepted that isostasy implied that such bridges could not sink into the ocean basins. Yet he was confident that "geophysicists would tell us how this occurred." Why wouldn't opponents grant Wegener this same benefit? Both Schuchert and Jeffreys emphasized the poor fit between the continents. Yet du Toit gave very precise fits that were later verified by Carey in the early 1950s and Bullard, Everett, and Smith in 1965. In sum, I think the best summary of the earlier debate is that opponents of drift were just as guilty as Wegener, if not more so, in their selective citation of the literature.

General Problems with the "Rationalist" Perspectives

There are a number of fundamental problems shared by all of the above perspectives. One of the major ones is the assumption that scientists will have little trouble reaching a consensus on certain key decisions about what are the "observational" data—the relevant "facts"—and how these are related to theory. Yet disagreements on these issues have plagued the entire history of continental drift and plate tectonics. For example, in the 1926 AAPG symposium discussed in Chapter 2, Schuchert denied there was a reasonable fit between the continents and used this to argue against drift theory. However, at the same meeting Bailey Willis accepted a reasonable fit between Africa and South America and used this to argue against continental drift because drift would deform the edges of the continents. More recent evidence for a close fit between the continents was given in the computer fit provided by Bullard, Everett, and Smith (1965) in Figure 4.14, but opponents (Meyerhoff and Meyerhoff, 1972) point out that this reconstruction excludes southern Mexico and all of Central America. In general, today's opponents tend to dismiss the geophysical evidence and emphasize geological or more "directly" observable "facts" (Meyerhoff and Meyerhoff, 1978). As Arthur Meyerhoff put it: "You will find a few geophysicists who are good mathematicians, but they are not geologists. The great majority have never actually hammered a rock. Now this is a bit shocking; it's the rocks that tell the story."

Even if there is agreement on what types of evidence is needed to test a theory, there may be the need to select the "best" data for the test. For example, many of the observed profiles of linear magnetic anomalies do not exhibit the predicted symmetry across ocean ridges, and even the most symmetrical ones vary in the degree they fit plate tectonic predictions. This point is stressed by many of the opponents of plate tectonics and even recognized by such proponents as Dan McKenzie:

> *McKenzie*: The Atlantic is an awful mess. You need to have a time scale; you need to pick your profiles; and you need not to be arguing about the truth of the hypothesis.
>
> *JAS*: By "not arguing about the truth of the hypothesis" do you mean basically looking for the profiles that give clear interpretations?
>
> *McKenzie*: Yes, you throw away most of them and pick up the ones that are good. Well, that never convinces anyone if they are questioning it.

Clearly, McKenzie's last point suggests that scientific research requires some decisions about what data will provide the best test of a theory and others may not agree with these decisions or feel that only the supportive data were used. As noted in Chapter 2, this same criticism was directed at Wegener by the opponents of drift theory.

Even if geoscientists can agree on the types of relevant "facts" and the selection of the best "facts" for testing a theory, they still might not agree on when the fit between the observed and predicted facts confirms the prediction. For example, there appears to be increasing symmetry and closer matching in the linear magnetic anomalies and theoretical models as one progresses from Vine and Matthews's (1963) to Vine and Wilson's (1965) and then Vine's (1966) papers on linear magnetic anomalies. However, none of the anomaly patterns were perfectly symmetrical or exactly replicated by the models. Geoscientists differed greatly at what point they considered the evidence persuasive. In 1974 Neil Opdyke, a Lamont geoscientist, recalled such an incident.

> I remember in 1966 [about the time of the Goddard conference] Bill Menard came up here [to Lamont] and spent the whole day looking at profiles—he never said a word. Joe Worzel, who was one of the chief opponents [at Lamont] . . . came up and said, "Bill, what do you think? Do you see this silly correlation; this lousy correlation [between magnetic profiles]?" Menard looked up and said, "Jesus Christ, how good do you want them to be?"

Conversely, Glen (1982: 335) describes several incidents where geoscientists agreed on the bilateral symmetry of the Eltanin-19 profile, but rejected it as "too perfect" to prove the Vine-Matthews hypothesis.

Another problem with these perspectives is the general neglect of non-empirical influences on the theory choice of scientists. Although many of the interviewed geologists mentioned the ability of plate tectonics to account for a number of facts, they also mentioned other reasons for their acceptance. For example, Allen Cox mentioned the quantitative nature of seafloor spreading as a major reason he was convinced by Vine's (1966) results.

> The difference between Holmes and Hess is that he [Hess] began to make the [seafloor spreading] ideas somewhat more quantitative: for example, the age of the sea floor and that sort of thing. He used data quite a bit, but it still wasn't firm enough for you to build your work on. There's a big difference in whether you build on something that's mushy or something that's firm. . . . for me the thing in the 1966 article was the quantitative nature of it. There was almost no chance that it was wrong.

Other geoscientists mentioned more aesthetic values. For example, Dalrymple's acceptance of the Vine-Matthews hypothesis was influenced in part because "it was a really neat idea. As a matter of fact it was such a good idea, it was so concise and so elegant that it almost had to be true because, in general, these very concise and elegant sorts of theories usually do turn out to be true and often the big arm-waving ones don't." Yet we have seen that other geoscientists do not share the same values or apply them in the same manner, as indicated by Fowler's (1972: 12) opinion that ". . . the terrestrial disharmonies of the New Global Tectonics [are] unbalanced, unconnected with the planetary motions, unmotivated, awkward, uneconomical of energy: in a word, ugly; and therefore false."

In summary, I find these various defenses of "rationalistic" accounts of the history of continental drift to be inadequate in important areas or impossible to apply by either the scientists involved or later historians. This is true for both the earlier debate about drift theory and the rise, acceptance, and development of plate tectonics. In particular, I think the earlier rejection of continental drift was not "rational" by any of the definitions offered here. Clearly the early evidence for drift was "inadequate" for a persuasive "proof," but I think it was at least adequate—even by the standards of the times—to allow drift theory to be entertained more widely as an acceptable "working hypothesis," especially after the contributions of du Toit and Holmes. It certainly did not deserve the extremely hostile rejection that it received, especially in North America. Menard suggests that du Toit's (1937) *Our Wandering Continents* should have caused continental drift to prevail ". . . had Wegener's integrated concept of continental drift not already been rejected. As it was, the war approached; the data were in another country; and besides, the controversy was dead" (Menard, 1986: 83). Finally, it is clear that to understand this earlier debate and even the subsequent acceptance of plate tectonics, we must understand how scientists make the crucial decisions mentioned in this section. To do so, we must adopt a broader perspective, and Thomas Kuhn's (1962) paradigm perspective provides a useful starting point.

The Paradigm Perspective

The discussion in Chapter 1 of Kuhn's perspective did not detail the components of a "paradigm," so the first subsection below illustrates this concept with some examples from plate tectonics research. We also find that Kuhn's description of "normal science" seems appropriate for current geoscience research. The second subsection considers whether *earlier* geoscience research was paradigm-directed normal science with subsequent stages of "crisis" and "revolutionary science" leading to the acceptance of plate tectonics. Although we only find limited support for these aspects of Kuhn's perspective, I suggest that Masterman's (1970) simple extension of Kuhn's perspective helps explain these "anomalous" results. I also defend the applicability of Kuhn's viewpoint from previous publications that argue it does not apply well. The last subsection describes what I see as the major problem with Kuhn's approach.

THE "NORMAL SCIENCE" NATURE OF CURRENT GEOSCIENCE RESEARCH

Kuhn (1970*a*) suggests that "normal science" research is conducted within a shared frame of reference provided by a paradigm, which he also calls a "disciplinary matrix." This is composed of several elements: symbolic generalizations, metaphysical models, values, preferred instrumental methods, and exemplars. These elements vary in their importance, function, and level of abstraction. The most abstract are the values used to evaluate research results. Typical values are simplicity, accuracy, quantitativeness, and the scope or range of applicability of the results. Such values, however, are seldom unique to the community of researchers who share a paradigm. They tend to characterize science as a whole, only playing a role in the choice between paradigms in revolutionary science. The metaphysical model and its related assumptions about the world are also relatively abstract, but are specific to individual paradigms. These models are the source of analogies developed in the course of normal science. The particle model of a gas as composed of tiny elastic particles in random motion is a typical model. The schematic figure of the basic plate tectonic model in Figure 4.19 illustrates the model providing the basis of current geoscience research.

Symbolic generalizations, such as $F = ma$, express some of the relationships between the different features of the model. However, symbolic generalizations may also serve *definitional* functions that are almost tautological, such as defining mass by the force necessary to accelerate it. The previously mentioned, $W_A = W_O \sin A$, is a symbolic generalization in plate tectonics, and it may be used to *define* the pole of spreading. Other symbolic generalizations may be only verbally formulated, such as "the direction of slippage in transform faults is the direction opposite of displacement" or "plate boundaries are always marked by earthquakes." All of the above components of a paradigm are abstract and would be difficult to relate to the results of specific instrumental output were it not for the critical role of exemplars.

Exemplars are specific examples of research that concretely illustrate how the above elements may be combined to solve specific research problems. This often requires developing specific techniques or assumptions so that the abstract model can be related to empirical data. For example, the Vine and Matthews (1963) paper connected Hess's (1962) seafloor spreading model to the observed data on linear magnetic anomalies by assuming that the seafloor layer recording the earth's magnetic reversals had a certain thickness and magnetic properties. These assumptions allowed them to generate predicted magnetic profiles that could be compared to the observed profiles. Later, Vine (1966) changed his model of the seafloor and added the assumption of a constant rate of spreading. Morgan's 1968 paper on relative plate movements provides another example of an exemplar. This paper was the first one that related the abstract model of crustal plates to data on transform faults, earthquakes, and linear magnetic anomalies by using the symbolic generalization of $W_A = W_O \text{ Sin } A$, but added the secondary assumption of rigid plates.

Exemplars also establish standards for how well theoretical predictions must correspond to empirical observations (Kuhn, 1977: 184–185). In the development of seafloor spreading, the tacit standards for a reasonable agreement between observed and predicted magnetic profiles were increased considerably by Vine and Wilson (1965) and Vine (1966). Another characteristic of exemplars is that they are widely recognized as important contributions by other researchers. Tables 4.2 and 4.3 in Chapter 4 show that these papers by Hess, Vine, Matthews, and Morgan were cited highly after publication, as were the other publications developing seafloor spreading and plate tectonics.

These comments illustrate the components of a paradigm or disciplinary matrix with plate tectonics examples, but they do not indicate the "normal science" nature of current geoscience research. To do this, we must show that plate tectonics is accepted widely, provides the basis of most research, and is not doubted when anomalies develop. The use of just one paradigm, instead of competition between paradigms, and the relative neglect of anomalies are two of the strongest contrasts between Kuhn's perspective and those of Popper, Lakatos, and Laudan.

Numerous lines of evidence indicate the general acceptance of plate tectonics. The citation counts to the key plate tectonics articles indicate the broad influence of these articles and a "cocitation analysis" in Chapter 7 will indicate that these key articles tend to be cited *together*. A 1977 survey of over 200 members of the Geological Society of America and the American Association of Petroleum Geologists showed that 87 percent of the geologists considered the theory as "essentially" or "fairly well established" (Nitecki et al., 1978: 662). The respondents showed a great deal of faith in the theory: only six percent thought the theory would still be in doubt by 1987. (If the survey had included geophysicists in the American Geophysical Union, then the results would have been even more positive.) Furthermore, plate tectonics is the only theory taught in schools, which particularly discouraged opponents of plate tectonics theory. Meyerhoff observed in a 1977 interview that "the great problem that I encounter in my lectures all over the United States, Canada, Western Europe, the Soviet Union, and Latin America is that they [the younger geologists] were never taught anything but plate tectonics; so they are unaware of or don't even know there is another hypothesis."

A paradigm must not only be accepted, but it must be used in such a way that research has a "puzzle solving" character. Kuhn suggests that determining fundamental constants, extending the basic model, and solving anomalies are typical puzzle solving aspects of normal science. Geoscience parallels for each of these include more accurate determination of the spreading poles between plates, applying the seafloor spreading model to other ocean basins or to earlier geological ages, showing how the model can explain such anomalies as "bent" linear magnetic anomalies or both tensional and compressional earthquakes at trenches, and elaborating aspects of the model to make it more relevant to other geological specialties, especially continental geology. Kuhn suggests that this aspect of normal science produces rapid growth in knowledge because researchers work within the framework of one basic model, focus on a narrow set of problems, pursue a restricted range of solutions, and can assume that others will share their major assumptions. Allègre (1988) provides numerous examples of how plate tectonics has evolved to integrate numerous areas of geoscience research. He also notes that geology has accomplished "more in ten years than it had in the previous one hundred years; it experienced an incredible acceleration in development, a sudden crystallization of ideas that had ripened slowly, often without being formulated, and a restructuring into new chapters" (Allègre, 1988: *xi*).

Finally, we have already seen that the existence of anomalies for the basic model has not caused geoscientists to abandon it. Old rocks at mid-ocean ridges, "bent" or non-symmetrical linear magnetic anomalies, extensional seafloor spreading near trenches where the crust should be in compression, vertical motions of large continental areas, and the lack of a causal mechanism are just some of the anomalies for plate tectonics. These become "problems" and solutions are proposed within the general plate tectonics framework. Wyllie (1974) provides an almost Kuhnian summary of how geoscientists should regard the new theory.

> I think that we should work the theory of the new global tectonics for all that it is worth—just as we should do with any theory or model, for it is through such intensive studies that we obtain large quantities of valuable information. We should reexamine the geologic data within this new conceptual framework, but we must not assume that we have arrived at the final solution—because geology just is not that simple. Wegman (1963) wrote: "Commonly the notions, concepts, and hypotheses control the selection of facts recorded by the observers. They are nets retaining some features as useful, letting others pass as of no immediate interest. The history of geology shows that a conceptual development in one sector is generally followed by a harvest of observations, since many geologists can only see what they are asked to record by their conceptual outfit." What we have to do now is use the new conceptual outfit but not be hampered by conceptual blinkers. (Wyllie, 1974: 14)

It might be difficult for geologists to avoid the "conceptual blinkers" because, if Kuhn is right, scientists must fully accept the new model to fully exploit it, and they may resist its overthrow as it becomes the foundation of their own research results and reputations.

THE TRANSITION FROM THE OLD TO THE NEW PARADIGM

The next issue is whether Kuhn's description of paradigm change seems to fit the earlier debate about continental drift and the eventual acceptance of seafloor spreading and plate tectonics. It is here that we find some, but not insurmountable, problems. To be consistent with Kuhn's analysis, we should expect to find an earlier period of normal science under a different paradigm. Next we should see the development of a "crisis" stage resulting from an increasing number of anomalies, especially some critical for the old paradigm. This should promote a stage of "revolutionary science" marked by competition between alternative paradigms. Eventually, we should see the acceptance of seafloor spreading and plate tectonics, not because they explain all of the old "facts" and some new ones, but because they provide better satisfaction of the widely shared values within the geoscience community. Furthermore, we should expect to see communication difficulties between proponents of the old and new paradigms and "gestalt" switches as opponents to drift accept the new viewpoint.

Although geology has a paradigm today, it is not apparent that plate tectonics replaced an older paradigm that previously provided the basis for "normal science" activities. Most of the interviewed geoscientists said there was an older paradigm, but their descriptions of it varied greatly. Some described it as contraction theory, others said it was geosyncline or tectogene theory, and still others said it was permanence of the oceans and continents coupled with continental accretion. After considering this diversity of earlier opinion, Hallam (1973: 107–108) has observed that "about the only common element that one can extract from such a hotchpotch of views is a belief in a 'stabilist' rather than a 'mobilist' earth, that the continents have remained fixed in position with respect to each other."

Furthermore, there was some disagreement among the interviewed geologists about how well the old paradigm(s) accounted for geological observations. For example, one proponent of plate tectonics mentioned the geosyncline theory and observed that "it was an absolutely beautiful hypothesis. Nothing wrong with it. There was no need, whatsoever, to bring in anything else." However, other comments by geoscientists indicated that geosyncline theory was inadequate, at least in retrospect. For example, during an interview in 1974 with a group of geologists at a special conference, I asked if geosyncline theory was the old paradigm, and got the following response.

> *Menard*: That was a model for how you got thick wedges of sediments, not for how you produced mountains.
>
> *JAS*: I guess I had a mistaken impression about it.
>
> *Menard*: Well, I think everybody, including me, at the time had an impression that it was a way of producing mountains.
>
> *Burke*: We had a vague feeling that subsequently these became mountains.

Menard went on to say that "there was a paradigm, but it didn't enter into anything very much. The paradigm was stability. . . ."

If we accept the older paradigm as the geosyncline theory or simply the assumption of stable continents, it is clear that it didn't have many well defined or widely shared implications for observational data. That is, geosyncline theory or the assumption of stable continents had little agreed upon scope, and this lack of agreement had several consequences. As Hallam (1973: 107) observed: "There was no consensus about which evidence was critical in deciding between rival hypotheses, if indeed there were any, and in consequence debate was frequently acrimonious." Another consequence was a reduced interest in research directed toward the "big picture" and more emphasis on concrete data collection. As Sykes observed in a 1974 interview,

> I think that there was a bit of a retreat in the sense of "Well, the big picture doesn't make sense, and so therefore there's enough to do in terms of data collection and analysis so I can retreat back to that." I didn't see it as a retreat, and other people didn't see it as a retreat at the time. It was just that you could see yourself making some progress there and you couldn't see yourself making progress in terms of the big picture.

Probably the best description of the philosophical state of affairs in geology before plate tectonics is provided by Masterman (1970) and endorsed by Kuhn (1970*b*: 272). She suggests that Kuhn neglected the important role of technology in science and that a limited form of "normal science" was possible if there was a consensus on the results of specific instruments. Thus, she suggested that a science could go from a "preparadigmatic state," where there is endless debate over fundamentals, to a "multiple-paradigm" science where

> . . . far from there being no paradigm, there are on the contrary too many. . . . Here, within the sub-field defined by each paradigmatic technique, technology can sometimes become quite advanced, and normal research puzzle-solving can progress. But each sub-field as defined by its technique is so obviously more trivial and narrow than the field as defined by intuition, and also the various operational definitions given by the techniques are so grossly discordant with one another, the discussion on fundamentals remains, and long-run progress (as opposed to local progress) fails to occur. This state of affairs is brought to an end when someone invents a deeper, though cruder paradigm, which gives a more central insight into the nature of the field, though restricting it and making research into it more rigid, esoteric, precise. This either by causing rival, more shallow paradigms to collapse, or alternatively, by attaching them somehow or other to itself, triumphs over the rest, so that advanced scientific work can set in, with only one total paradigm. Thus multiple-paradigm science is full science, on Kuhn's own criteria; with the proviso that these criteria have to be applied by treating each sub-field as a separate field. (Masterman, 1970: 74)

Masterman noted that in retrospect multiparadigm-science might appear to be less than full science.

In a multiple-paradigm science the major level of social and cognitive organization is based on specific instruments or narrow specialty interests. This seems like the best description of the geosciences before the acceptance of plate tectonics. The seismologists had a solid earth model with a layered structure that they were refining with the study of seismic waves and their reflection and refraction from different layers. Experimental petrologists identified how different patterns of melting and crystallization produced different rock types from different materials. Structural geologists focused on the application and elaboration of the geosyncline model to specific mountain ranges. Oceanographers emphasized the collection of a broad range of basic geophysical data with each scientist tending to specialize in a limited number of techniques, but here at least there was some effort to integrate the results of different techniques.

Numerous comments support this view. For example, Cox has suggested that before plate tectonics "... the earth sciences were characterized by increasing specialization and divergence. Paleontologists, seismologists, geomagneticists, geologists, and marine geophysicists became better and better at what they were doing, but at the same time they had less and less to say to each other" (Cox, 1973: 2). The contrast between geologists and geophysicists was particularly striking. Heezen suggested that in 1955 the geophysicists at Lamont "... were thinking of the earth as a physical and mathematical model. It didn't have rocks in it; it had things of different velocities. It didn't have history; it didn't have an age. They were on a different plane: it was the difference between talking to a historical geologist and a solid-state physicist" (quoted from Glen [1982: 314]). Wilson has remarked that the major geology departments of Princeton, Yale, and MIT in the 1930s "were teaching a few good techniques and a lot of pedantic nonsense" (Wilson, 1974).

Stewart (1987*b*) provides quantitative evidence that the geosciences were divided into isolated specialties before plate tectonics. When he examined the referencing patterns in random samples of geoscience articles published in 1963 and 1970, he found that the references cited together ("cocited") in the 1963 articles formed clusters related to specialty topics, such as geochronology, marine studies, and isotopic analyses, without any "theory" cluster. In contrast, a "plate tectonics theory" cluster was the most prominent cluster among the cocited references in the 1970 articles.

As noted by Masterman, the impact of the new paradigm may cause a reevaluation of past research. In retrospect, earlier work may not appear as "scientific" as more recent research.

> *Burke*: Well, in fact, a lot of what we were doing was fairly unscientific, wasn't it. It was describing the fossils from a particular bore-hole or something.
>
> *Menard*: Yes, I would say that's right. A man could work for a lifetime simply redefining terms in stratigraphy and the like. Never doing anything except resetting the boundary. At least the work didn't hurt anybody.

A final piece of evidence for the fragmented state of earlier geoscience research is the increased sense of integration provided by the plate tectonics model. All the histories of this revolution emphasize this point, so only a few illustrative quotations will be given here. Roger Larson, an oceanographer at the University of Rhode Island, made the following comment after reviewing an earlier draft of this book.

> [E]arth science is no longer a compartmentalized science, it is a unified science that is glued together by the theory of plate tectonics. This is most vividly displayed on the [deepsea] drillship Glomar Challenger where paleontologists, geochemists, sedimentologists, petrologists, and geophysicists all work together on sediments and rocks recovered from beneath the deep sea floor. Without the theory of plate tectonics as a background framework, this interaction would be almost impossible.

Similarly, Oliver noted the personal sense of integration provided by plate tectonics. When asked whether it helped integrate the geological sciences, he responded:

> Yes, incredibly so. I found that for me the day after our hypothesis [interpreting the Benioff zone as a slab of descending oceanic lithosphere (Oliver and Isacks, 1967)] came out that I could read a lot of papers in geological journals and start to see something important and significant in them. Before that I just read the ones in my specialty, I didn't even understand what the others were about. But suddenly—really almost overnight—I could see significance in these things that I had never paid any attention to before.

The suggestion that there was not a single, established, global paradigm needs two qualifications. First, some of the geologists suggested that there may have been a consensus on a paradigm among the elite scientists, but the average geologist was not concerned with such issues. Second, even if geologists shared the assumption of fixed continents, it was a "paradigm" of little shared scope that had few implications or constraints on the theories and models used at the specialty level of research. Most geologists could proceed with their research interests without much concern over whether drift theory was right or wrong. Before plate tectonics the broadest level of active intellectual integration appears to have been at the specialty level. Given this diversity of interests, techniques, and theories before plate tectonics, it is not surprising that the interviewed geoscientists reported a variety of transitions to plate tectonics.

One's initial reaction upon reading about the exciting, post-war developments in oceanography might be that geology should have experienced a full "crisis." These data disclosed the extensive mid-ocean ridge system—the most prominent feature on the earth's surface—with its central rift valley, the large fault zones extending for hundreds of miles, and the lack of the expected thick sediments in the ocean basins. Furthermore, the unexplained linear magnetic anomalies—well-named features from Kuhn's point of view—in the oceans suggested that the ocean floor had extensive displacements along the fault zones, and the new technique of paleomagnetism provided evidence that the continents had moved over the surface of the earth. However, among those interviewed, only Bruce Heezen seemed to feel a sense of "crisis" because of these results.

There seem to be several reasons why most geologists did not experience a sense of crisis with these new data. First, the assumption of fixed continents, although widely and rather strongly held, did not have strongly shared implications for oceanographic data. Second, when there were expectations about these data, there were many supporting assumptions that also could be modified. Finally, the fragmented nature of research training and interests insured that any one individual experienced only a few of the anomalies. We can see all of these points in the following dialogue among the conference geologists; note that Menard and Burke are geologists and Grommé is a geophysicist. I asked them how they explained the lack of extensive sediments in the oceans.

Menard: There was no problem disposing with that because we didn't know the rate of sedimentation in the ancient past. Kuenen made predictions that there would be about five kilometers or so of sediments out there and we went out and measured it and there was not much sediment. Then we started getting rates of sedimentation and they turned out to be very slow. The better the measurements were—and you had your choice of people you wanted to believe—the slower the rates were . . . it was easy to get around that. [We might note here that these "better" measures of sedimentation rate must have been wrong since the seafloor is now believed to be less than 200 million years old instead of the billions expected in the 1950s.]

JAS: How about the linear magnetic anomalies and the long faults they disclosed?

Menard: Until somebody explained the magnetic anomalies there was no problem at all with those either because you didn't know what they were.

Burke: You just said they were remarkable objects off the coast of California.

Menard: That's right.

Grommé: I remember reading those articles and being really astonished by the maps of the magnetic anomalies. They were unbelievable. How much time did you stay awake at night wondering how those things formed?

Menard: I didn't really . . . because I was really trained as sort of a general purpose geologist and nobody had ever told me that there was any reason why I had to worry about magnetics.

Burke: But what about those faults . . . because they came to the continents and didn't come in?

Menard: Oh, I worried about those faults. Well, you know when I first interpreted them, I thought they came in. They can run right into the Transverse Range you see. . . . So I thought they did and so did everybody else.

As geologists, Menard and Burke were not very concerned with magnetics, and only focused on interpreting the faults disclosed by the magnetics, whereas Grommé, a geophysicist, was particularly impressed by the magnetic anomalies. Even when the assumption of stable continents had implications for the amount of sediments that should be present in the ocean basins, the relative lack of sediments did not cause geoscientists to doubt the stability assumption because they could doubt instead other assumptions about the rate of sedimentation. The "multiparadigm" nature of the geosciences before plate tectonics helps to explain the different responses to these anomalies.

Another controversial aspects of Kuhn's perspective is that acceptance of a new paradigm does not depend simply on which paradigm can explain the most facts. He suggests that a scientist's values play an important role in paradigm choice. We already have seen that geoscientists varied in the value basis for their acceptance of the Vine-Matthew hypothesis. Allan Cox accepted it when Vine (1966) illustrated its "quantitative nature," whereas Dalrymple was impressed because it was so "elegant and concise." Yet Fowler found plate tectonics "ugly and therefore false." Many other factors were important determinants of when specific geoscientists accepted the new paradigm. One of these was the specific research interests of the geoscientist, which would be expected in a multiparadigm discipline.

Since the quantitative aspects of seafloor spreading developed first in marine magnetics and later in seismology, we might expect that the time of a specific person's acceptance could be predicted from their specialty interests. Peter Molnar's experience illustrates this point. During the middle 1960s he was a graduate student majoring in seismology at Columbia University. In an interview he mentioned that he had heard an in-house talk by Heirtzler on the new magnetic evidence for seafloor spreading, but remained skeptical because Heirtzler's talk was "low key" and because he [Molnar] "didn't really quite understand that kind of data." Then he heard an in-house talk by Sykes on the seismological evidence for Wilson's transform faults, and thought ". . . it seemed such a reasonable idea, it explained the data so well there seemed to be no alternative. So I just bit it right there on the spot. I was free; my mind was the original tabula rasa if there ever was one." Molnar's openness to new ideas fits Kuhn's suggestion that younger scientists will find it easier to adopt a new paradigm.

Numerous geologists suggested that researchers were convinced when they could see the relevance of the new theory to their own research interests. Menard mentioned a particularly striking example.

> Reidel came to me at some time in 1967 or so and said, "Look my data can't be compatible with seafloor spreading." He collects cores and dates radiolaria [microfossils] and he knew the stratigraphy of the sea floor before the JOIDES [ocean drilling] stuff. He said that his data don't agree with this, but he didn't know how to run a decent test. Since he knew I had the magnetics data that existed, he asked if we could work together and I would tell him how old the sea floor is from the magnetics and he would tell me how old it really was. . . . I got out the maps and did the best I could to give him the ages, and he plotted this with his samples. He didn't destroy the idea; they were in perfect agreement. So he says "Ye Gods!" and writes a paper about the confirmation of seafloor spreading.

Not only does this illustrate the importance of seeing the relevance of the new theory to one's own interests, but it suggests that one's acceptance of the new theory will also be more striking—more of a "gestalt switch"—when the data are more relevant and one is initially very skeptical about the theory.

A number of the interviewed geologists experienced a dramatic gestalt switch, but not always in the manner suggested by Kuhn. Generally, there were several stages of acceptance ranging from an intellectual consideration of the theory, to an intellectual belief in the theory and maybe even an application of it, and finally to an emotional commitment to use the theory as much as possible to explain empirical phenomena. This aspect of the switch to the new paradigm seems to fit Laudan's (1977) model better than Kuhn's model.

Even those without a strong commitment to an older paradigm still experienced a dramatic switch. Allan Cox's acceptance of seafloor spreading suggests that even those doubting the stability of continents could still experience a dramatic event when accepting seafloor spreading. In an interview Cox recalled that when he was a graduate student in the mid-fifties he and another student formed a Geology Club to talk about the "big picture" because the departmental presentations seemed "absolutely dull and pointless" and expressed the departmental attitude that there had been "enough BS [on the global picture] and [one] should work on a very narrow subject and do it very well." The club discussed continental drift theory and voted on it; "When the department heard that we had voted on a scientific issue [drift], they were just furious." Cox was in the majority favoring drift, but his later work on magnetic reversals seemed unrelated to drift theory, even though he was familiar with the Vine-Matthews hypothesis and had read the Vine and Wilson (1965) paper.

> I read the article [Vine and Wilson, 1965] and just didn't believe it. I don't know why now. Then Vine was at the [1966 AGU meeting] and . . . showed me the preprint of his 1966 article [including the Eltanin-19 profile]. When I saw that profile, everything that we had ever found was all there and some things that we thought might be there were there. . . . I remember really having chills looking at it and seeing that it had to be true. That was it for me. It was just amazing to find everything that you had worked on for five years laid out in one data profile.

The stage of emotional acceptance is usually the most dramatic because it is a commitment to see the world through one paradigm. The scientist no longer expends effort deciding how to relate different models to his data; all effort is spent on elaborating a single model as much as possible. When Cox saw the Eltanin-19 profile he switched from "believing something to really believing it with a lot more conviction and enthusiasm." It became something that "you could just use to build on like we use chemistry—you don't have to worry about whether the composition of water is H_2O—you start playing your games on top of that."

Tanya Atwater described a similar experience after she and Menard had published an article showing some of the geometrical relationships expected in linear magnetic anomalies when there was a change in the direction of sea floor spreading. "I was utterly amazed when we got some new lines near the great magnetic bight [the bend in the linear magnetic anomalies shown in Figure 4.20] and the pattern was there, just as predicted. That day I was converted from a person playing a game to a believer" (from Cox, 1973: 410). A number of geologists mentioned that Vine was not totally convinced of the validity of his hypothesis until he saw the Eltanin-19 profile (Glen, 1983: 358). Apparently even the proponents of a new idea can still experience a greater level of commitment to their idea.

Jack Oliver reported a similar dramatic moment when he and Bryan Isacks suddenly perceived that their seismic data on the Tonga Trench (Oliver and Isacks, 1967) could be explained by a descending slab of lithosphere. "So as soon as we saw it, both of us were convinced that the general idea of seafloor spreading was right. Once it fits your own data then you are sold on it. . . . It was because I was so confident in our data, I guess, and not so much in anybody else's." Not only did the fragmented nature of the organization of geology require that acceptance came easier when plate tectonics became relevant to one's research interests, but as Oliver noted, it might cause one to distrust others' data and results.

Two other factors help to explain the timing of the acceptance of drift theory: access to relevant information and professional age. Being tied into informal communication channels helped many of the geoscientist to hear about relevant data before many others. The previous comments by Cox and Molnar indicated that both were persuaded when exposed to informal talks or preprints. Being present at special conferences and informal talks was especially important for many of these geoscientists. Scientists at Lamont presented the magnetic and seismic evidence for plate tectonics in many informal talks—called the "roadshow"—at different universities.

Oliver thought the major generalization about who accepted plate tectonics first "was communication of information. To me it wasn't how bright somebody was, but it was mostly a matter of when the right information got to them." When asked if the younger geoscientists were likely to accept plate tectonics sooner than older ones, he noted that maybe this was true for the "masses," but pointed out that both Hess and Wilson were in their sixties. Other geoscientists also mentioned that these two originators were older, but felt that youth was an important factor. Cox mentioned that this was especially true in the Soviet Union. "A few years ago when we were there at the [International Union of Geophysics and Geodesy] meeting, they had to reschedule the session on plate tectonics three times because the young people over there wanted to go to it and it got so crowded that they couldn't go on."

Kuhn suggests that younger people are more receptive to new ideas because they are not committed fully to the established paradigm, while those who have taken a stand against the new paradigm will resist it more than others. Many of the Lamont scientists seemed to have had this difficulty. Walter Pitman had graduated in physics from Lehigh University and worked in the electronics business, but grew restless and returned to graduate school at Columbia in oceanography and geology.

> I had a completely open mind, and I had no fixed ideas about geology or the history of the earth. When the idea came along, I was perfectly willing to take a look at it. That's a great advantage of being a graduate student—when ideas begin to develop you have a clean slate. You aren't filled up with any long-term prejudices. . . . For a lot of people at Lamont it was a real, real hard corner to turn. Maurice Ewing—he's a really smart fellow—fought the thing off and on for a couple of years.

Pitman also noted that others at Lamont, especially those who had written papers against drift, had trouble switching over to plate tectonics. Glen (1982) provides some vivid illustrations of their difficulties.

Other factors may have influenced when a geoscientist accepted plate tectonics, such as the country one was trained in. Kevin Burke, who was trained in England, noted that compared to North Americans many of the English geoscientists were more likely to take a neutral position toward drift. Oreskes (1988) gives a number of reasons for this difference. These included (a) the high importance placed on geophysical techniques, which caused Americans to emphasize the geophysical evidence against drift and to have less experience with the traditional geological evidence used by Wegener, (b) the relative neglect of theoretical issues in preference for the pragmatic uses of geological findings disclosed with further mapping of the West, (c) the lack of much experience in the southern continents, and (d) the fact that the contraction model was more popular in Britain and more directly threatened by the evidence for isostasy.

Although these factors help explain the acceptance of the new paradigm by different individuals, there were still many geoscientists who simply "jumped on the bandwagon" at some point. Many of the geoscientists mentioned this "bandwagon" phenomenon and it alarmed them because many of the new followers did not really understand plate tectonics or the evidence for it. In their analysis of a 1977 survey of AAPG and GSA members, Nitecki et al. (1978) found that many of the geologists who accepted the theory of plate tectonics in the late 1960s were no more familiar with the relevant literature than those who accepted it earlier or later. They concluded that it was plausible that in the sixties there occurred a " 'chain reaction' or other general shift in opinions toward the theory; this more or less uniformly altered the attitude of the majority of the profession as a group and was not, at least in most cases, the result of individual judgments of the accumulating evidence and arguments for and against the theory" (Nitecki et al., 1978: 664). As might be expected, opponents of plate tectonics find the bandwagon effect particularly offensive.

Most of the interviewed geoscientists switched to the new paradigm well ahead of the majority. Consequently, they had numerous opportunities to talk to those who still resisted the new theory. As Kuhn suggests, these conversations were often frustrating for all concerned. For example, Molnar mentioned an incident during a trip in the Soviet Union.

> In some places people were very sympathetic to plate tectonics and in others they argued vehemently. I had one very strange argument where this lady was completely convinced I was wrong and I was just as convinced she was wrong, and someone was telling her to not waste her time on me because I was so completely wrong that it was not worth it, and I was thinking how could I be wasting my time with her. . . . Most [of those we met] were fairly quiet and not interested.

After a few such experiences, some of the early followers simply gave up on discussions with opponents.

Menard: Did you ever talk to an anti-drifter in a serious vein?

Burke: No, they were just holding up the whole operation. There was nothing they could say that was of any interest to me.

Menard: I mean, it was a curiosity and they were nice fellows, but there was no organized opposition in the sense of against Darwin or even Einstein. I always thought the opposition was basically ludicrous.

Menard went on to say that a "serious person with their own data," who wanted to test those data against the theory, was a different matter, and he provided the Reidel example given earlier. The discussion in Chapter 4 of the "debate" between Beloussov and Sengör and Burke suggests that communication problems still exist between proponents and opponents.

These communication problems arise from different emphases on data sources, instruments, language use, and assumptions. For example, a Harvard "skeptic" mentioned that

> when I came to this country, I actually had great hopes to do lots of cooperation with marine geologists and marine physicists, but I very soon realized that they talk such strange language and have such a different philosophy that there was not much sense of talking together since the best they might do was to try and brainwash me.

This skeptic also mentioned that there seemed to be two approaches in geology.

> The "classical" approach starts with "documents of the past"—rock specimens or fossils—and fits these into a regional context. . . . You can see the basic thing is very crucial: factual observation. . . . Now it is a very different approach if you work with geophysics. Geophysics is the present day condition of the earth, which you try to figure out by way of remote sensing. . . . If you are working with this sort of thing you need some sort of instrumentation, and in order to even design your instrumentation, you first of all have to have mathematical models. . . . you have to start with models and with theories.

Opponents tend to distrust geophysical procedures because their interpretation requires hypothetical models of underlying structures, not actual observation of these structures. As Meyerhoff noted in an earlier quotation: "It's the rocks that tell the story."

Thus the new paradigm has emphasized different techniques of research and types of data, i.e., different "facts." As Menard put it: "In geology you are always confronted with a vast array of facts, as well as in geophysics. Some of these facts are inconsistent and in order to produce a unified hypothesis you have to reject some of the facts as being 'erroneous facts,' that is, as 'non-facts,' or not the most important facts. Any of these simplifying hypotheses must do that." Geophysical techniques and data provided the initial foundation of plate tectonics, whereas much geological data, which appeared to be explained in the earlier theories, was less directly related to the new global model. Even the proponents recognize that plate tectonics needs more development before it is tied closely to the extensive geological data from the continents. As indicated in Chapter 4, the application of plate tectonics to continental geology is now one of the primary research areas of proponents.

With the new paradigm new concepts may be invented, such as "subduction" and "exotic terranes," and old concepts may be given new meanings. There may be a tendency to modify or abandon old concepts. Robert Dott (1974) provided a clear statement of this problem. He noted that "causality very early became a part of the geosynclinal concept especially in North America. It was an article of faith that 'thick sediments must invariably lead to mountains–in fact are a prerequisite–and mountain building leads to enlargement of continents' (Dott, 1974: 7). Within the new paradigm of plate tectonics, such wedges of sediments do not necessarily lead to mountains. Instead, they must be caught between colliding continents. Thus, Dott observed that

> . . . as a new general model is discovered, the bases for classification will change, which results in an important language shift. In short, changing concepts require changing language! . . . So many of the premises associated with the older taxonomy are incompatible with the new idiom that, if perpetuated, such terminology would certainly act as a mental straitjacket. Conversely, exploitation of the new idiom's truly revolutionary implications could be greatly accelerated by a clean break with the old language. (Dott, 1974: 11)

Similarly, Coney (1970: 742) suggests that continued use of many of the concepts associated with the geosyncline model has created "confusion" and that these concepts "have lost their meanings and dependence on models which no long apply. I would suggest we need new terminology and a new model."

It appears that Kuhn's general perspective applies quite well to the history of plate tectonics, especially if we accept that geology was in a multiparadigm stage of development before plate tectonics. This addition resolves some problems identified by previous efforts to apply the Kuhnian perspective to the history of continental drift. In particular, Rachel Laudan (1981), Ruse (1981), and Frankel (1981*b*) all suggest that this history does not show either of the two possible transitions identified by Kuhn (1962): old paradigm to crisis and revolution to new paradigm *or* "pre-paradigmatic" science with rampant disagreement on fundamentals to *first* paradigm science. However, the "multiparadigm" state of pre-plate tectonics geosciences is consistent with their observations about the existence of earlier paradigms and the lack of a "crisis." Ruse (1981) suggests that the psychological and sociological aspects of Kuhn's perspective are supported by this history, but not the "epistemological" and "ontological" aspects. That is, he does not see new epistemological aspects dealing with emphases on new data sources, methods of research, or rules of reasoning in geoscience research. However, the comments in this section clearly indicate that geophysical data gained an increased importance in terms of evaluating the best global model with a reduction in importance of concrete geological "facts." Hence there was a change in data sources and methods of research (Wood, 1985). Furthermore, the previous American and British emphasis on working from facts to theories was replaced with a stronger emphasis on letting theory determine the relevant facts. Allègre (1988) even suggests that geologists will increasingly have to use statistical methods to decipher the geological complexities produced by plate tectonic processes. Finally, Ruse argues that the earth seen by pre- and post-plate tectonics geoscientists consists of the same "ontological" entities, such as continents and ocean basins. Yet we have seen that some changes were made. The older "crust" of the earth, defined by the Moho, is now less important than the "lithosphere" defined by the low-velocity zone. The "plates" of the lithosphere may have both continental and oceanic surfaces. There are now "subduction zones," "triple junctions," and "transform faults," which did not "exist" before. (In the next chapter we will try to show how there has been a fundamental change in the classification of basic earth structures.) Furthermore, Ruse's assertion that there were no communication problems is not supported by quotations given in this section.

Finally, all of these critiques of Kuhn's perspective have tried to show that the transition from continental drift to plate tectonics seems to be simply the result of the geoscientists responding to increased information about the properties of the earth. Like Lakatos, L. Laudan, and Popper, these critics of Kuhn assume that there is a set of shared *rules* for rational decisions about theory choice. Thus they cannot explain the diversity of individual decisions about what data are relevant, when predictions are "adequate," and the communication difficulties between geoscientists using different paradigms. In contrast, Kuhn's perspective covers a broader range of scientific decisions by embedding scientists in a paradigm or disciplinary matrix: a "culture" that sets limits on decisions and actions, but does not formally specify all decisions. Consequently, the diversity of individual actions is not a "problem" for Kuhn, but one of the "virtues" of his perspective. Yet Kuhn does not describe fully the social processes occurring within these "cultural" communities. He simply suggests that "persuasion" process using shared values is the mechanism of paradigm change. The next section considers this particular problem with Kuhn's perspective.

THE "MECHANISM" PROBLEM WITH THE PARADIGM PERSPECTIVE

Although Kuhn places his scientists in a social culture, he employs a psychological model for the behavior of scientists and neglects more sociological processes. This is apparent in his emphases upon such psychological processes as the personal sense of "crisis" that scientists have during a revolutionary period and the "Gestalt switch" occurring with the acceptance of a new paradigm. Kuhn argues that he differs from Lakatos and Popper because they assume that scientists have "special minds" or a special capacity for logical thought, whereas he emphasizes the decision processes in a diverse community of "normal minds" with "special ideals," which include "the value system," the "motivations," and the "ideology" current in their disciplines (Kuhn, 1970*b*: 237–238).

However, we need to consider more than just values and motivations in our attempts to understand how a consensus is reached in a scientific community. For example, a values explanation cannot explain completely why Lysenko's theories dominated previous Soviet biology (Graham, 1972). Kuhn suggests that Lysenko gained control because Soviet scientists rejected "values which its members ordinarily share" (Kuhn, 1970*b*: 263). More is needed here. We also need to consider the structure of the Soviet scientific community and its degree of autonomy from the political system. Lysenko gained control of Soviet biology because his theories were compatible with Soviet political ideology and he was allowed to influence promotions and demotions among Soviet biologists. In other words, we must also consider the structure of authority and influence in a scientific community and its degree of autonomy from outside influence.

Sociological processes not only deal with the relationships between the scientific community and other social institutions, but they also apply to processes occurring within the scientific community itself. Lysenko may have gained dominance because those who liked his theories were outside of the scientific community, but even when the evaluators are within the scientific community itself, there can still be an "entrenched hierarchy" that influences the reception of certain ideas. Numerous geoscientists have suggested continental drift was rejected because established geoscientists tended to oppose it (Hallam, 1973; Wood, 1985; Allègre, 1988).

Some observers have given a similar explanation for the extended resistance of Soviet geologists to plate tectonics.

> Soviet reluctance to accept plate tectonics is often attributed to isolation from western research results, national pride (they obviously were not first), a lack of oceanic data of their own, and the entrenched hierarchy of Soviet science. Vladimir Beloussov, the dominant fixist and author of the Soviet geology textbook on the subject, headed that hierarchy until his retirement 3 years ago. (Kerr, 1978*b*: 283)

In other words, Kuhn's perspective needs some elaborations of the "persuasion" mechanisms used by scientists as they reach their consensus on the best theories or paradigms. Appealing to the use of shared values is an incomplete analysis (Barnes, 1982). We must consider other factors influencing the decisions of scientists, including their personal interests and intellectual "investments," and characteristics of the broader scientific community, such as its degree of autonomy from other social institutions and its internal reward and stratification systems. Describing the character of such social systems is one of the goals of the sociological perspectives discussed in the next chapter.

These perspectives emphasize the "negotiation" and "persuasion" processes creating a scientific consensus, but such processes occur within a specific intellectual context and aspects of this intellectual context may themselves be powerful persuasion resources. This is true even if these "intellectual" aspects are simply "social conventions" established in prior "negotiations." Thus the remainder of this chapter examines some of the general relationships among the major intellectual elements of a paradigm: the metaphysical model and assumptions and the related symbolic generalizations. This will help us understand the process of scientific discovery, the development of scientific arguments, the role of tacit knowledge in scientific thought, *and* indicate how we might proceed in the sociological analyses of scientific thought.

Models and Analogies in Science

One of the important elements in a paradigm is the metaphysical or visual model conveying a gestalt image about how the world operates. A plate tectonics example is the visual model in Figure 5.1 from Isacks, Oliver, and Sykes (1968). Nearly all introductory geology texts reproduce this model because it conveys the basic plate tectonic imagery. As Drummond Matthews described it, "that little block diagram of the crust being formed is so perfect that it just carries conviction by itself." This section briefly describes the models and analogies perspective in the philosophy of science, shows how this perspective relates empirical data to the abstract elements in the disciplinary matrix, and illustrates this perspective with examples from the history of plate tectonics.

Many philosophers of science emphasize the fundamental role of models, analogies, and metaphors in scientific thought (Harré 1970, 1976, 1986; Hesse, 1966, 1974, 1976; Ziman, 1978; Giere, 1988; Leatherdale, 1974). Some use this perspective to criticize both the logical empiricist's emphasis on the logical nature of scientific theories and the apparent relativism of the paradigm perspective. Harré (1976) and Hesse (1976) also assert that the emphasis on models reduces the apparent differences in how natural and social scientists produce knowledge: both are seen as using the same fundamental process of reasoning from models.

Harré notes the two major tasks of scientists: "They have tried to sort out the non-random patterns in nature from the enormous multiplicity of happen-

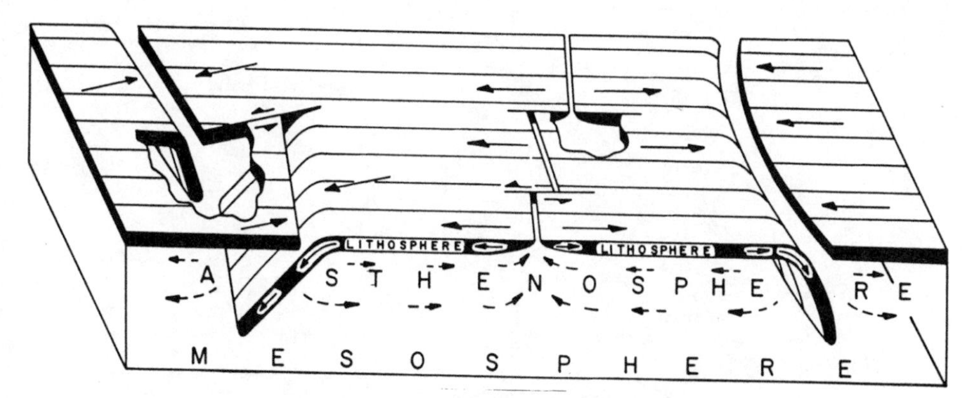

Figure 5.1 The plate tectonics block diagram as an iconic model for basic features of the earth's outer layers. (From Isacks, Oliver, and Sykes, "Seismology and the New Global Tectonics," *Journal of Geophysical Research* 73 [1968]: 5855–5899. Copyright © 1968 by the American Geophysical Union.)

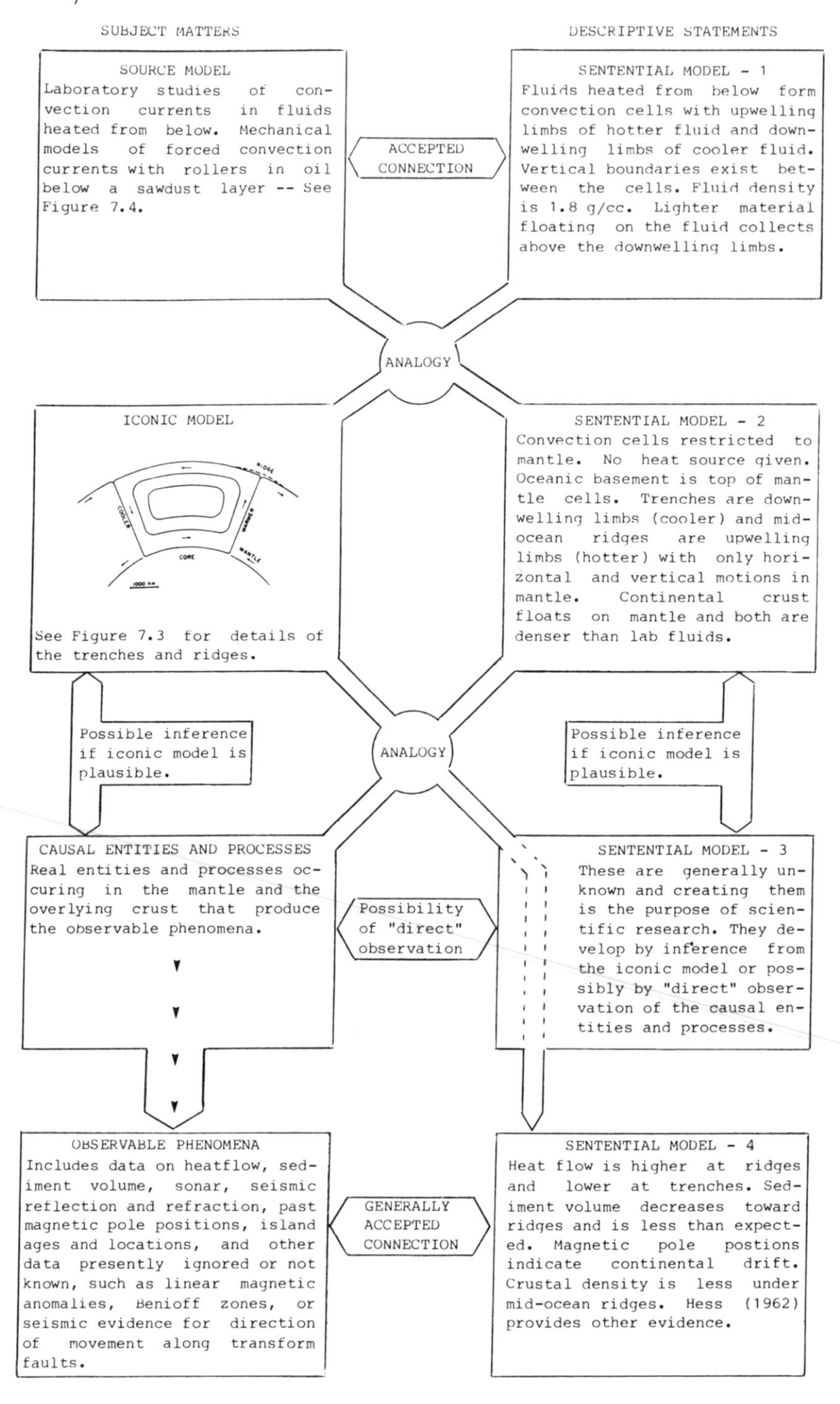

SUBJECT MATTERS
DESCRIPTIVE STATEMENTS
SOURCE MODEL
Laboratory studies of convection currents in fluids heated from below. Mechanical models of forced convection currents with rollers in oil below a sawdust layer -- See Figure 7.4.
ACCEPTED CONNECTION
SENTENTIAL MODEL - 1
Fluids heated from below form convection cells with upwelling limbs of hotter fluid and downwelling limbs of cooler fluid. Vertical boundaries exist between the cells. Fluid density is 1.8 g/cc. Lighter material floating on the fluid collects above the downwelling limbs.
ANALOGY
ICONIC MODEL
RIDGE
COOLER
WARMER
CORE
MANTLE
1000 KM
See Figure 7.3 for details of the trenches and ridges.
SENTENTIAL MODEL - 2
Convection cells restricted to mantle. No heat source given. Oceanic basement is top of mantle cells. Trenches are downwelling limbs (cooler) and mid-ocean ridges are upwelling limbs (hotter) with only horizontal and vertical motions in mantle. Continental crust floats on mantle and both are denser than lab fluids.
Possible inference if iconic model is plausible.
ANALOGY
Possible inference if iconic model is plausible.
CAUSAL ENTITIES AND PROCESSES
Real entities and processes occuring in the mantle and the overlying crust that produce the observable phenomena.
Possibility of "direct" observation
SENTENTIAL MODEL - 3
These are generally unknown and creating them is the purpose of scientific research. They develop by inference from the iconic model or possibly by "direct" observation of the causal entities and processes.
OBSERVABLE PHENOMENA
Includes data on heatflow, sediment volume, sonar, seismic reflection and refraction, past magnetic pole positions, island ages and locations, and other data presently ignored or not known, such as linear magnetic anomalies, Benioff zones, or seismic evidence for direction of movement along transform faults.
GENERALLY ACCEPTED CONNECTION
SENTENTIAL MODEL - 4
Heat flow is higher at ridges and lower at trenches. Sediment volume decreases toward ridges and is less than expected. Magnetic pole postions indicate continental drift. Crustal density is less under mid-ocean ridges. Hess (1962) provides other evidence.

ings and phenomena, and they have tried to explain these patterns" (Harré, 1976: 20). The first task of finding the nonrandom patterns is certainly a critical part of the process and will be discussed later, but Harré has focused on the process of explanation. He suggests that explanation is accomplished by a "statement-picture complex,"[1] which is composed of a series of statements describing a parallel series of subject matters, such as visual models or observed phenomena. Figure 5.2 diagrams the relationships between the statement/descriptive parts and the related pictorial/subject matter parts of Hess's theory of seafloor spreading. It shows the four possible subject matters—source model, iconic model, empirical "reality," and observed data—and the statements or "sentential models" possibly associated with each subject matter.

The starting point of a scientific theory is some "source model," which provides a source of analogies for the construction of a particular "iconic model." Both of these models are described by a set of associated "sentential models" that describe their major characteristics. For example, Harré (1976) suggests that the "source model" for the particle theory of gas behavior is our shared knowledge about the behavior of macro-sized objects, and the associated sentential model includes statements about the characteristics and behavior of such objects, e.g., about color, texture, elasticity, their behavior in a vacuum, and their manner of interactions, including such formalizations as "$F = ma$." The "iconic model" is based on the source model by *analogy*. Thus some features are carried over, perhaps idealized, and others are not. The iconic model of a gas assumes that gases are composed of microscopic, perfectly elastic particles in random motion and following the same laws of mechanics that apply to macroscopic objects. Only some of the many possible descriptive statements associated with the source model are carried over to the set of statements associated with the iconic model, e.g., there is no concept of color for the iconic model's particles, but perfect elasticity is assumed and the same laws of motion apply.

The same approach provides some insights into how Hess developed his seafloor spreading model. Figure 5.3 is a composite of four features of Hess's iconic or metaphysical model for the formation of the ocean basins. Part A of Figure 5.3 is his model for how the lighter continental material separated from the denser mantle and core materials during the first convection cycle occurring early in the history of the earth. Part B suggests that after the formation of the

Figure 5.2 A representation of the crucial features in scientific reasoning according to the models and analogies perspective, as applied to Hess's (1962) model of seafloor spreading. The two basic components are four subject matters and the four sentential models describing them. The initial ideas come from source models from which analogies are used to construct an iconic model and its description. When the description of the observed data is analogous to the description of the iconic model's behavior and the model is based upon plausible assumptions, then the iconic model is taken as an accurate portrayal of the real causal entities and processes.

core subsequent convection cells were confined to the mantle. Parts C and D show more details of the upwelling and downwelling parts of the later convection cells. These are hypothetical or metaphysical models for processes that cannot be directly observed in the same way that gas particles cannot be directly observed. These models are based by analogy upon directly observed and well described processes or source models, which raises the question, what was Hess's source model?

Laboratory studies of convection currents in fluids are a possible source model. These studies indicated that heated liquids form convection cells where the upwelling limbs of the cells are hotter than the downwelling limbs. However, these studies usually consist only of a single fluid without an analogy for the continental crust floating on the convecting medium. A further clue to Hess's source model is his reference to Griggs's (1939) article on mantle convection, in which Griggs portrayed an operating mechanical model composed of sawdust floating on oil that was caused to convect by a pair of mechanical rollers. When the rollers were turned, they forced convection cells to form in the surrounding oil with a site of downwelling currents between the rollers.

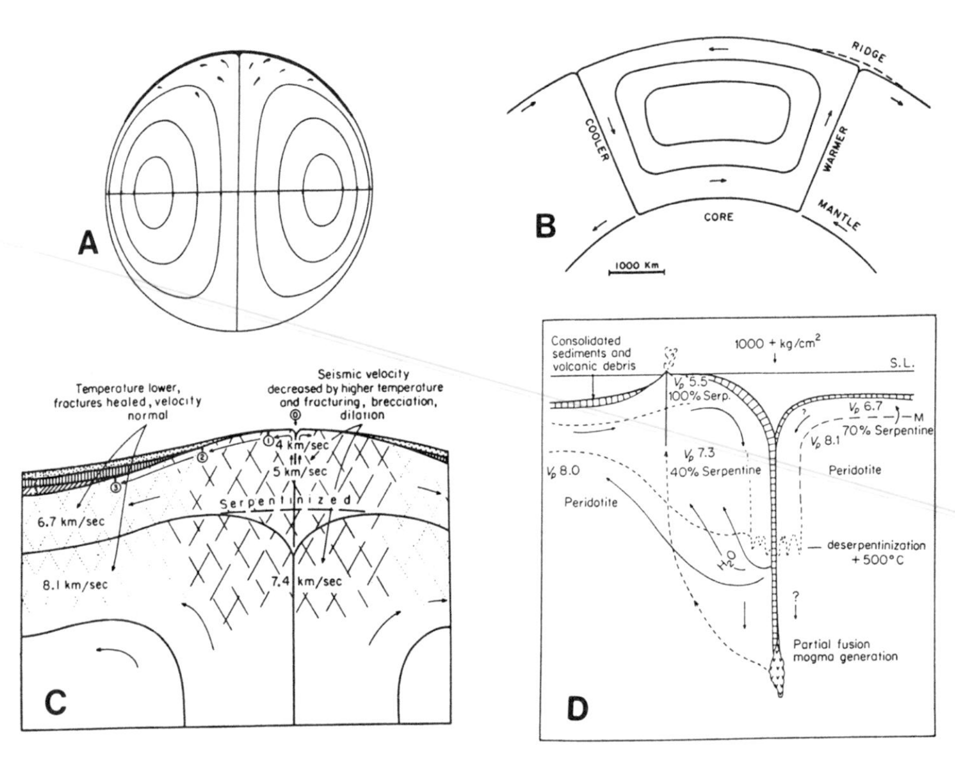

Figure 5.3 Various aspects of Hess's (1962) model of seafloor spreading. See the legend for Figure 4.1 for an explanation of these diagrams and their sources.

The floating sawdust tended to pile up over the downwelling site. During an interview with Arthur Buddington, who was Hess's professor and colleague at Princeton, he recalled that Hess had observed this mechanical model in action: ". . . one demonstration that I am sure influenced Harry was given by David Griggs . . . he demonstrated [convection currents] with an oil and sawdust type of thing. I am sure that made an impression on him."

Hess's seafloor spreading model represents an iconic model that was based by analogy on actual laboratory studies of convection currents and a mechanical model—the source models. The sentential models corresponding to each of these models are related by analogy. Some of the statements that describe the source models are carried over to the iconic models. For example, statements describing the laboratory results would include "a heated fluid forms convection cells," "the upwelling arms of these cells are hotter than the downwelling arms," and "the density of the fluid was 1.8 grams per cubic centimeter." Only the first two statements would be considered applicable to an iconic model for convection in the earth's mantle. In the same manner, only some of the statements describing Griggs's model would be carried over to the iconic model. There would be no analogies for Griggs's mechanical rollers, but the piling up of the sawdust at the downwelling arms would have analogical statements in the description of the iconic model. Hess proposed that continents were less dense materials floating on a convecting mantle, that the mid-ocean ridges corresponded to the upwelling limbs of the mantle convection cells, and that the ocean trenches corresponded to the downwelling limbs.

This example illustrates two of the subject matters (the source and iconic models) in a scientific theory and the two sentential models that describe them. Harré (1976) suggests there are two other subject matters and corresponding sentential models that describe them. The third subject matter is the "real causal entities and processes" that produce the "observed phenomena," which are the fourth subject matter. For the scientist seeking the explanation of the observed phenomena, an important goal is developing the sentential models describing the unseen causal entities and processes. Although these statements are unknown initially, Harré (1976: 23) suggests that these statements may develop from later "direct" observations or by the gradual extension of the statements describing the iconic model as this model becomes more "plausible." The plausibility of the iconic model depends on the degree of correspondence between the statements describing its behavior and the statements describing the observed phenomena. In addition, plausibility depends on whether the source model is compatible with widely held metaphysical assumptions and "reasonable" analogies were used to construct the iconic model.

Hess tries to establish the plausibility of his seafloor spreading model in the same manner as outlined by Harré. Although he notes that mantle convection is a "radical hypothesis," Hess suggests that "whole realms of previously unrelated facts fall into a regular pattern, which suggests that close approach to satisfactory theory is being attained" (Hess, 1962: 606). Among the correspondences between his iconic model and observed phenomena, Hess noted the following: (a) the high heat flow and lack of sediments near the mid-ocean ridges, (b) the gravitational anomaly over trenches, (c) the paleomagnetic evidence for continental movements, and (d) the topographic elevation of the mid-ocean ridges due to the rise of hotter, less dense material below them. These descriptions of observed phenomena correspond to the descriptions of the proposed behavior of the iconic model, which increase the plausibility of the iconic model as an analogy for the actual causal processes. Furthermore, the role of heat in the source model is compatible with the common assumption that heat is important in geological processes.

Harré also mentions that technological changes may allow "direct" observation of the proposed entities or processes, such as using the electron microscope to see viruses. Similar developments are occurring in the geosciences. Geoscientists now use submarines and underwater cameras to directly observe lava eruptions at mid-ocean ridges and "seismic tomography" now allows geophysicists to "see" convection currents in the mantle.

In contrast to the philosophies of science emphasizing the logical characteristics of the sentential models, Harré argues that the iconic model or pictorial part of the statement-picture complex is the essential element in scientific reasoning. The important, creative process is the use of analogies to suggest relationships among the entities and processes producing the observed phenomena. He suggests that the three sentential models associated with the source and iconic models and the observed phenomena may individually "be organized in a deductive manner but since the relations between the subject matter of the three sets is only one of analogy it is impossible to complete, a priori, a set of transformation rules which would enable deductive connections to be established [vertically] in [Figure 5.2]" (Harré, 1976: 23). Of course, scientists sometimes do develop a logical, deductive system relating the various sentential models and it has great heuristic value for new predictions and elaborations of the theory, but it is not essential for scientific progress, nor does it constitute the essential feature of scientific reasoning.

Analogical reasoning is the key process and it cannot be formalized into a set of "rules." Hence, "[t]he selection of a model is based upon real resemblances and differences between the model and what it represents, and *decisions* as to what are the proper degrees of likeness and unlikeness can be the subject of argument" (my emphasis; Harré, 1976: 16). Thus it seems that scientists must still confront such key decisions as: what source model to use, what parts are used to construct the iconic model, and how well the statements about the iconic model correspond to those for the observed phenomena. In many cases scientists may regard the source model and analogies used by others as absurd and violating their own "commonsense" understandings.

Harré discusses more thoroughly than Kuhn the relationships between models (Kuhn's metaphysical models) and statements about models (Kuhn's symbolic generalizations), but both Harré (1976) and Hesse (1976) use the model perspective to attack the logical empiricist position *and* the paradigm perspective. Hesse considers the paradigm perspective to be implausible when applied to the natural sciences. She suggests that science can be viewed as a "learning machine," where raw input data are transformed by the chosen model into "initial data," which is then compared to model predictions, and then followed by adjustments to the model. In her view the paradigm perspective only states that it is always possible to adjust a model so that it is consistent with the input data, so knowledge is always relative. In contrast, she argues that scientists collect new input data to test the modified model. Hesse, however, tends to ignore one of Kuhn's major points: to a large extent the model determines which data are important and worth collecting. Harré explicitly recognizes the directive role of models: "the work of imagining an iconic model leads to the postulation of novel processes, hitherto unobserved structures and even novel kinds of things, perhaps with novel properties" (Harré, 1976: 23). This process aids the development of new ideas, but it also constrains what one will observe and what one expects to find, as noted in a previous quotation from Wyllie. Harré argues that the real causal processes and entities also constrain the degree of correspondence between the sentential models for the iconic model and the observed phenomena, so that science provides an increasing understanding of the causal processes occurring in the real world. Hesse argues that such an assertion is a matter of faith, but that scientific change, even revolutionary change, always shows an advance in our capacity to manipulate the natural world.

Kuhn, Harré, and Hesse each perceive the research process as a continuing interplay between models and empirical data, but two of the major differences between them concern the relative importance of models and data in their mutual interaction and the status of the resulting knowledge. Harré and Hesse appear to place highest importance on the data's influence and assume that the evolving model increasingly approximates an external reality or at least our capacity to manipulate it. Kuhn emphasizes the importance of the model and only assumes some improved satisfaction of broadly shared values. A further examination of Hess's seafloor spreading paper illustrates how the source model can constrain the choice of relevant data.

Since his source model was convection cells in a fluid, where the flow in the cells reached the surface of the fluid with vertical movements at the limbs of the cells, Hess's iconic model had these features. Thus he proposed that the "basement" of the seafloor below the sediments was the upper surface of a mantle convection cell. Consequently, there was no separate crustal surface with a composition different from the mantle. Yet seismic studies showed a density discontinuity at the Moho; Hess accounted for this by suggesting it was due to a change in crystal structure, not a change in composition. Furthermore, the vertical motions in the source model do not include analogies for the *inclined* plane of earthquakes—the Benioff zone—that are always associated with trenches. Although Hess's specialties included the study of ocean trenches, his iconic model had no analogies for the Benioff zones. These examples illustrate how one's initial imagery or source model may constrain the type of data emphasized, which is one of Kuhn's major points.

Although Hess's initial model was not very quantitative, other researchers added elements to it that increased its scope and quantitative capabilities. For example, Wilson (1963*c*) modified the model by placing a rigid layer over the convection currents, thereby providing analogies for the Benioff zones. He also compared island ages to their distance from the mid-ocean ridges. Other researchers attached other source models to seafloor spreading. For example, the Vine-Matthews hypothesis indicated how seafloor spreading, when combined with reversals in the earth's magnetic field, provided quantitative predictions for the patterns in the linear magnetic anomalies. In order to generate these predictions, they had to borrow models and symbolic generalizations from studies in magnetics to create an iconic model for the geophysical structure of the sea floor, which was illustrated previously in Figure 4.8.

Vine's 1966 article persuaded many geoscientists that the real causal explanation for the magnetic anomalies was the hypothesized blocks of alternately magnetized material. This acceptance was based on the degree of correspondence between the descriptions of the linear magnetic anomalies and those generated from the iconic model. Note, however, that these blocks had not been directly examined by ocean drilling to see if they had the proposed magnetic properties, which was a serious problem according to the opponents of plate tectonics.[2] Others could object because they felt the Vine-Matthews iconic model, which had frequent volcanism constrained to the mid-ocean ridge crest, was inconsistent with the source model provided by the observed sporadic and diffuse volcanism on the continents (Glen, 1982: 303). Thus, some geoscientists could examine the symmetry in the Eltanin-19 magnetic profile and conclude it was "too perfect" to support the Vine-Matthews hypothesis. Clearly the interplay between models and data is quite complicated and neither completely determines the characteristics of the other. Kuhn's perspective permits a more complex interplay between models and data than that allowed by Harré's or Hesse's emphasis on how data constrain plausible models.

Another excellent example of use of models and analogies is the study of surface waves in seismology. These are particular sinusoidal wave patterns that occur after the arrival of primary and secondary waves in the seismic record of an earthquake. They were seldom studied until Maurice Ewing and Frank Press (1955) showed how they provided information about the thickness of the earth's crust. In an interview with Jack Oliver, who was a student of Ewing's at the time of this work, I asked how Ewing discovered the interpretation of surface waves. Oliver, who said that he knew that story "from start to finish," noted that Ewing in the 1930s had studied the dispersion of sound waves in the ice on the surface of a lake.

> In World War II he [Ewing] got to study the sound waves in shallow water over rock. That gave a form of dispersion too, and that showed a very characteristic signal—a long sinusoidal signal. Because of the ice he was able to understand the waves in the water. After the war it was logical to just keep extrapolating and just go up to bigger problems still, so he used the same theory as for shallow water, but

just scaled it up to look at the oceans. With that theory, instead of just looking at the beginning of the signal [the typical use of the primary and secondary waves that arrive before the surface waves], you could explain every little wiggle of this long train of waves. Ewing and Press showed what kind of structures you needed to provide the particular pattern in the wave that was observed.

This example shows how a model for sound dispersion in ice was the source model for sound dispersion in shallow water and then a model for seismic wave dispersion in the crust of the earth, which enabled Ewing and Press to use the previously ignored surface wave data. It is a good illustration of scientific discovery and reasoning from the use of models and analogies, as well as how available models influence what "facts" are important.

J. Tuzo Wilson's (1965*a*) discovery of the transform fault provides a final illustration of the role of models and analogies in the development of plate tectonics. Wilson said that he discovered the transform fault concept while he was "playing with a paper model" (one similar to Figure 4.7). If this mechanical model is manipulated as indicated in Figure 4.7, it predicts a certain pattern of movement along the transform faults that periodically offset the mid-ocean ridges. In general, such mechanical models may convey considerable tacit knowledge and a visual gestalt about the basic causal entities and processes.[3]

The models and analogies perspective of Harré and Hesse provide a valuable supplement to Kuhn's general perspective. First, they show how Kuhn's metaphysical model, assumptions, and symbolic generalizations are interrelated, and how an exemplar combines these components with other procedures and applies the combination to an empirical problem. This accomplishment establishes standards of scope, accuracy, and quantitativeness that subsequent researchers must try to meet. Second, we have seen how the source model constrains the resulting theories and appropriate data sources, as well as providing important analogies necessary for scientific research. There is a continuing interaction between models and data, and analogies may play an important part in the tacit knowledge that characterizes scientific thought. Third, the study of the origins of source models may help us understand the process of scientific discovery, as illustrated by the study of surface waves. Finally, the fundamental role of the source model suggests how paradigms are based on different imagery; switching paradigms requires "seeing the world through a different set of glasses."

There are, however, some aspects of scientific reasoning left unanalyzed by the models and analogies perspective. As Harré (1976: 16) notes, there may still be disagreements among scientists about which analogies from the source model should be carried over to the iconic model and about how well the statements describing the iconic model's behavior correspond to the statements describing the observed properties. Furthermore, we still must explain the initial choice of a source model. These represent crucial decisions, but advocates of the models and analogies perspective do not provide answers that account for the diversity of individual decisions on such topics.

Although the models and analogies perspective fills out some of the details of scientific reasoning, it does not remove the requirement for many decisions by scientists as they conduct their research. Source models must be chosen, analogies emphasized, data procedures invented or borrowed, and judgments must be made about the adequacy of the correspondence between the statements about the iconic model's behavior and those related to the observed phenomena. In short, we still need to consider other factors before we can understand how a consensus is achieved within a scientific community. The next chapter considers possible sociological contributions to our understanding of this process.

Conclusion and Overview: The Paradigm "Paradigm"

Most philosophers of science have serious reservations about the paradigm approach (Lakatos and Musgrave, 1970), but Harold Brown (1977) adopts it in a fully reflective sense. He not only uses it as an epistemology for the development of scientific knowledge, but he also argues that it applies to the developments in epistemology itself. He argues that epistemology is an empirical science because its theories can be tested against studies from the history of science, so he suggests that the developments in epistemology from logical empiricism to Popper's falsificationism to Kuhn's paradigm approach can be seen as changes in epistemological "paradigms."

If the paradigm approach to the history of science constitutes a new epistemological "paradigm" in the philosophy of science, then we should expect to see some of the features of "revolutionary science" as we contrast the different epistemological "paradigms" described in this chapter. For example, Kuhn's "paradigm" starts from an entirely different "source model" for understanding scientific behavior. Theories in the social sciences, especially psychology, provide the basic models and analogies for constructing the iconic model that Kuhn uses to describe scientific behavior. He especially emphasizes a Gestalt psychology model of perception and a decision making model on the basis of shared values. The paradigm "paradigm" places the study of scientific thought squarely within the domain of the social sciences.

Consistent with any change in paradigms, this new epistemological "paradigm" has different meanings for traditional concepts, emphasizes new empirical phenomena, and uses different methods of research. For example, the concept of "rationality" has changed. Rational thought is not equivalent to the rule directed decisions of deductive logic or the minimization of conceptual problems coupled with the maximization of solved empirical problems, but is the reasoned judgment of *informed individuals* about the coherence between models and evidence. "It is the trained scientist who must make these decisions, and it is the scientists, not the rules they wield, that provide the locus of scientific rationality" (Brown, 1977: 149).

> A central characteristic of our new model of rationality is that it recognizes that different thinkers can analyze the same problem situation and come to contrary conclusions without any of them being irrational. But the fact that a theory is arrived

> at rationally is not sufficient to make it a part of the body of science; that requires not an individual but a group decision. No thesis becomes a part of the body of scientific knowledge unless it has been put before and accepted by the community of scientists who make up the relevant discipline. (Brown, 1977: 150)

Thus this new "paradigm" shifts our central concern from an emphasis on the nature of scientific "rationality" to an *empirical* inquiry into the nature of the consensus formation process within a scientific community. We need to focus on the actual processes of socialization, persuasion, negotiation, or even coercion used by scientists to reach and maintain a consensus. Much of this occurs outside of the scientific literature, so we have a much broader area of empirical inquiry and, correspondingly, will need a broader range of research methods, such as interviews, psychological tests, surveys, and participant observation.

The major conclusion in this chapter is that scientific knowledge is the result of a consensus formation process within a community of specially trained individuals using normal cognitive processes shared by all humans. Our next task is to study how this consensus formation process occurs. This task will require proceeding in the same fashion as scientists themselves proceed: by developing an "iconic model" from a reasonable "source model" and showing that the iconic model "behaves" in a way similar to the observed behavior of scientists. Thus the next chapter examines some of the models used within the social sciences to understand the behavior of scientists.

CHAPTER SIX

Social Perspectives on Decision-Making in Science

The previous chapter left us in a dilemma. It suggested that the interactions between empirical evidence, theory, and logic cannot completely explain how scientists reach a consensus on different theories or paradigms, nor could Kuhn's "value-satisfaction" approach. However, the chapter mentioned some possible components of a solution to this dilemma. First, all of the perspectives mentioned in Chapter 5 emphasized the key role of various *decisions* in the work of scientists. The more we can understand how scientists make these decisions, the more we will understand how a consensus develops (and breaks down) in science. Second, these decisions are made by individuals in a community sharing a "cultural tradition" that includes a complex cluster of models, data, techniques, assumptions, values, and exemplars. Thus the theories and methods of social scientists may help us understand how this community creates and maintains its traditions. Since human reasoning is even more complex than geological processes, we too will need simplifying models and measuring "instruments" to guide the empirical study of how scientists reach a consensus. Finally, scientific theories generally start from source models that provide analogies for constructing an iconic model or basic imagery of the "real" processes producing the observed behavior. In our case we need a model for how scientists make decisions and an "instrument" that will measure important aspects of scientific decision-making.

This chapter explores the possible uses of a basic decision-making source model that can be applied to many aspects of scientific research. The first section reviews some of the typical "intellectual" decisions involved in a research project and briefly contrasts the different perspectives mentioned in the previous chapter. The second section describes a general source model for studying human decision-making and suggests that recent approaches in cognitive psychology and artificial intelligence use iconic models derived from this general model. The third section places the decision-making model in a broader social context by briefly summarizing a few aspects of a classic analysis of the decision-making process in social organizations. The fourth section contrasts different sociological perspectives on decision-making in science. The final section develops a quantitative procedure for measuring the relative importance of "social" and "intellectual" factors involved in a key decision influencing the consensus in scientific research: whose research will provide the basis of one's current research. This procedure provides an "instrument" for the study of some of the "global" characteristics of scientific decision-making. It is based upon a simple iconic model, combines aspects of the different sociological perspectives, and can be related to other studies of social organizations. The next chapter applies this quantitative model to some data sets related to the history of plate tectonics.

Decisions in a Typical Research Project

The examination of a simple example of "normal science" research will illustrate some of the many decisions occurring during a research project and the differences among the perspectives in the previous chapter. Suppose we wish to see how well the plate tectonics model fits magnetics data to be collected along a segment of a mid-ocean ridge between two plates. Previous studies provided us with some essential information. First, earlier studies of earthquake locations had mapped the approximate location of the ridge in our planned study area. Second, the study of transform faults and fracture zones along another part of the ridge had estimated the location of the spreading pole for the movement of the two plates. Unfortunately the magnetometer was broken during this latter survey, so the researchers could not estimate the rate of spreading between the two plates. However, other studies of the spreading and subduction rates at other plate edges on a "great circle" passing through the planned study area predicted that the spreading rate should be 4.0 cm per year. (The reader may recall that on any "great circle" around the earth the spreading and subduction rates should sum to zero, provided the earth is not expanding.) Our primary task is to test this prediction by collecting magnetic data on this unsurveyed section of the ridge. This is clearly "normal science" research in that we are filling in details by applying the plate tectonic paradigm to a new area.

Some of our first "intellectual" decisions involve planning the traverses across the ridge crest. Suppose we decide that we need to study crust that is at least five million years old so that we will have enough magnetic anomaly peaks to test the predicted rate of spreading. At 4.0 cm per year this means we must extend our survey to at least 200 kilometers on each side of the ridge. Furthermore, we want to minimize the chances of crossing a fault zone, which could distort the anomaly pattern with respect to the ridge, so we plan the traverses to follow lines of "latitude" from the assumed spreading pole. In both of these decisions, the model we are "testing" actually influences our basic data collection decisions.

We next collect magnetic field data from five traverses across the ridge that extend to 200 kilometers on each side. These raw magnetics data are "meaningless" with respect to our test without further work and decisions. Some of the necessary decisions include (a) identifying the ridge "crest," which require analysis of sonar, seismic, gravity, magnetic data for a ridge's distinctive features, (b) removing the effects of the regional magnetic field and submarine topography so that just the "pure" anomalies are apparent—the anomalies are only a few percent variation in the total field, and (c) connecting the magnetic anomaly profiles to distances from the ridge crest along lines of latitude from the proposed spreading pole.

Several aspects of these data manipulation decisions are important. First, many data manipulations are now so formalized that computer programs make these "decisions" automatically, such as connecting distances and magnetics data or correcting for seafloor topography. Second, the final data are not theory free, but depend on the theories built into the designs of the magnetometer, sonar, radar, and other instruments. Third, the theories behind these instruments seem "independent" of the theory we are testing, but this is less true of the decisions about the ship's course. However even this "independence" of the chosen instruments is based upon conventional assumptions about what data (mainly magnetics data) provide the best test of the theory. For example, gravity, seismic reflection, air and water temperature, and seafloor sample data are not seen as particularly "relevant." The "decisions" to emphasize certain types of instrumental data are now so accepted that they are unconscious and similar to the routine "decisions" in the computer programs.

After many other research "decisions" we are ready to calculate a spreading rate for each profile, but this requires other decisions. We first have to accept some standard magnetic reversal chronology for the last five million years. We then have to make some other key assumptions, such as: (a) the properties of the ocean floor whose natural magnetic remanence records the magnetic reversals as new seafloor is created at the ridge, (b) that the rate of spreading is constant over the last five million years, (c) that the plates are rigid, and (d) that new seafloor is created only in a very narrow zone at the ridge crest rather than in a diffuse area. These are easy, maybe even unconscious, "decisions" because a standard chronology is in the scientific literature and geoscientists share conventions about constant spreading, narrow zones of creation, and the magnetic properties of the seafloor. Such assumptions are essential for drawing the implications of the abstract model for our specific empirical results, so they might be used even though there are reasons to doubt them. For example, drilling samples from the basaltic (sima) layers below the sediments, do not support the assumption of a uniformly magnetized layer used in typical tests of seafloor spreading (Kennett, 1982).

Now we are ready to generate predicted profiles at different rates, compare the predictions to the observed profiles, and visually pick the best matches, and hence the implied spreading rates. We might be more "sophisticated" and use a statistical analysis to determine the spreading rates and, perhaps, an estimated "standard error" for the rates. Finally, we have a crucial decision. Do the results fit the predictions? Five spreading rates of "about" 4.0 cm per year would provide additional support for plate tectonics theory. Five rates that are "too high" could raise a number of questions. Is there a transform fault or a fault zone displacing the magnetic anomalies? Were the locations of the ship properly calculated? Were the measuring devices working properly? Is the estimated spreading pole location wrong? Is the reversal chronology wrong? Did the previous studies estimating the spreading rate contain errors? Is plate tectonics wrong? Is the earth expanding? None of these possible questions would be raised if the four observed spreading rates were "close enough" to the predicted rate, even though some of the assumptions might have been wrong.

The more likely outcome of the study will be some variability in the estimated rates around the predicted 4.0 value. We would have to decide which observed profile to use and whether its estimated rate was "close enough" to the predicted rate. Deciding whether the prediction is adequate requires considerable experience with the results from previous studies (exemplars) and the tacit knowledge learned from those studies about what provides an "acceptable" prediction. Clearly any research result is embedded in an immense "network" of assumptions, models, instrumental preferences, and previous results, but the previous decisions establishing these elements would only be questioned when unexpected results occur.

Other decisions occurring both before and after the project are more "consciously" made and equally important. For example, the proposed study must first be approved, probably by a committee, because the project will use valuable resources, such as ship and staff time. Numerous factors would affect the negotiations about the project, such as the "importance" of the study, the characteristics of the chief scientist, and the degree the study fits with the "interests" of the other researchers involved with the expedition. Other key decisions follow the study and include (a) whether the researcher presents or writes a report, (b) whether she submits it for publication—in some institutions submissions must be approved internally, (c) whether the journal editors and reviewers accept the submitted article, and (d) whether other researchers read the article and incorporate the reported results into their own research activity. Without positive responses to these other decisions, a research result will not influence the consensus that forms within a community of scientists. All of these decisions may be influenced by both "intellectual" and "social" factors.

This simple example illustrates some of the differences between the perspectives discussed in the previous chapter. Each one neglects some of the key decisions made in this slice of normal science. For example, the logical empiricists would emphasize the Euclidian logic used to locate the spreading pole or predict the expected spreading rate, but ignore the many decisions and assumptions required to produce the observed spreading rates. Popper's falsificationism incorporates these elements, but still emphasizes deductive reasoning and its capacity to "disprove" theories once there is agreement on the empirical facts. Yet it is certain that if the research results in this hypothetical study had not been consistent with plate tectonics, few geoscientists would have started doubting plate tectonics. Instead, they would have followed Lakatos and Laudan and sought modifications in the "auxiliary belt" surrounding the "hard core" of plate tectonics. For example, they might have assumed that fault zones had disrupted the predicted pattern of anomalies in the profiles with a poor fit. To follow Lakatos they should seek further evidence for these faults, but more likely the "bad" profiles would just be ignored if some profiles are "adequate."

None of these perspectives specify, for example, how scientists negotiate the planned research, select their instruments, interpret their data, decide when theoretical predictions adequately match the observed results, or how other scientists evaluate and use the results reported in the scientific literature. Kuhn's perspective comes the closest to considering these additional decisions by arguing that the research process should be seen as the result of the activities of a group of individuals sharing a cultural whole: a paradigm that includes favored models, analogies, symbolic generalizations, instruments, interpretive procedures, assumptions, value criteria, and exemplars providing the tacit knowledge about these components and their interrelationships. Kuhn emphasizes the intrinsically "social" nature of this process, but tends to neglect sociological approaches and theories.

This simple example illustrates the importance of numerous types of "decisions" in the consensus formation process and the need to consider decisions within a broader social context to understand this process. For example, we need to consider how scientists choose where to gain their education, who to study under, who to let study under them or to hire in their department, what grants to support, what articles to read, and whose research will provide a basis for their own work. These represent some of the decisions that will determine how a consensus forms in science and are some of the topics studied by sociologists. To empirically study how these decisions are made, we must explore possible source models that can represent *both* the "intellectual" and "social" decisions made by scientists. This is the subject of the next section.

Source and Iconic Models for Studying Decision-Making in Science

The idea of a decision is a very general concept that has been applied to the analysis of many aspects of human behavior. "The fact that this concept has been seized upon independently in such a variety of contexts to provide a framework for a theory of behavior suggests that it represents the real core of the new behavioral *Zeitgeist*" (Simon, 1957: xxix). As we shall see, the variety of applications of this concept is immense and includes important aspects of the behavior of scientists (Simon, 1977).

According to the models and analogies section in the previous chapter, most scientific research is premised upon a basic visual gestalt image or source model. Analogical reasoning from the source model provides an iconic model, whose behavior is compared to actual observed behavior. The study of human thought and decision-making proceeds in the same fashion. We will start with a simple source model of decision-making and elaborate it.

The typical visual representation of a decision-making situation is in terms of a "decision-tree," where a series of choices branch out from a common decision point and each choice may lead to another decision point. The choice at each decision point is based upon some test criteria and a choice of action or a conclusion results at the end of the decision process. Economics and statistical decision-making theory have formalized the characteristics and operation of this simple model. In the simplest, "decision under certainty" model it is assumed that the decision-maker knows all the possible alternatives, the precise outcome of each choice, and the utility associated with each choice. The "rational" choice is defined as that which maximizes the desired utility (March and Simon, 1958). Note that "rational" here is defined in terms of "efficient goal attainment," rather than the use of deductive logic.

There are many elaborations on this simple model, such as adding "uncertainty" to the possible outcomes or incorporating the effects of new information, but we only need to note that this source model is similar to the iconic models used by many researchers in artificial intelligence and cognitive psychology. Some researchers in these subjects try to produce computer programs that behave in a manner similar to observed human behavior. Since computer programs are an elaborate set of branching decision-points with "choices" based upon some criteria, this approach to human thought processes is compatible with the simple source model of decision-making. Furthermore, researchers in cognitive psychology follow the procedure outlined in the models and analogies section in the previous chapter. When the computer program solves a problem in a fashion analogous to actual human subjects, the program is assumed to model the real cognitive processes occurring in the human subject (Newell and Simon, 1972).

Computer programs can discover proofs in symbolic logic, solve puzzles, play championship chess, manipulate objects, mimic human image processing, and even learn from experience (Newell and Simon, 1972; Hofstadter, 1979; Waltz, 1982; Kosslyn, 1983). The most successful computer programs for modeling human reasoning have been the "expert systems" that model the reasoning used by *scientists* in very specific applications (Duda and Shortliffe, 1983). Some of these programs are particularly relevant to the topics in this book, such as the BACON program that discovered Black's Law of temperature equilibrium by a series of steps similar to the original discovery (Bradshaw, Langley, and Simon, 1983), the PROSPECTOR program that has successfully predicted the existence and size of a new ore deposit (Campbell, et al., 1982), and LITHO, which combines numeric and symbolic data from oil well drillings to produce descriptions of the penetrated rocks (Ganascia, 1986).

Within this cognitive science perspective all aspects of scientific research—from measurement to thinking—can be related to this decision-making or computer program model. As Barnes (1974) notes, "in many ways it is apposite to employ it [the computer program model]. It is increasingly being taken up by natural scientists in their attempts to understand man, so it is interesting to press their perspective as hard as possible, and discover what it implies for the knowledge and activity of science itself" (Barnes, 1974: 82). Others have used the general decision-tree or computer program imagery to describe the behavior or thought processes of scientists (Hesse, 1974; Barnes, 1981; Werner, 1970; Simon, 1977), but DeMey (1980, 1982) provides its most thorough application to the analysis of how scientists learn and develop scientific theories.

DeMey (1982) adopts the cognitive sciences' assumption that any person is an information processing system and must have an "*internal model* or *represen-*

tation of the environment in which it operates" (DeMey, 1982: xv). These "cognitive structures" are the basis of all human processes involved in scientific research from perception to thought and communication. He proposes that the heart of a paradigm is a cognitive structure that forms the basis of scientific research. This structure not only provides for the cognitive integration by directing scientific observation, but it promotes social integration by providing a framework for scientific communication. Without this cognitive element, social processes alone cannot create a community of scholars sharing a paradigm.

In particular, these cognitive structures are represented as a network of interconnected "frames," each of which is related to a particular concept. Frames do not provide a formal definition of a concept, but a list of the *typical* features possessed by an object illustrating the concept. DeMey argues that these cognitive structures are tacitly learned by the study of exemplars, which not only communicate the basic cognitive structure, but also provide weak "default" values (the typical features) for what should be seen when the world is studied with this structure. That is, one's model of the world influences what one expects to observe in the world. In any scientific research only a few of these expectations are tested empirically.

DeMey argues that people, including scientists, have a multitude of cognitive structures composed of loosely organized frames that provide expectations for a variety of settings. In fact, he suggests that the mind be viewed as a "community of interacting experts" (Minsky, 1985). Each "expert" is a scheme or subframe appropriate for a general setting or task and these are organized together by higher level frame systems to produce a hierarchical structure.

Changes in these cognitive structures can occur at any level. A scientific revolution occurs when changes occur at a more global level so that the web of concepts in lower level frames are reorganized and have new "meanings" because the meaning of a concept depends on its interconnections to other concepts. Gradual cognitive change constantly occurs because these cognitive structures are idealized abstractions that need modifications, called "debugging," in any specific applications. Debugging involves slight modifications of cognitive structures by "hooking" different frames onto the cognitive structure so that it can be applied to the specific problem. Analogy provides the basis of deciding which frames will be added to the cognitive structure, and the analogy is based upon the exemplars from which the different frames were learned. Thus cognitive structures evolve and change as they are applied to more topics, but their basic structure always reflects the problems they were meant to solve. For example, the Vine-Matthews hypothesis combined the seafloor spreading model with magnetic field reversals to solve the problem of the linear magnetic anomalies.

This perspective is quite compatible with the model and analogy approaches of Harré (1970, 1976, 1986) and Hesse (1970, 1976), but allows the model or cognitive structure to more directly influence the data collection process. Furthermore, Masterman (1980) illustrates how cognitive structures *and* analogies between these structures can be represented by decision-tree structures. Thus this simple source model can represent both cognitive structures and the analogical reasoning process. However, DeMey's presentation is abstract with only simple examples of cognitive structures and has an emphasis on cognitive processes in *individuals* without much consideration of the effects of the social context on the individuals. To make this approach more concrete, we will consider some specific examples of cognitive structures in the geosciences.

If this cognitive structure approach is appropriate for the analysis of the thoughts and behavior of geoscientists, then it should be able to represent the drastic changes occurring with the acceptance of plate tectonics. Figures 6.1 and 6.2 provide simple attempts to show the changes in the way geoscientists classified the earth's features and components before and after the acceptance of plate tectonics. Each of these figures is a decision-tree representation of cognitive or classification structures assumed to be shared by geoscientists. A visual image or iconic model is included in each figure to show how the respective classification schemes are related to a gestalt image about the important features of the earth. Figure 6.1 represents the "stablist" cognitive structure before the acceptance of plate tectonics. The earth is first divided into a crust and an interior and each of these is subdivided further into very broad categories. The figure illustrates the possibility of more detailed analysis by incorporating Heezen, Tharp, and Ewing's (1959) classification of the features of submarine topography.

Figure 6.2 represents the plate tectonics scheme for the important features of the earth. Now the initial division of the earth's features is into lithospheric plates and an interior. The lithosphere is much thicker—about sixty miles—and includes the old concept of the crust defined by the depth to the Moho. Furthermore, the major features of the lithosphere are based upon the plate tectonic distinction between plates and their three types of margins. Figure 6.2 also includes a visual diagram showing the basic imagery of plate tectonics. The comparison of the old and new cognitive structures in Figures 6.1 and 6.2 suggests that plate tectonics represents a conceptual reorganization at a "high" level in the cognitive structures used by geoscientists to classify features of the earth.

These are very general classification schemes that will be modified as researchers try to solve particular problems. For example, Figure 6.3 is actually from the geoscience literature. It is Pearce and Cann's (1973) scheme for classifying basaltic rocks. Comparing it to the general scheme in Figure 6.2 shows how it simplifies the more general cognitive structure into one more appropriate for petrologists studying the origins of volcanic rocks.

In a 1974 interview Marshall Kay provided an illustration of how "default" expectations are only checked at a few points. He suggested that the evidence initially supporting plate tectonics was more "objective" because it was not gathered in an effort to support or apply the theory. He gave the following example of how things are different now.

> Dewey interpreted this data [I'm working on] as a subduction zone. I say, "Okay,

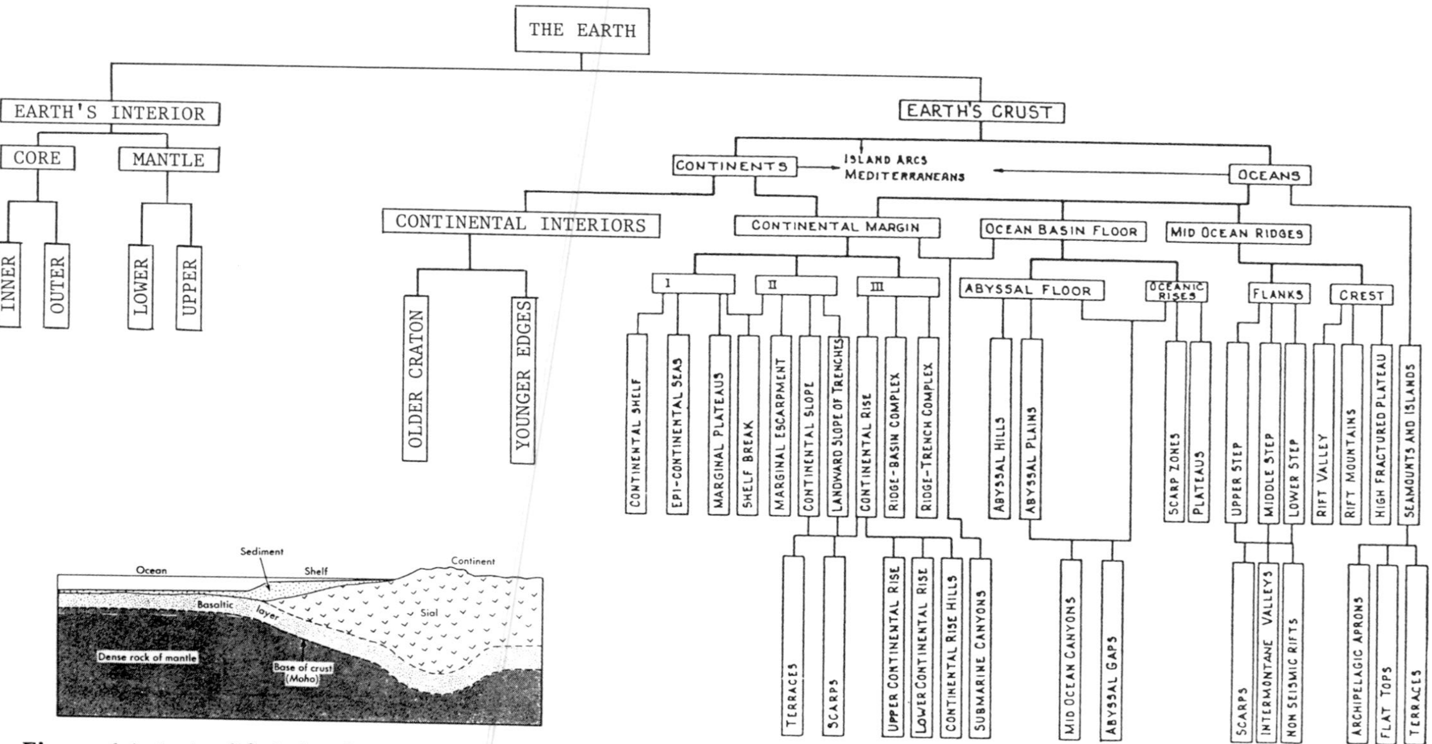

Figure 6.1 A simplified classification scheme for the major structural components of the earth under the "stabilist" model for the earth. The oceanic component of the scheme illustrates that considerable detail is possible. Note that in the visual model the crust is defined as the depth to the Moho. (The classification scheme for the oceanic crust is from Heezen, Tharp, and Ewing, *The Floors of the Oceans, I, The North Atlantic*, Special Paper 65 [1959], with permission, Geological Society of America; reproduced by permission of Marie Tharp, 1 Washington Ave. South Nyack, NY 10960. The visual model is from Longwell and Flint, *Introduction to Physical Geology*, with permission, copyright © 1962 by John Wiley & Sons.)

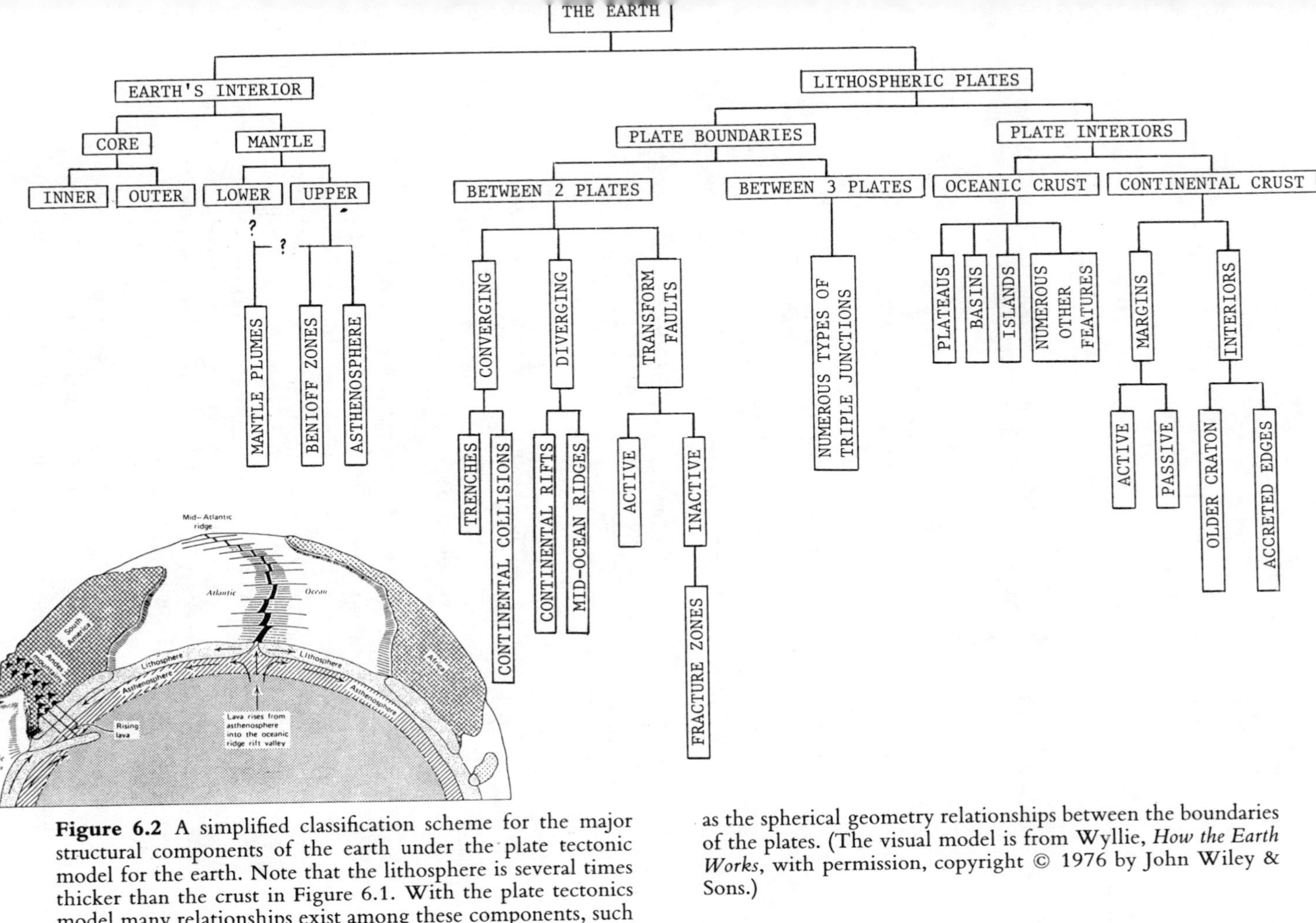

Figure 6.2 A simplified classification scheme for the major structural components of the earth under the plate tectonic model for the earth. Note that the lithosphere is several times thicker than the crust in Figure 6.1. With the plate tectonics model many relationships exist among these components, such as the spherical geometry relationships between the boundaries of the plates. (The visual model is from Wyllie, *How the Earth Works*, with permission, copyright © 1976 by John Wiley & Sons.)

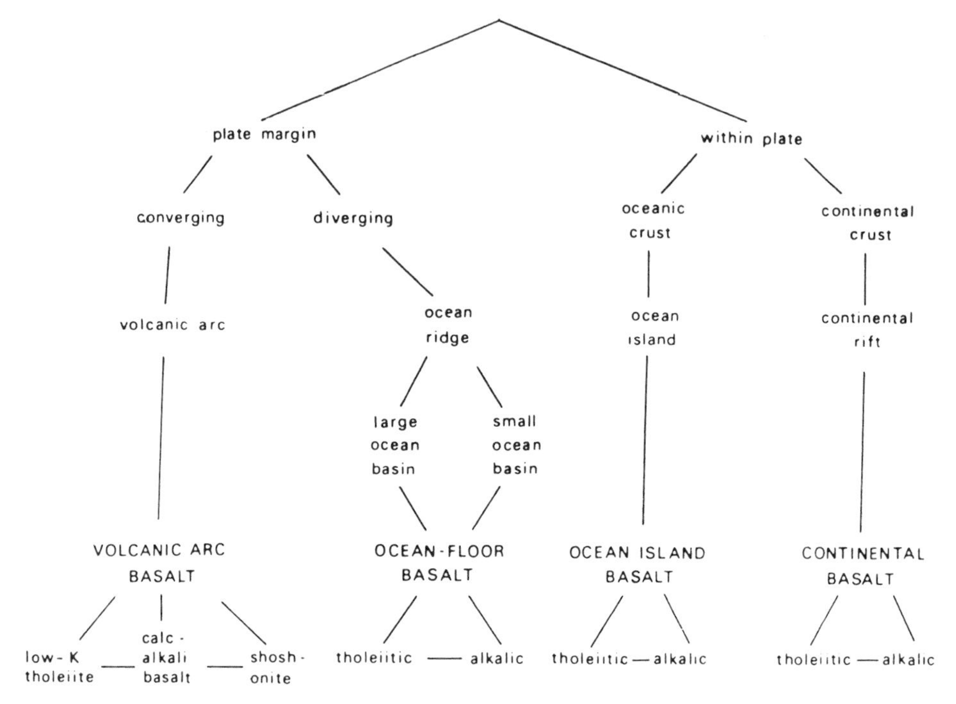

Figure 6.3 An example of a "decision-tree" representing a possible classification scheme for basaltic volcanic rocks based on the tectonic setting of their origin. (From Pearce and Cann [1973: 290–300], with permission, Elsevier Science Publishers.)

> if this is a subduction zone, then there ought to be ocean crust to the south." I go south and find oceanic crust. But this doesn't prove that the ocean crust belongs there. I'm making certain assumptions . . . that there's no transcurrent fault in between that brought two things together that don't belong together.

Other "default values" might include assumptions about rigid plates, the magnetic reversal chronology, and the structure and properties of the seafloor.

Several features of this general approach to scientific thought and behavior are worth noting. First, as mentioned before, these cognitive structures can be seen as analogies based on the more general imagery of decision-trees. Second, these structures have two important components: (a) a general *structure* given by the set of interconnected decision points and a set of choices at each point and (b) a set of decision-making *procedures and criteria* directing how "choices" are made at each decision point. The former is similar to the "declarative knowledge" and the latter is similar to the "procedural knowledge" in artificial intelligence research. Finally, it is assumed that all aspects of the typical research project discussed in the previous section can be modelled by this decision-making approach. For example, choosing the length of traverse, adjusting the magnetics data, and comparing the observed and predicted magnetic profiles are easily seen as typical decisions, some of which are so formalized that computer programs may make the "decisions."

Other aspects of the research process may seem less clearly based upon this decision-making imagery. For example, it is difficult to represent more complex cognitive structures dealing with *processes*. Simple taxonomies or classification schemes are represented easily and are crucial for the development of more abstract theory (Ziman, 1978), but it is not clear how a set of decision points and choices can represent such dynamic imagery as seafloor spreading, magnetic field reversals, or plate motions. However, the ability of cognitive psychologists to write computer programs that mimic aspects of scientific thought *processes* suggests that the basic decision-making imagery may still be a useful analogy.[1]

One of the critical elements in research is the measurement process, such as the magnetometer output in the simple example at the start of this chapter. Although the magnetometer can be seen as a decision-making structure consisting of a set of structured decision points and alternative choices, human decisions do not appear to influence the actual output from the magnetometer. Instead it is "Nature" that seems to make the "decisions" about what will be recorded by the instrument. To illustrate that this view is too simple, we can examine in more detail the Pearce and Cann (1973) classification scheme for igneous rocks in Figure 6.3.

This scheme is a set of structured decision points and alternative choices and the choices at each point are determined by the results of technical or empirical procedures. The final result is the "proper" classification of a rock sample into the classification scheme. That is, scientists develop the technical procedures, but "Nature" seems to provide the information that causes the ultimate classification. However, there are two problems with this analysis. First, the general classification structure of basic rock types comes from "outside" of this analysis. So "Nature" has to work within this general structure. Second, when we examine the origin and use of the technical procedures providing the empirical data used at each "decision" point in the general structure, we find the effects of many human decisions.

Fortunately for our purposes, Pierce and Cann detail these technical procedures. Part A of Figure 6.4 shows the "flow diagram" of technical procedures and decisions needed to use the general classification scheme given in Figure 6.1. Parts B, C, and D in Figure 6.4 present the results of using the flow diagram rules for some rock samples from the Troodos Massif of Cyprus, which at that time was accepted as an example of ocean floor basalt. Chemical analyses for different elements in the rock specimens are specified in Part A of Figure 6.4 and the concentrations of these elements allow the samples to be plotted on different discrimination diagrams, which are shown in Parts B, C, and D. Ocean floor basalts should all fall into the discrimination diagram areas marked by arrow heads in Figure 6.4.

Now we can identify some of the "human" decisions involved in these procedures. First, only some of the many possible elements are used in the

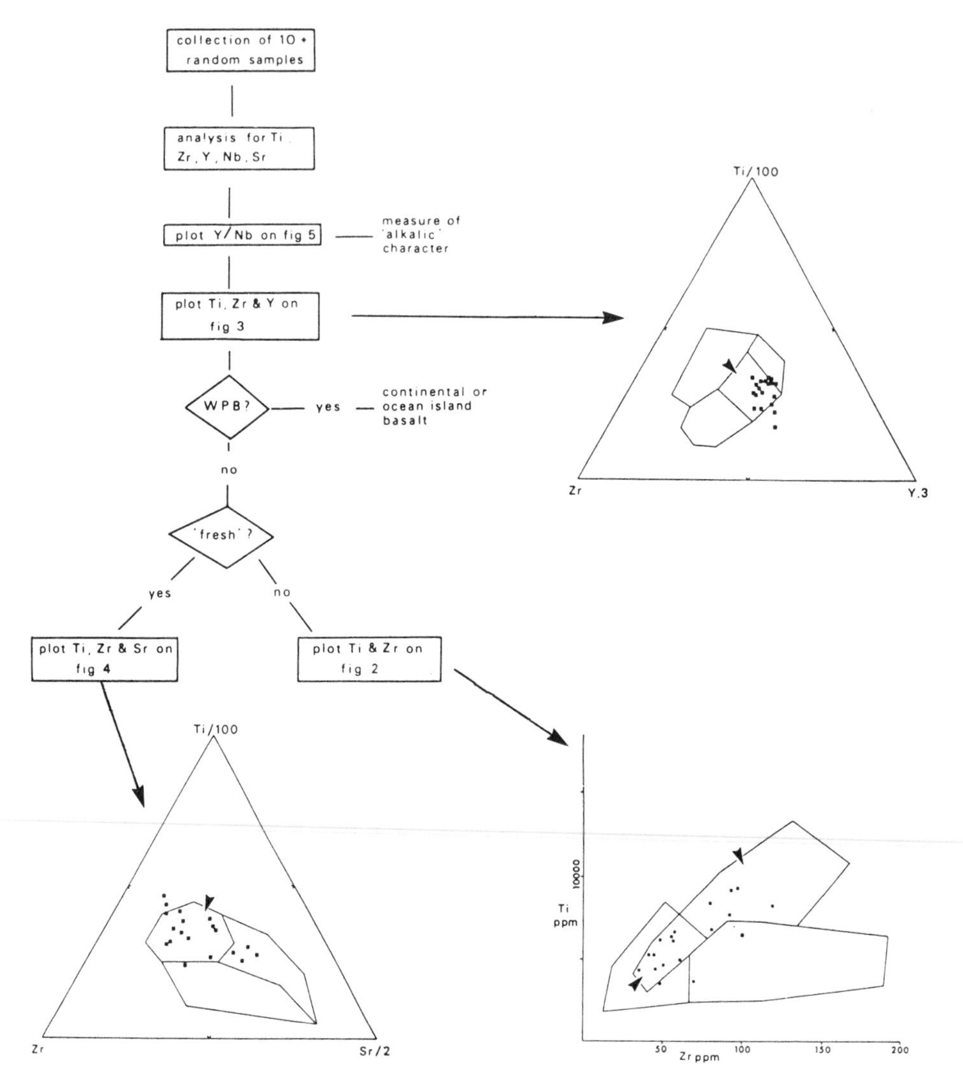

Figure 6.4 Pearce and Cann's flow diagram of the decision procedures used to classify basalts according to the tectonic setting in which they formed. The results for 19 rocks from the Troodos Massif of Cyprus are shown in three different discrimination diagrams. The Troodos was taken as an example of an ocean-floor basalt, so all the data points should fall within the areas marked with arrow heads. (From Pearce and Cann, [1973: 290–300], with permission, Elsevier Science Publishers.)

chemical procedures and the chosen ones are based upon widespread convections about which will discriminate between the origins of rocks. Second, the chemical analyses required to use the flow diagram involve numerous tacit assumptions, procedures, and decisions, as does the decision about "fresh" rock samples. Third, the scientist using these procedures must decide when the plotted results in Parts B, C, and D provide an unambiguous classification of the rock samples. Finally, the areas marked on the discrimination diagram, which are used to discriminate the different possible origins of the rocks, were themselves defined by executing the prescribed analysis on many rock specimens from different tectonic environments. Human decisions were used to provide the initial classifications of these rocks and to define the marked areas on the discrimination diagrams. In fact, the Troodos example, which was used by Pearce and Cann to illustrate how mid-ocean ridge basalts are "properly" classified by their procedures, is now considered by many to have formed in an island arc environment rather than at a mid-ocean ridge (McCulloch and Cameron, 1983). Clearly, "Nature" influences the final results, but human decisions permeate the whole measurement process.

When such decision processes become widely accepted and tacit knowledge, the measurement process becomes a "blackbox" (Latour, 1987) whose output is not questioned and the human decisions involved in the construction and use of the procedures are forgotten. The same process applies to assumptions needed to interpret the output of the magnetometer "blackbox," such as constant spreading rates, rigid plates, or seafloor properties. In other words, various components of the network of cognitive structures and processes used in a discipline are no longer considered problematic or the result of human decisions. Instead, they are taken to be characteristics of the "real world," the only "proper" measurement procedures, and the most appropriate models and analogies to use in scientific reasoning. Yet the close examination of the origins of these cognitive structures and procedures indicate that human decision-making is always involved in their construction. Thus to a large extent they are "cultural conventions" accepted by the community of researchers working within a shared paradigm (Barnes, 1982).

The above examples illustrate that the decision-making approach not only provides a basic source model for many aspects of the "intellectual" structures and reasoning processes used by scientists, but we can use the same source model for the human decision-making involved in the creation of these structures and processes. The latter decision-making is more conscious and analyzable in terms of the "classic" decision-making model that considers the conscious "utilities" sought by a decision-maker. Thus we must examine scientists' motives and goals in order to understand how a consensus is reached in science. Furthermore, we have neglected the influence of the broader social context of decision-making in science by focusing on the most individual and intellectual aspects of a research project. As mentioned earlier, research projects occur in a broader social context given by scientists' socialization into the discipline, their research experiences, and interactions with others whose decisions determine the acceptance of grants, research proposals, and articles. Since these interacting individuals differ in their knowledge, status, and power over important resources, these factors will influence the consensus that develops in a scientific community. So before considering the motivations of individual scientists, we should first examine how decision-making is influenced by the broader social context.

Decision-Making in Social Organizations

The previous examples of decision-making in science have neglected the fact that most decisions are made by individuals with limited time, energy, and cognitive capacity, who are involved in a web of relationships with other scientists whom they influence and who seek to influence them. These features have been considered most explicitly in studies of decision-making in "bureaucratic" organizations. Although studies of bureaucracies might seem irrelevant to the study of scientists, such studies seek principles of behavior that apply to "social organizations" in a more general sense, with the traditional bureaucracy representing a more "formalized" extreme. Thus studies of organizational decision-making may provide useful insights into the study of science, despite their obvious differences in content and style.[2]

One might justify this approach by noting how a scientific specialty has many of the essential features of a "social organization."

1. Scientific research on a particular topic is conducted by a relatively well defined group of "organizational members," whose activities contribute to common or closely related "organizational goals."
2. These members share an "organizational culture" or belief system composed of shared, often tacit, assumptions, models of reality, and exemplars of behavior and thought. This culture provides a basis of communication and evaluation among members and is used to develop and defend an "organizational technology" that creates the organization's "products." In science, the "products" are ideas, or more concretely, scientific articles containing ideas, and scientific instruments or other observation procedures are the "technology."
3. There is some degree of role specialization and an "organizational hierarchy" of authority among members. That is, some scientists have more esteem, influence, and power over valued resources than others and use these to maintain social control and coordination of the activities of the individual members in their different roles. For example, teachers teach the basic paradigm to students, research supervisors direct the tasks of subordinates, grant and article referees determine others' access to research support and communication channels.
4. There are established formal and informal channels of communication between members. In science, major channels of communication occur through scientific journals, conferences, university classrooms, and personal networks.
5. These features have a relatively continuous existence and stable enough re-

lationships with other "organizations" in their "environment" (such as journals, universities, and funding agencies) to insure an adequate supply of needed resources (such as people and money) so that the organization and its cultural system is maintained.

If the social organization perspective provides a reasonable approach for understanding the decision-making of scientists, we should be able to find plausible analogies between studies of organizational decision-making and studies of scientists' decision-making. A brief review of March and Simon's (1958) classic study of decision-making in organizations illustrates that such analogies exist.

March and Simon (1958) begin with the simple economic decision-making model, where a person has a clearly defined set of choices, a set of consequences associated with each choice, and a "utility function" that ranks the choices in terms of their consequences. The "optimal" or rational choice is the one leading to a preferred set of consequences. One could translate this simple model directly into Laudan's "problem-solving" perspective by making different theories the choices and problem-solving ability the utility to be maximized. The general difficulties in his approach are apparent as March and Simon modify this simple model by embedding it in a social context and reflexively applying it to itself so that each step is itself a product of decisions by humans with limited time, energy, cognitive abilities, and membership in a organization with an authority system. The resulting description of decision-making has many parallels to Kuhn's conception of normal science.

Let us first consider their modifications of the steps in the simple economic model. In typical situations all alternatives are not known, so some searching and screening efforts are required to find additional alternatives, but these efforts are limited by finite human capacities. "A theory of choice without a theory of search is inadequate" (March and Simon, 1958: 174). However, where to search and how to evaluate the consequences of each considered alternative require a "model" of reality because the "objective situation is far too complex to be handled in detail . . . [so] rational behavior involves substituting for the complex reality a model of reality that is sufficiently simple to be handled by problem solving processes" (March and Simon, 1958: 151). Even the preference orderings are uncertain because they are dependent on this simplified model. Consequently, these modifications imply that the rule for choosing the alternatives cannot be one of maximization in any "objective" sense because not all possible alternatives are known and the consequences and preferences are model dependent. The rule is one of "*satisficing*" in which some minimum criteria are established and the chosen alternative is the first one that satisfies these criteria. If none of the considered alternatives are satisfactory, then more alternatives are sought or the minimum criteria are reduced.

The whole decision-making process is influenced by the model of reality. It also provides the basic vocabulary used for communication within the organization. This vocabulary provides "a set of concepts that can be used in analyzing and in communicating about [the organization's] problems. Anything that is easily described and discussed in terms of these concepts can be communicated readily in the organization; anything that does not fit the system of concepts is communicated only with difficulty. . . . The particular categories and schemes of classification [the organization] employs are reified, and become . . . attributes of the world, rather than mere conventions" (March and Simon, 1958: 164–165).

Since organizational decision-making is embedded in the context of the organization's model of reality, the concept of rationality has changed: "from a phenomenological viewpoint we can only speak of rationality relative to a frame of reference . . ." (March and Simon, 1958: 138). The task of a good administrator is to provide the premises and framework for their subordinates' models so their "rational" behavior will promote the organization's goals. Consequently, there are many levels of decision-making in an organization and the "rational" choices at each level are dependent on the models used by the decision-makers. In March and Simon's terms, there are only areas of "bounded rationality" in complex organizations.

There is a potential for infinite regress here. If each aspect of the traditional economic model is the subject of a decision process, then we could also have decisions about when to make decisions and so on. March and Simon avoid this possibility by suggesting that there are three basic levels of decision-making "programs" in organizations. "Task programs" are established, which are sets of responses requiring little discretion on the part of individuals using them. "Switching rules" direct the choice of which task programs are used in particular situations. Finally, "procedure programs" direct the modification of old task programs or the development of new ones. The switching rules and procedure programs provide the relatively stable aspects of organizations and are based upon the simplified model of reality. Thus the behavior of the organization's members consists to a large extent of many established "task programs" that are held together and interrelated by "higher level" decision procedures (switching rules and procedure programs) based upon the organization's model of reality.

The analogies between decision-making in organizations, DeMey's analysis of cognitive structures, and the behavior of scientists should be apparent. To a large extent scientific research consists of using established instrumental and interpretative procedures that are coordinated by theories or models of reality. These cognitive elements are shared by a specific set of socially related individuals, but they do not yield a set of optimal decisions for the description of the external world, only a set that is satisfactory within the tacit expectations provided by the current paradigm's exemplars. Decision-making in science has the character of "bounded rationality" in that numerous assumptions, beliefs, models, etc., are necessary for the more rigorous applications of logic or mathematics. In other words, March and Simon have described decision-making processes in Kuhn's normal science.[3]

However, a crucial element is missing in this organizational approach. Perrow (1972) suggests the decision-making model provides a fuller view of the *cognitive* processes in organizations, but the model tends to ignore conflict. In Perrow's view conflict will always be present in organizations because "there is a never-ending struggle for values that are dear to participants—security, power, survival, discretion, and autonomy—and a host of rewards" (Perrow, 1972: 159). For example, do the premises for the "rational" behavior of subordinates facilitate the "organization's" or the administrator's goals? Does a scientist with a theory preference encourage his students to work on any theory or his preferred theory? In other words, we must consider some of the values, goals, and motivations that influence the decision-making activities of scientists. These are discussed in the next section.

The Motivations of Scientists and the Context of Research

A key component in the analysis of the decisions made by scientists is the goals or motivations they are trying to achieve through their decisions. In this section, we will consider some of these motivations, how they might influence typical career decisions, and some of the possible consequences of these decisions.

Since science is now one among many possible professions, a person can pursue it simply because it is a means of earning a living. For some individuals geology offers a means for making a comfortable living that involves outdoor field work. Financial rewards may be considerable, especially for geologists working for mining and petroleum companies. One of my informants mentioned that he was a millionaire because of his role in the discovery of several oil deposits. Although this is true of some geologists, the discussion here will emphasize the motivations of those in academic or professional settings that place the greatest emphasis on the "pursuit of science for its sake alone."

Within this traditional perspective the only "legitimate" motivation is the thrill associated with scientific discovery and the advancement of knowledge. Some of those involved with the development of plate tectonics mentioned this excitement. Tanya Atwater describes one of these experiences.

> It is a wondrous thing to have the random facts in one's head suddenly fall into the slots of an orderly framework. It is like an explosion inside. This is what happened to me that night [when McKenzie explained the geometry of plate tectonics to her] and that is what I often felt happened to me and to others as I was working out (and talking out) the geometry of the western U.S. I took my ideas to John Crowell one Thanksgiving day. I crept in feeling very self-conscious and embarrassed that I was trying to tell him about land geology starting from ocean geometry, using papers and scissors. He was very patient with my bumbling, but near the end he got terribly excited and I could feel the explosion in his head. He suddenly stopped me and rushed into the other room to show me a map of when and where he had evidence of activity on the San Andreas system. The predicted pattern was all right there. We just stood and stared, stunned. [From Cox, 1973: 536]

In a 1974 interview Walter Pitman described a similar experience.

> The biggest thrill I had was when I was [on Leg 3 of the DSDP project when they tested the Lamont extrapolation of the magnetic reversal chronology to 80 million years to predict the age of the South Atlantic sea floor; see Figure 4.21]. . . . I went over on their ship and they were really excited. They showed me all of this [Figure 4.21]. I couldn't believe it. . . . That was a tremendous experience.

A few years earlier (in late 1965 and early 1966), after Pitman worked all night to get the first tracings of the Eltanin-19, 20, and 21 magnetic anomaly profiles, he pinned them on Opdyke's door and went home. "When I came back the guy was beside himself! He knew that we'd proved seafloor spreading! It was the first time that you could see the total similarity between the profiles—the correlation, anomaly by anomaly. The bilateral symmetry of Eltanin-19 was the absolute crucial thing. Once Opdyke saw that he said, 'That's it—you've got it!' " (from Glen, 1982: 334–335). Both Glen (1982) and Wertenbaker (1974) describe numerous examples of the excitement among the geoscientists involved in the testing and development of seafloor spreading and plate tectonics.

Undoubtedly these were intense and rewarding experiences, but scientific "discovery" occurs in a social and intellectual context, which "sets the stage" for these experiences. For example, others at Lamont were much slower to accept seafloor spreading on the basis of the Eltanin-19 profile (Glen, 1982; Menard, 1986; Wood, 1985). The social and intellectual context of an individual scientist is the result of numerous decisions by the scientist and by others. For example, decisions concerning graduate schools, specializations, supervising professors, and post-graduate appointments will influence the eventual social and intellectual climate of one's research experience. Before considering specific examples of these decisions, we need to consider some other motivations that will influence such decisions.

As mentioned in Chapter 1, a key component of a sociological analysis is that scientists desire "recognition" from their peers (Merton, 1957). This is also an important aspect of the "thrill of discovery"—knowing that your peers will recognize it. Although recognition will have some *extrinsic* benefits because it aids promotion, tenure, or increased access to research support, it also appears to be *intrinsically* rewarding to scientists. A few examples illustrate this aspect.

A scientific revolution offers both exciting discoveries and recognition from one's peers—provided the revolution is successful. Some geoscientists saw this possibility and responded quickly to it. For example, when John Sclater heard a seminar by McKenzie, which gave the McKenzie and Parker's (1967) explanation of the northeast Pacific in terms of plate tectonics, he was convinced that plate tectonics was important:

> I knew exactly what to do. . . . We [Sclater and Francheteau] knew we had to try this thing for the whole world. We thought about this for a bit, but by January we found out that LePichon was doing it. . . . I was just sitting there and I can remember saying to myself: "My God, here you are, you understand this thing, it's very easy to do, it's obviously going to explain [most geological features in the oceans], and

the whole thing is going to pass you by without you doing anything." I went through a crisis!

Sclater clearly saw the potential for recognition in the new theory and that it wouldn't be given for repeating LePichon's analysis. Instead he related heat flow and ocean depth to plate tectonics: see Chapter 4. He also had an advantage over others because of his early exposure to plate tectonics through the seminar. Early adopters of a new theory will be able to "skim the cream" off a new area by expressing its most obvious implications (Hagstrom, 1965).

Hagstrom (1965) notes that recognition can take several forms. Formal recognition is represented by giving professional medals and awards, the naming of physical processes or constants after their discoverers, and explicit references to a person's contributions in scientific articles. All of these forms of recognition have been bestowed upon major figures in the development of plate tectonics. For example, there is now an official Wilson Medal, named after and first given to J. Tuzo Wilson for his contributions to plate tectonics. He also received the Penrose Medal, as did Hess. Hess was also the first North American to win the Feltrinelli Prize, worth $32,000. The Balzam Prize was awarded to McKenzie, Vine, and Matthews. Doell, Cox, and Runcorn were awarded the Vetleson Prize, one of the esteemed prizes in geology. The cycle of opening and closing ocean basins is called the "Wilson Cycle"; there are mountain ranges in Antarctica named after Wegener and Wilson; there is the "Hess Rise" in the Pacific; some of the craters on the moon are named after Wegener; the relationship between ocean depth and seafloor age is often called the "Sclater Curve," and, of course, there is the "Vine-Matthews hypothesis." The tables in Chapter 4 for the citation counts to major articles indicate that these contributions were widely recognized. Figure 6.5 shows that plate tectonics articles were increasingly cited during the late sixties and early seventies.

It appears, then, that those who worked on plate tectonics during the sixties have received extensive formal recognition for their contributions to the geosciences, but scientists also value less formal recognition given in personal conversations and feedback from presentations. In contrast to the delay in formal recognition, informal recognition is more immediate and personal. In summary, both forms of recognition may be desired for several reasons. First, they are intrinsically rewarding to scientists. Second, other institutional rewards, such as promotions and salaries, may be tied to the recognition received by a scientist. Finally, Storer (1966) argues that scientific research is a creative process without clear criteria for the validity of the results, so peer recognition validates a scientist's "discovery."

Hagstrom (1965) notes that since recognition is based upon peer assessment of the merits of one's contributions, it provides a powerful means of social control in science. For example, he found that scientists would switch research topics or even specialties if they perceived greater potential for recognition in the new areas. Of course, the *lack of recognition* for one's contributions might decrease a scientist's motivation to do research.

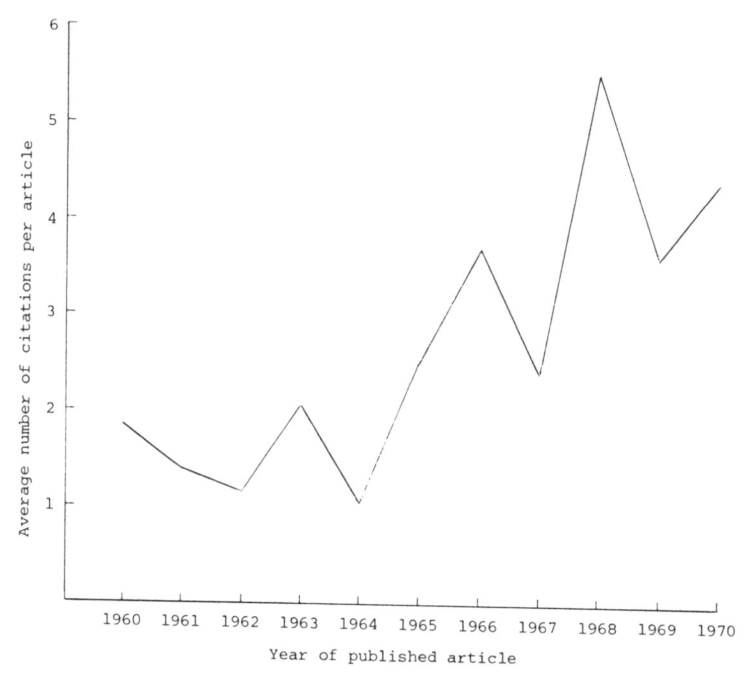

Figure 6.5 Average number of citations to articles related to continental drift and published between 1960 and 1970. Citation counts are based upon those occurring in the second year after publication.

The lack of recognition for early proponents of drift or seafloor spreading was discouraging for many of them. About a year after Vine and Matthews (1963) published their hypothesis, Vine said he "was getting pretty discouraged and beginning to lose faith myself. It went over like a lead balloon; in some ways there was no response. People just sort of turned away" (from Glen, 1982: 281). Opdyke had a more personal rejection during a talk at Lamont on the paleomagnetic evidence for the drift of Africa. "I got not one question after I finished that talk. Not one question—zero! It was the most disheartening experience I've ever had in my professional life." Patrick Hurley described the reaction of a Harvard audience to Hess's presentation of his seafloor spreading model: after the lecture "somebody asked a stupid question and everybody laughed; nobody shook his hand or anything else . . . he was almost laughed out of the hall."

It is apparent that recognition by one's peers is a strong motivation for scientists and it can explain many aspects of their behavior, such as the desire to publish ahead of competitors and to control access to their data. Both of these concerns are illustrated in the following comment by Walter Pitman who explained how Fred Vine obtained the Eltanin-19 profile from Lamont.

Vine saw the Eltanin 19 profile in February 1966 and went away. Heirtzler and I began to write our paper up in about May or June of 1966; we had it pretty much in the shape in which it appeared in *Science* [Dec. 2, 1966]. Vine was coming up to Lamont from Princeton to give a Friday afternoon colloquium talk in early May or June 1966 after the April 1966 A.G.U. meeting; Heirtzler was away, and I was a little worried that Vine might possibly scoop us. So I finished off a redraft of the paper Heirtzler and I were writing and sent it to Vine by special delivery, so that he got a copy of the paper before he came up here to talk—which was a smart thing, because when he gave the talk, there it was, the paper that he eventually published in *Science* [Dec. 16, 1966]. I had not known if Heirtzler had given Vine the Eltanin 19 profile, which he saw early in the year, or not. I went back and looked through the files and found a letter from Heirtzler to Vine giving him the data, he was very vague about it. Heirtzler said, "here is the data, but please keep in mind that several of my students are using it in their theses." Vine was obviously preparing a large paper based on the Eltanin 19 profile.

So there we were, up a tree. Heirtzler was coming back very shortly, so I said to Vine, "Maybe we ought to think of publishing together," but he declined. Very shortly after that, we went down to Princeton to confer with Vine about how we were going to settle all of this. I think basically Vine didn't want to be a coauthor with us for a lot of good reasons. His paper went into much more depth about the implications of this than we did; and, again, it was because the guy had been thinking about it all these years. If [the Vine-Matthews hypothesis] were true, what did this [Eltanin 19 profile] mean? Well, God, he knew what it meant—why should he share that with anybody? After some arm-wrestling with Vine and with *Science*, it was agreed that we would publish [two weeks] before he did; his would be a major article. *Science* wanted to publish both papers together in one issue, which would have been very logical from a purely editorial point of view. But on the other hand, the guys that run *Science* understand geopolitics, and once that was explained they said "We don't like it, but we'll do it."

As a general rule in this business, you never give data away, and you don't have to apologize to people. The Eltanin 19 was the whole ball of wax. You can't give an intelligent man a piece of data and tell him not to work on it. You just can't do it, particularly if the guy is a scientist; particularly a young guy like Vine, who had his brains beaten out by that time [by opponents to the Vine-Matthews hypothesis]. . . . I must say that I learned something from it: Data is either given away with no strings or not given at all. (From Glen, 1982: 336–337)

To understand this comment, we must assume that scientists are motivated by more than just the desire to discover properties of the natural world. Recognition from their colleagues is essential and this recognition flows to those who present new and convincing results. Hence, Pitman and Vine were concerned with the order in which they published their results. Other scientists control this recognition and can use it to encourage or discourage intellectual developments, as indicated by the "beating" that Vine had received from other geoscientists. Finally, the desire for recognition helps explain why scientists try to control the use of their data until they have a chance to publish their results.

These examples suggest, then, that scientists seek not only the thrill of discovering properties of the natural world, but are motivated by a desire for recognition from their colleagues. This implies that scientists might also seek to "protect" the recognition received for their previous research. That is, they may resist new theories that would challenge the theories providing the basis of their previous publications. Edward Bullard has suggested that this motivation may account for the early resistance and rejection of drift theory.

> There is always a strong inclination for a body of professionals to oppose an unorthodox view. Such a group has a considerable investment in orthodoxy; they have learned to interpret a large body of data in terms of the old view, and they have prepared lectures and perhaps written books with the old background. To think the whole subject through again when one is no longer young is not easy and involves admitting a partially misspent youth. (Bullard, 1975*a*: 5)

This desire to increase and/or protect one's reputation can help explain Kuhn's observation that younger scientists are more likely to develop or quickly adopt a new paradigm than are older scientists. Kuhn's emphasis was on the greater "cognitive freedom" of younger scientists, but a new paradigm also offers faster routes to recognition because there will be an abundance of new "puzzles" that have not been tackled by anybody and the newcomer will not have to learn as much before reaching the "research front."

Intellectual investments build up over the course of a scientist's career and result from the many decisions made by the scientist along the way. One of the first decisions a prospective scientist must make is where to pursue her educational training. Assuming she is accepted in the department, which requires a decision by others, the choice of schools may affect strongly the assumptions, theories, and methods that she will adopt in later research. For example, choosing to study at Columbia University and do oceanographic research at Lamont in the fifties or early sixties would expose her to a fixist point of view. As Ellen Herron, a graduate student at Columbia in the mid-1960s noted, "Doc Ewing's philosophy that the oceans were permanent features was the party line at Lamont; seafloor-spreading was anathema" (from Glen, 1982: 313). In 1974 Peter Molnar, who graduated from Columbia and worked at Scripps, contrasted the two schools. "Lamont has always worked pretty much as an institution with very little dissent—as far as I can tell. Dissent was somehow weeded out; Ewing got rid of it or something. But Scripps is full of it. Scripps is just a bunch of entrepreneurs who do their own thing—prima donnas, really. I don't think you'll find a Scripps opinion that holds." One possible explanation of these differences is given by the percentage of the faculty members that are graduates from the same department. In 1975 46.7 percent of the Columbia/ Lamont faculty were "inbred," whereas only 17.9 percent of the San Diego/ Scripps faculty were inbred.

In 1974 I asked John Sclater about institutional differences; he mentioned that "Harvard [the geology department] is having terrible trouble. Harvard has had it and is essentially destroying itself [as a major research department]. They still give seminars on anti-drift tectonics up there; but they are changing now." Princeton now is very much pro-drift; to such an extent that Oxburgh, who graduated from Princeton in 1960, was alarmed by the uncritical attitude present when he participated in a Ph.D. exam there in the late 1960s.

> I was horrified—absolutely horrified—at my experience in that examination: by the totally uncritical assumptions . . . that the candidate was allowed to make regarding plate tectonics. It was absolutely assumed! I went back a couple of years later and found exactly the same thing about plumes. It was never a question of whether plumes existed—that wasn't open at all for discussion—the question was where this plume had been at that time or exactly how this plume had moved.

These comments indicate that the seemingly innocent decision about which school to attend can have a major impact on the ideas that one will use in future research. These choices can either promote or hinder the acceptance of new ideas. However, as noted in Molnar's characterization of Scripps, within some schools there may be a wide diversity of opinions among the faculty. Consequently, the graduate student's choice of a dissertation professor may be extremely important because the dissertation research often lays the foundation for the student's future research program. A contrast in the styles of research between Ewing and Wilson illustrates the importance of this factor. A USGS geologist contrasted his own approach to Wilson's. "My philosophy is to get things down on a map and to find the facts. I don't jump ahead like some people do and spin a gorgeous theory. I like Tuzo's leaping ahead of things. I have told him many times that half the time I don't believe a god-damned word of it, but I would like to listen to him all day. Some other people are simply annoying. Tuzo is delightful. . . ."

Walter Pitman also noted this aspect of Wilson's style by contrasting it to a person who had said, "I just never want to publish anything unless I'm absolutely one hundred percent certain that I'm right." Pitman suggested that this attitude is "the total antithesis of scientists like Wilson who could not care less. He was not afraid of being wrong. The people who want to be one hundred percent certain they're right, never make a breakthrough in science—never!"

In contrast is the description of Ewing's research style, as recalled by Lynn Sykes, a research associate at Lamont and a Columbia graduate.

> His [Ewing's] whole philosophy was that you didn't take a great leap into fantasy, that you built things up very carefully and documented them piece by piece. I think that he tended to feel that all the other hypotheses were arm-waving, but out of really solid data collection would emerge the fundamental hypothesis. When essentially the version put forth by Hess or when the idea of magnetic strips of Vine and Matthews came along, he tended to dismiss those, feeling that it was something else that was going to emerge. . . . There is the tragedy of the person who very early recognized that the ocean would be the clue to understanding the big scale of things, but when the correct hypothesis came along, he resisted it.

Ewing was a physicist by training and apparently developed his geological beliefs from Walter Bucher of Columbia University, who was a strong opponent of continental drift (Glen, 1982: 313), whereas Wilson had a dual major in geology and physics (Wood, 1985). Such differences among individuals in research styles are likely to have major influences on those they train.

Mentors in science not only influence styles of research, but also preferences for particular theories and research subjects. Bullard (1975*a*) and Wood (1985) note that Richard Field played a major role in the development of marine geology because of his feeling that geology was incomplete without knowledge of the ocean basins. "With the burning zeal of an Old Testament prophet" (Bullard, 1975*a*: 9), he recruited Bullard, Hess, Wilson, and Ewing for this purpose and encouraged them to complement their specialty training with theories and methods from each side of the chasm between geology and geophysics. This "role hybridization" (Ben-David, 1971) helped these four scientists lay the empirical and theoretical foundations of plate tectonics. Other early proponents of drift theory or plate tectonics had this characteristic, e.g., Blackett, Runcorn, Morgan, McKenzie, Menard, Dietz, Carey, Holmes, and Wegener.

Choices about schools, majors, and professors represent decisions that scientists must make during their educational careers, but none is an individual decision that can be made alone. They also require the consent of others. That is, the school and department must decide to admit the student and the professor must allow the student to study under him or her. Each step in the student's educational process is subject to decisions by others and these steps cumulate in the departmental decision to grant the Ph.D. This important evaluation by others controls what types of ideas and scientists will have legitimate access to a scientific community. According to Wilson (1974) some departments were reluctant to grant degrees to students writing dissertations on continental drift. "At Cambridge poor Ted Irving became so excited about the new [paleomagnetic] results that he devoted his thesis to them, so the conservative examiners refused to award him his Ph.D. and he left for Australia."

This decision is only one of many that are made by scientists in their role as "gatekeeper" (Zuckerman and Merton, 1971) for the scientific community. Another major one is deciding whom to hire. As might be expected, some of those who advocated continental drift may have had difficulty finding jobs. Heezen mentioned the example of a distinguished European mathematician who came to Lamont, helped write a classic text in seismology, and who also believed in continental drift.

> When he came to this country in 1941 or 1942, he wrote another [paper on drift] and that finished him. To me, if you had a man that bright anywhere in the country now, he'd be a full professor in any department. He couldn't get a job except as a physics lab instructor at Manhattan College because he did the unforgivable—he believed in drift.

Hallam also thought that "until quite recently in some institutions an open adherence to this doctrine would have put at serious risk the attainment of tenure by junior faculty members, while their more secure senior colleagues would have been all but drummed out of their invisible college. At the present time, only a few years later, almost the reverse situation holds" (Hallam, 1973: 105).

Once one has a degree and an institutional position in a scientific community, one is a "legitimate" member of a professional group. Yet this does not free one from being influenced by colleague decisions because they control access to needed research resources, such as grant money, students, data sets, and instruments, not to mention recognition for one's efforts. Unpopular topics may not receive support or recognition. Arthur Meyerhoff thought one of the primary reasons that plate tectonics is so popular is that the National Science Foundation (NSF) and other funding agencies would not fund research opposed to plate tectonics. He mentioned that William Benson at NSF had written him stating that they would not fund his research because he was just trying to disprove plate tectonics.

> I don't mind seeing you put this in print; he [Benson] and I are bitter enemies . . . he is not a scientific person. To me a scientifically honest person is a person who will take both sides and weigh "them" and will say, "Look, I don't have enough facts or it looks as though such and such might be true, but we need more data. . . ." His [Benson's] attitude is, "Look, everybody accepts it, therefore it must be true." . . . To my knowledge an anti-drift grant has never been funded by NSF since the hypothesis became popular; that means since 1968.

Even without grants it is often possible to do some research provided you have access to the available resources and instruments, which may be controlled by the decisions of others. During an interview Allan Cox mentioned that his work on the magnetic reversal chronology required access to potassium-argon (K-Ar) dating instruments. Even though his professors at Berkeley helped develop this instrument, they would not date his samples.

> Two of my professors there were world leaders in [K-Ar dating]. I tried to get them to date the rocks I was working on, but they were working on other things and didn't think it was an important problem to work on. . . . I knew it was important for geology, [but] I just couldn't talk them into doing it. . . . What we did was to start doing paleomagnetism [reversal dating] on stuff that people had collected for other reasons.

Glen (1982) suggests that Cox's professors were overwhelmed with requests for K-Ar dating of samples. Consequently, they had to decide which ones were most important to them and they made their choices on the basis of which would gain them the most recognition. Magnetic reversal studies were not considered very important and in geology more recognition was given for studies of rocks much older than those studied by Cox (Glenn, 1982: 93–94, 162).

Finally, after getting the approval of others for degrees, jobs, and research funds, the scientist can start the research project and proceed with the decisions outlined in the typical project in the first section of this chapter. However, not just any project can be worked on; rather, it must be similar to what was described in the grant proposal, which was approved by others. Thus even the actual research process is not completely under the researcher's control. Even after the research is finished and written up, decisions by other scientists continued to have a major influence. The scientist is "free" to interpret the data in any manner, but the results are more likely to influence others only if the paper is accepted for publication in a scientific journal. Thus journal editors and referees control what becomes publicly available to the wider scientific community. A particularly important example of journal control of ideas is provided by Lawrence Morely's anticipation of the Vine-Matthews hypothesis.

Morley not only outlined the Vine-Matthews hypothesis, but he more explicitly noted that the spreading rates could be calculated with better knowledge of the magnetic reversal chronology. (Excerpts of Morley's original article may be found in Cox [1973: 224–225], Glen [1982: 299–300], and Lear [1967].) Morley did not use data to support his suggestion, but he did submit it as a letter to *Nature* several months ahead of the Vine and Matthews article. About two months later, the editor said they "did not have room to print" his letter, whereupon he submitted it to the *Journal of Geophysical Research*. Five months later, after Vine and Matthews were published, the editor of JGR rejected the letter saying that an anonymous referee argued that "such speculation makes interesting talk at cocktail parties, but it is not the sort of thing that ought to be published under serious scientific aegis" (from Glen, 1982: 299). Morely's letter has been described as "probably the most significant paper in the earth sciences ever to be denied publication" (from Glen, 1982: 302). Some geoscientists writing on the history of plate tectonics have sought to redress this denial of recognition by reference to the "Vine-Matthews-Morley" hypothesis. Today the situation is different in that articles opposed to plate tectonics may have a more difficult time getting accepted. Arthur Meyerhoff mentioned that several journal editors had said to him that they would not consider any articles by him because of his known opposition to plate tectonics.

In some institutions there are internal reviews of articles before they are submitted to journals. This was apparently the case at Lamont, where Ewing would review papers before they were submitted to journals. A Lamont scientist recalled an incident where he had submitted an article without Ewing's approval.

> *A*: Ewing was trying to stop me from publishing the paper, and he called Maddox [the editor of *Nature*] and tried to stop it, but it went to press the next week: it came right out. . . . Maddox just wouldn't put up with pressure from the establishment to stop something.
>
> *JAS*: Ewing was trying to stop your publication?
>
> *A*: Right—several times.
>
> *JAS*: That's backbiting—trying to cut you off at the pass.
>
> *A*: Yes, well, we had a situation where you had to submit manuscripts to him for review before you could send them, and then he'd act like he wasn't doing anything. But if he knew where you sent it to and didn't want it published, he'd call the editor and just stop it. He could stop things in *Science* and stop things in other journals, usually long enough so he could get something lined up himself or some

friend of his could. He was a great man, but there were a lot of things about him that were pretty nasty.

Menard (1986: 225) mentioned that some geologists in the U.S. Geological Survey had their pro-drift publications delayed for over a year because of internal opposition to drift.

Finally, there is another key decision made by other scientists—which research in the scientific journals is worth reading and using as part of the foundation for one's own research. A published research result cannot affect the consensus in a scientific discipline unless others read and use it. Since use of a research paper usually results in citations to the paper, the citation counts in Chapter 4 indicate considerable differences among articles in how much influence these articles had on other researchers, especially the relative neglect of articles opposed to plate tectonics. The decision about whose research will provide the basis of one's own research is a particularly critical decision and the factors influencing this process are considered in more detail in the remaining sections of this chapter.

Sociological Perspectives on Science

These previous examples indicate that the consensus that develops within a scientific community cannot be understood by just examining the decisions made during a research project, which is the narrow focus of most philosophies of science. These perspectives ignore the fact that scientists are members of a social system in which there is competition for recognition and a multitude of interrelated decisions among its members that result in (a) the socialization of new members into the social system, (b) the control of such rewards as degrees, jobs, research money and facilities, tenure, publications, recognition, and power or authority, and, ultimately, (c) the very beliefs that form the basis of research activity in a scientific community. The different sociological perspectives try to describe how these decisions are made.

The traditional sociological perspective on science, Merton's functionalism, emphasizes the rules or norms directing the decisions related to the *social context* of research. The newer "constructivist" viewpoints include a focus on the research process itself. This section briefly reviews these diverse perspectives, discusses some of their problems, and illustrates them with examples from the history of plate tectonics.

THE FUNCTIONALIST PERSPECTIVE

As mentioned in Chapter 1, Merton (1942) suggested that the function of the scientific community for the broader society is the production of "certified knowledge." Merton did not analyze the "scientific method" itself, but suggested that it would operate more effectively in a social environment where special norms constrained the relationships between scientists and helped them avoid biasing processes. Two important norms were "universalism" and the acknowledgement of others' contributions to your own research. Universalism required scientists to evaluate the merits of a knowledge claim solely on the basis of intellectual criteria and ignore such intellectually irrelevant criteria as the race, religion, personality, or nationality of the person making the knowledge claim. The acknowledgement of others' contributions was crucial because the desire for recognition was held to be one of the major motivations of scientists (Merton, 1957). It is apparent that these norms pertain more to decisions regarding the social relationships between scientists than to decisions made in the research process itself.

The desire for recognition helps us understand some important characteristics of the social system of science. We have already seen examples of how scientists can grant or withhold it as a means of social control within science. Furthermore, the autonomy of the scientific community is protected by the intervening role of peer recognition in the distribution of other social rewards. For example, academic scientists' access to such social rewards as job security or higher salaries is based to a large extent on their contributions to the scientific community and the recognition earned by these contributions. This arrangement helps to protect scientists from outside political and social pressures.

However, competition for recognition may have some negative consequences for science by encouraging violations of the functionalist norms. For example, Hagstrom (1965) found evidence that competition for recognition caused some scientists to keep their results secret, which violates the "communality" norm to share ideas and information. Such violations, however, may still be functional for overall scientific progress. If secrecy allows a scientist to more fully develop and publish a coherent argument for a new idea, then it will not only give the scientist more recognition, but may also advance scientific knowledge more rapidly. Similarly some violations of the universalism norm may be functional for scientific advance. Hagstrom (1965: 24) noted that when two or more scientists reported the same discovery, the most famous of the scientists is likely to receive most of the recognition, which violates universalism and the norm to acknowledge the sources of one's ideas. Merton (1968) called this process the "Matthews Effect" and suggested that it may be functional for science because it makes important discoveries more visible to other scientists, even though it is not functional for the motivations of the neglected scientist.

Some examples from the history of plate tectonics indicate that some of these norms exist within the geoscience community. There are clear statements about the norm to acknowledge others' contributions: "To fail to cite known work is dishonest" (Menard, 1986: 5), but even better evidence of the existence of this norm is indicated by the penalties extracted for its violation. When discussing the history of plate tectonics with one geoscientist, I mentioned Dietz's (1961) publication on seafloor spreading. This prompted a strong response: "He stole it from Harry Hess! Don't give him credit. I was there [with Dietz at a Princeton meeting where Hess gave an informal talk after the 1960 AGU meeting] and he went home and wrote it up and sent it to *Nature*. . . . He's the only plagiarist I have ever met in the geological field in fifty years."

Clearly Dietz was seen as having broken a norm and the denial of recognition was needed as a sanction for this violation.

There is, however, a possible alternative explanation: Dietz's "terrible memory." Several geologists noted this characteristic of Dietz.

> *Menard*: Dietz and I wrote about five papers together and he simply had a memory—he was really quite capable of forgetting completely that he had talked to Hess.
>
> *Burke*: I think that's realistic; that you assimilate things that you got from other people and then you say something to somebody and they say, "Well, I said that to you six months ago."
>
> *Menard*: That's exactly what Peter Molnar said to me at the beginning of this meeting. I was telling him about this exciting new idea that I had been developing and he said to me, "You bastard! I knew that you would forget about that." He had come in and we had a discussion and after that I had written a little note that said "idea" on it and stuck it up on the wall . . . he hadn't signed it.

Although Merton (1963) has suggested that such "cryptomnesia" may be quite common in science, its occurrence can still disturb those, such as Molnar, who were denied recognition for their contribution.

Menard (1986) documents a particularly dramatic example of a violation of this norm and its consequences. As noted in Chapter 4, in 1966 two Lamont groups had found persuasive new evidence for seafloor spreading and reversals of the earth's magnetic field: Pitman's analysis of the Eltanin magnetic anomaly profiles and Opdyke's analysis of the pattern of reversals in the deep-sea cores, but it was Heezen and his group who moved the fastest to explore the analysis of the cores. Opdyke objected to Ewing about the infusion of Heezen's students into his lab and Ewing limited Heezen to one graduate student's work. To allow Opdyke priority in publishing his results, "Heezen was expressly forbidden [by Ewing] to talk on the subject of magnetism in sediments at the [upcoming] International Oceanographic Congress in Moscow" (Menard, 1986: 269). Heezen discussed these results anyway and was censured, but his tenure protected his employment, so other penalties were taken. "At age 42 the internationally famous oceanographer Bruce Charles Heezen was forbidden access to the data of his home institution, denied the right to have Lamont as a sponsoring agent for his requests for grants, and denied access to Lamont ships" (Menard, 1986: 272). This is a particularly clear example of punishment being enacted for a violation of a social norm, which indicates the existence and importance of the norm and the general desire for recognition.

The desire for recognition is recognized in a number of the comments made by these geologists and it may cause other behaviors that conflict with the norm of "open communication" of research results, such as control of data or secrecy. Being "scooped" by others can deprive one of recognition and encourage secrecy (Hagstrom, 1965). For example, McKenzie and Parker (1967) gave the first *published* presentation of the basic plate tectonics model, but Morgan's (1968) more extensive paper was actually submitted earlier. Several geoscientists *emphasized* this during interviews and some noted Morgan was more secretive in his development of his first mantle plumes paper (Morgan, 1971). Pitman's comments in the previous section about the desire to establish publication order with Vine and to control data also indicate the importance of recognition.

Geoscientists recognize such dangers from the competition for recognition and have developed some mechanisms to avoid this conflict. One of these is the Penrose conferences, which are modeled after the Gordon conferences in chemistry. Allen Cox helped develop the Penrose conferences for the Geological Society of America. He noted that

> . . . people are much less defensive at that kind of meeting. They are more willing to express their doubts and they see each other for a couple of days. We structured the Penrose so that there were no published reports. . . .
> *JAS*: Does the idea of not publishing volumes permit less inhibited discussion?
> *Cox*: That's right. It's also conducive to people bringing their data, showing it, and not defending a position they know they will have to write up. They are free to change their views while they are there.

Recognition can be denied because of the "Matthews Effect," where several scientists become associated with a particular result and the more famous gets the credit. Menard initially published with Dietz, but one of his first separate publications described the fracture zones in the Pacific. He had sent a preprint to Hess for comments and received a "highly complementary" reply and his "joy at this recognition from the master was only slightly muted by the fact that it was addressed to Howard S." instead of Henry William Menard (Menard, 1986: 64). Clearly Menard valued this recognition, but others often credited the better-known Dietz with this discovery (Menard, 1986: 65). Similarly, the computer generated fit by Bullard, Everett, and Smith (1965) is commonly referred to as the "Bullard fit."

Scientists are aware of this danger and try to avoid it. The previous example of the conflict between Heezen and Opdyke over how to divide up the study of reversals in sea cores caused Opdyke to tell Heezen ". . . that he could not be an author on the first paper, which caused a big blow-up, because I'd been doing paleomagnetism—it's my business—and his very presence on the paper would have been enough to lead people to think he had been the prime mover . . ." (from Glen, 1982: 330).

These examples strongly support Merton's suggestion that scientists desire recognition and that there is a norm to acknowledge contributions from others. However there is less support for the importance of the other norms because there are numerous violations of them without much censoring of the violators. For example, one can find examples of violations of universalism, but without much in the way of repercussions. In the earlier debate on continental drift, Wegener's status as a "meteorologist" was cited frequently as a reason to doubt the competence of his conclusions. Opponents noted that some journal editors refused to even consider their submitted articles because of their (the opponents') opposition to plate tectonics. Bruce Heezen suggested that there used to be a tendency for editors of *Nature* to regard speculation from Cambridge and Oxford as "intelligence, [whereas] speculation from a redbrick university in the United States was bullshit."

Ian McDougall, an Australian, thought that continental drift theory was not accepted earlier because North American geologists tended to distrust, ignore, or reinterpret results published by geologists from the Southern Hemisphere. Some comments by a few geologists at a conference support this conclusion.

> *Menard*: The evidence down there is overwhelming; they believed in the glacial evidence.
>
> *JAS*: But I take it you weren't familiar with the evidence or didn't believe it, as a North American geologists?
>
> *Menard*: You didn't have to pay much attention to it.
>
> *Burke*: You knew in general terms that there were tillites [from glaciers] in the southern continents. . . .
>
> *Menard*: Oh, come on now. Why should I believe there are tillites in the southern continents, when my learned colleagues who looked at these things tell me that it looks like a mudflow? I mean there was just an overwhelming skepticism among geologists.

Many of the proponents and especially the opponents to plate tectonics were bothered by the "bandwagon" effect. As one put it: "You get on the wagon and ride it. There are a lot of people who follow the plate tectonic model and if you asked them for the evidence, they wouldn't be able to give it to you." This violates the norm of individualism (also called "organized skepticism") in that scientists are responsible for critically questioning the previous work used as the basis for their present research. Scientists are also supposed to be emotionally neutral about their own and others' beliefs. As noted earlier, Bullard (1975*a*) suggests that the violation of this norm helps to account for early resistance to Wegener's theory of continental drift. "The whole story of the fits is an illustration of the sloppy way in which new ideas can be treated by very able men when their only object is to refute them" (Bullard, 1975*a*: 6). These selected examples suggest that the norm to acknowledge intellectual debt does exist in the geosciences, but the evidence for the other norms is less persuasive and subject to alternative interpretations. Chapter 1 included references to relevant studies.

Before considering the constructivist perspectives, some characteristics of the functionalist studies merit emphasis because they highlight the differences between these two general approaches. In particular, studies testing the functionalist norms tend to use the quantitative technique of "regression analysis" to examine how some "reward"—such as recognition, rank, or assessed quality of research articles or proposals—can be predicted by various "resources"—such as productivity, previous recognition, quality of graduate education, or gender. Universalism implies that some "resources," such as sex, should not be significant predictors of these rewards. Thus functionalist studies have emphasized *quantitative* types of statements about the structure of rewards in *large aggregates* of individuals (typically, whole disciplines) and *neglect the research process itself*.

The constructivists differ in all of these features. After considering the constructivist perspectives on science, I will suggest how a synthesis can combine some of the strengths of each approach and avoid some of their problems.

CONSTRUCTIVIST PERSPECTIVES

As noted in Chapter 1, there are several different approaches among the constructivists, but they tend to share several common features, including a skepticism about the utility of the functionalist analysis of science. For example, functionalist norms are viewed as too common among academics to characterize science itself and too vague to really constrain behavior. Instead, constructivists suggest that "cognitive norms" about the "proper" methods, assumptions, and theories are more useful for understanding the behavior of scientists, so they study the research process itself to see how these norms or "conventions" become established.

Another common feature is a general acceptance of the "strong programme" in the sociology of scientific knowledge. As mentioned earlier, this programme holds that four principles should be used in the explanation of scientific beliefs: (a) all beliefs have causes, (b) causes should be sought for both "true" and "false" beliefs, (c) the same types of causes explain both types of beliefs, and (d) the same processes identified *in science* apply to the constructivist's beliefs *about science*. Finally, rather than viewing science as a special institution, they emphasize the study of science because the nature of "knowledge construction" is more visible than in other social institutions (Barnes, 1974; Collins, 1982). For example, they would argue that the procedures for labeling a rock as an "ocean-floor basalt" are more explicit, but no different in character, than the social procedures for labeling a person as "deviant."

Clearly philosophers and scientists would regard these as "radical" assumptions that would seem to undermine our "faith" in science. How can they be defended? First, constructivists would note what we have already seen in this and the previous chapter: the "under-determination of theory choice." That is, empirical evidence and deductive logic cannot specify the correct interpretation of empirical evidence or dictate the proper choice of theories or paradigms. A host of additional assumptions and analogies are needed and scientists can still disagree about when there is "acceptable" agreement between predictions and observations. Second, despite these "logical" problems and disagreements, scientists eventually manage to reach a general consensus on certain beliefs, so it is at least plausible that "social" factors may influence this process. Finally, constructivists argue that, unlike the philosophers' desire to *prescribe* the proper scientific method, they only seek to *empirically describe* how science is actually done and determine whether social factors can be identified (Bloor, 1981; Barnes, 1982).

The Relativists: Scientific Activity as a Process of "Negotiation"

The relativists elaborate Kuhn's suggestion that persuasion on the basis of shared values creates the consensus on a new paradigm during "revolutionary science." First, they emphasize the *interaction* between scientists and suggest that "negotiations" about assumptions and beliefs is a better description of the process. Second, the negotiating "resources" are more concrete than Kuhn's shared values and include both "social" resources, such as the scientist's reputation, institutional prestige, style of presentation, and personality, and "scientific" resources, such as finding predicted results or using "standard" instruments or assumptions. All of these resources will help convince others to accept a scientist's ideas. Finally, this same causal process occurs during both "normal" and "revolutionary science," even though the types of the resources may be different (Barnes, 1982).

By studying current controversies in science, relativists find that scientists evaluate others' results by using some of the "social" characteristics mentioned above. In fact, they have to use these because each other's empirical results always rest upon a host of assumptions that can always be challenged in order to undermine the reported results. Eventually, a consensus is reached about the proper assumptions or the existence of an empirical phenomenon and anybody who continues to question them is defined as "irrational" by other scientists. Although relativists recognize that both social and scientific resources are important in this consensus formation process, they suggest that as a *methodological* rule one should try to explain scientific beliefs as much as possible with social factors and allow "Nature" to explain the residual (Collins, 1981).

How well does the relativist approach apply to plate tectonics? The Oxburgh quote in Chapter 1 indicated the basic problem identified by the relativists: the inability of a theory (plate tectonics) to provide definitive implications for some concrete data (regional geology). This has not changed. "It is impossible to write a rational, integrated and ordered account that quantitatively or even qualitatively explains the geological history of the British Isles, or anywhere, in a plate tectonic framework" (Dewey, 1982: 372).

Other comments by the interviewed geoscientists support the conclusion that nonscientific factors influenced the opinions of geoscientists in the earlier debates about continental drift and seafloor spreading. For example, Heezen suggested that Blackett's Nobel Prize may have helped some geoscientists to accept his paleomagnetic evidence for continental drift.

> The big guys . . . don't read the literature because they are directing labs, promoting the money, . . . and working on their own research. . . .
> *JAS*: So how do they change their minds?
> *Heezen*: When one of the prize winners comes up, like Blackett, and says this is right; they don't look at and test all his theories and all his diagrams or go and try to repeat it. It's good enough . . . so his authority comes in. . . . they couldn't go through everything, it's impossible.

Heezen's comment that it is impossible for scientists to "go through everything" before accepting the results represents a basic thesis of Kuhn and the relativists: any research result depends on a host of tacit decisions and assumptions. Some element of trust is required and it is more likely to be extended to esteemed scientists (Hagstrom, 1965).

Trust is also more likely to be given to those who are seen as having little reason to give "biased" results. For example, one interviewed geoscientist mentioned that when he evaluated someone's publication, he also considered "whether the author's marriage or career were going well." Several geologists suggested that Meyerhoff and Beloussov continued to oppose plate tectonics because their "egos" were involved in their opposition. This was given as a reason for doubting their conclusions, which is similar to earlier rejections of Wegener's evidence for drift because of his "bias."

Hess also felt a person's reputation affected the response to innovative ideas. In a personal communication to E. Moores, Hess attributed the lack of interest in one of his early attempts to synthesize to the fact that ". . . at the time of the original article he was a young scientist, as yet without much of a reputation trying to synthesize. He said that he had learned a hard if valuable lesson from the experience, that is, if one wants to say something controversial or to synthesize, then one must first make one's name in an established field" (Moores, 1982: 737). Hess did this before he wrote his "geopoetry" synthesis.

Another example of how previous recognition increases the reception of an idea was mentioned by Peter Molnar to account for the great popularity of mantle plumes in the early seventies. In 1974 he felt that the plume concept was very difficult to test empirically. "I don't think that plumes, as originally suggested . . . actually exist, but I don't think I can disprove it. I don't think that as many people would have accepted the suggestion had it not been Jason Morgan who made it. He's a creditable scientist, whom people took seriously."

Heezen mentioned that using certain methods or instruments might present a more convincing argument by contrasting the continental fits presented by Carey (1958) and Bullard, Everett, and Smith (1965), see Figures 3.7 and 4.14. "Carey's reconstruction of the South Atlantic was just as convincing to me, or more so, than Bullard's, but Bullard said he did it with a computer. That's great—put a little holy water on it, but the fact that Carey took a one meter globe and took the two sides and fit it together was a lot more conclusive to me because you knew nobody made a mistake because you could see it." Menard also emphasized the persuasive importance of the computer technique and Bullard's demands that large scale fold-outs be used to present the results because "a new idea must be advertised and sold" (Menard, 1986: 234). Additional examples illustrate the effects of scientific authority and style of presentation.

Another negotiation resource in getting one's conclusions accepted is the rhetorical style of presentation. Many of those writing on the history of plate tectonics noted that Hess (1962) introduced his paper on seafloor spreading as an essay in "geopoetry." This paper appeared in a volume honoring Arthur Buddington, a Princeton colleague and former teacher of Hess. In 1974 Buddington recalled the effect of this rhetorical device on his first reading of a preprint of Hess's paper.

> One of the first things that he said was that this was going to be an essay in geopoetry. But you know if he had not said that, so help me I would have told him to go

throw the damn thing in the wastebasket. But on account of that, he disarmed me completely and I read the thing with an equitable frame of mind, neither against it or for it, and I was rather intrigued by it as a result. But I am sure that there was enough in there that I would have reacted against if it had not been for that first lubricant that he put in.

Heezen also mentioned that Hess called his essay "geopoetry" and asked, "What was he worried about? He was worried about the same damn thing I was worried about [when he adopted the expanding earth hypothesis]. If we had come out for continental drift, [we] would have been thrown out of the fraternity and wouldn't be able to get funding so easily. So he put a little frosting on it [saying in effect], 'Well, it's not really serious, we're just playing around—let's just try this.' "

The style of oral presentations may also be an important factor, as indicated by Oxburgh's recollections of the differences between Blackett and Runcorn.

Blackett is a speaker in a thousand. I mean, he's a brilliant speaker—he's a careful person. . . . He's a guy whose presentations are so good and so careful that he probably makes more impact than he should. Whereas Runcorn . . . at that time wasn't . . . a very good speaker. He tended to mutter, he tended to look at his notes, and he tended to deliver his lectures with his elbows on the table.

If the relativists were to look at the history of plate tectonics they might suggest that the established reputations of Blackett, Hess, Bullard, and Wilson were important resources in getting other geoscientists to consider the subject of continental drift or seafloor spreading, especially when compared to the lesser resources of Runcorn and Carey. However, negotiating resources can include more "scientific" criteria. Simply having data to present and discuss is important. Several geoscientists emphasized that at least Vine and Matthews presented new data, whereas Morley's rejected paper had no data, even though he presented the same ideas. Quantitative data tend to be more persuasive than qualitative data. An earlier quote by Cox indicated that the quantitative arguments of Vine's papers were more persuasive than the earlier qualitative arguments in Hess's seafloor spreading paper. Such data allow better tests of the accuracy of predictions, even though these can always be "deconstructed" by challenging other aspects of the study, such as the assumed model of the seafloor or the selection of the "best" magnetic profiles.

Once a theory is established and tacit knowledge provides key assumptions, analogies, and shared expectations about results, simply finding results within the range of expectations will be a good negotiating resource for having others accept the results. It is still possible to undermine results by challenging tacit assumptions, as Beloussov (1979) did in his questioning of the assumed seafloor model, but these are not potent arguments any more because the assumed properties are conventional beliefs. To question them may cause others to label one as "irrational." This process can extend to general theories. Jason Morgan commented in 1974 that he thought journal editors and referees demanded more explicit reasoning in articles claiming incompatibility with plate tectonics than in articles making vague claims of compatibility.

Although relativists would accept that these more "scientific" negotiating resources, such as finding expected results or appealing to quantitative values, influence the negotiation process, they would emphasize that this does not free scientists from social influences because social factors helped establish these shared expectations and values in earlier negotiations. Consequently, some relativists argue that the logical gap between theory and data is ultimately bridged only by social factors, which may be directly involved in current debates or indirectly involved as the earlier basis for widely shared assumptions, beliefs, procedures, and tacit knowledge that are now seen as "scientific" criteria for theory choice. Before considering some of the problems with this approach to the study of science, we should examine the complementary "interests" approach among the constructivists because both share similar problems.

Interests, Conflict, and Competition

Other constructivists emphasize the "interests" that scientists have in different theories (Barnes, 1982; Bloor, 1978; Dean, 1979; Pickering, 1980; Shapin, 1982). The conception of "interests" is very general and can range from a preference for quantitative theories to the compatibility of specific theories with political ideologies (MacKenzie, 1978) to the "interests" of the scientific elite in maintaining the theoretical foundations of their previous research.

As noted earlier, Bullard explained the early resistance to drift theory by suggesting that "there is always a strong inclination for a body of professionals to oppose an unorthodox view. Such a group has a considerable investment in orthodoxy" (Bullard, 1975*a*: 5). This clearly expresses an interests viewpoint, but interests proponents would make an important addition. In line with the strong programme's emphasis on symmetry, they would argue that the beliefs currently accepted as true must also be explained by the interests of their proponents. For example, geophysicists might prefer the quantitative nature of the predictions of seafloor spreading to the more qualitative ones provided by geosyncline theory or Beloussov's "oceanization" theory. Furthermore, they will prefer theories that explain geophysical data rather than geological data. Younger scientists will have an interest in new paradigms because, if successful, they will alter the older cognitive and social structures and thereby offer more exciting intellectual problems and faster routes to recognition.

Clearly, the interests approach is simply looking at the other end of the negotiation process in science—the central focus of the relativists. However, the interests approach more explicitly adds an element of conflict in the negotiation process within science. This conflict arises from the different interests that scientists have in different theories, models, and instrumental techniques. Many of the interviews with geoscientists contained illustrations of different interests and conflict.

Allen Cox mentioned how the avoidance of vested interests by a rhetorical device probably caused the evidence for magnetic field reversals to be accepted earlier than it might have been. He commented that by the early 1960s

> . . . the data just spoke for itself. . . . Also it didn't affect anyone else's vested interests because nobody cared that much. I think that's why plate tectonics caused more of a hassle—because it was related to continental drift and the age of the ocean basins and those were old ideas with a great set of vested interests. We decided at the beginning to not try to relate reversals to the age of the Pleistocene, [which] is an ancient, famous fight. We decided [to use] neutral names. We named them [the reversal periods] after famous people instead of type of localities [which is the means used to date the Pleistocene] so we wouldn't get into any hassles. . . . We did this very deliberately. . . . So when the data got good enough, people accepted it.

Constructivists would argue that data can never speak "for itself" unless there are a host of other shared assumptions, but the rest of the quotation illustrates the importance of conflicting interests and how these can be minimized by the style of presentation—a point mentioned by the relativists.

Conflict can arise between those who collect empirical data and those who develop theories about the data, especially when most of the recognition goes to the theorists (Hagstrom, 1965). This problem was strongly stated by Heezen, who had helped collect much of Lamont's vast bank of oceanic data.

> In many ways we considered Hess and Dietz as just sort-of little gadflies that ran around and talked bullshit because they weren't collecting any data; they weren't applying themselves to the problem. We were the ones going to sea; we were the ones taking the instruments and going out and getting data, and trying to solve it, and doing the serious work. They were—Tuzo Wilson and these people—just trying to live off of us. This sort of antagonizes us because they end up in the literature as being the ones who did it, and the people who really did it are hardly mentioned.

A related conflict is the difference in the techniques and explanatory goals of geologists and geophysicists, which was illustrated by comments in Chapter 4 indicating that geologists distrust evidence obtained by remote sensing instead of "direct" examination and understanding of the rocks themselves. Furthermore, Verhoogen suggests that some of the "mistrust" between geologists and geophysicists occurs because the latter aim at producing "general" results based upon simplifying assumptions, whereas geologists are interested in understanding the particular and unique (Verhoogen, 1983: 6–7).

Some geologists suggested that outside social factors helped to develop plate tectonics.

> It's basically after World War II [that the oceans were studied], and I think here again part of it has been activated by defense contracts. . . . So I think there was a need—the drive did not come from the geologists—it was political and military needs that made the exploration of the oceans of vital interest. I think in the exploration of the oceans . . . that a completely new group of workers became involved. And I think most of them barely saw rock before, so you could hardly call them in any

> way geologists. Mostly they were physicists, . . . for many it was just an opportunity because there was so much research money available. . . . I think for them, the earth was like another machine—they were trying to find out about the design. And then comes, of course, what I mentioned at the very beginning, most of what these people do is remote sensing; they're not at all in touch with rock material.

Here we see a variety of interests and conflicts in action: military interests provide funds; physicists study oceans because that's where the money is; geophysicists emphasize abstract or global models using instruments measuring global properties; and geologists dislike the results because they do not pertain directly to their interests in explaining the actual rocks observed in the field.

Wood (1985) develops this general thesis in more detail. He suggests that, in opposition to the old-style, hard rock geology of the past, an alternative "organization" of the "Earth Sciences" had been developing before the arrival of plate tectonics. This alternative emphasized the global views of geophysics and geochemistry, which had always been relegated to the sidelines by geologists, and argued that "equal status" be given to the geophysical and geochemical data on properties of the earth below its surface and the oceans. Several factors increased the "power" of this new group. First, the increased commercial importance of geophysical techniques in oil and mineral exploration caused a rapid growth in the number of geophysicists. Although the American Geophysical Union only began around 1930, by 1950 it had over twice as many members as the American Geological Society (Stewart, 1979: 535). Second, their research was becoming increasingly relevant to geology. In 1949 the *Geophysical Abstracts* incorporated geological subjects in its subject index. Third, the success of the first International Geophysical Year in 1957 increased the visibility of this group. Finally, geophysicists received most of the huge influx of defense department funds related to the exploration of the ocean basins and to monitoring the test ban treaty. Consequently, even before plate tectonics arrived, the balance of power between these groups had shifted: new academic departments of "Earth Sciences" were being merged from older separated departments. "In the late 1950s the concept of Earth Sciences was created but still lacked a model for the behavior of the whole-Earth that would turn an ideal into a coherent structure" (Wood, 1985: 192). Plate tectonics provided this model. "Yet the core of this revolution was neither moving continents, nor a convecting Mantle, but the Idea of the Earth. In Kuhnian terms there had been a paradigm shift from Geology's little earth to the whole Earth of the Earth Sciences. . . . That this parallel structure existed was more important to the success of [the] revolution than the eventual trigger: the announcement of plate tectonics" (Wood, 1985: 193).

A number of studies suggest that "interests" in the broader society may influence scientists' decisions to adopt different theories (e.g., MacKenzie, 1978). Both Wood (1985) and Carozzi (1985) suggest that anti-German feelings may have influenced some of the reactions to Wegener's drift theory in Europe. Later in Germany the Nazi Party explicitly endorsed some geological theories

(Wood, 1985: 100), but in North America the opposition to "German" theories was not overtly expressed as a reason to dismiss Wegener's drift (Menard, 1986: 27). There seems little evidence that such factors played a role in the acceptance of plate tectonics by Western geoscientists. Certainly defense department funding helped collect the data, but this funding was not tied to support for any particular theory. If there was any "outside" influence, it was of a very general nature. For example, academic geologists are the most politically conservative scientific discipline (Mazur, 1987: 271). Perhaps this conservatism extended to the consideration of new scientific theories.

The same cannot be said for the reception of plate tectonics in the Soviet Union. Wood (1985) describes the direct impact of Marxist ideology on geological thought in the Soviet Union and the formation of a strong geological "gerontocracy," which is strongly opposed to the new emphasis on geophysical data. "To replace the Soviet hierarchy of Geology will require gradual demolition and reconstruction over many years; an opposition to plate tectonics will probably be sustained into the 21st century" (Wood, 1985: 222). Meyerhoff, who has traveled extensively in the Soviet Union as a petroleum consultant, noted that even a shift in the official "party line" has not caused a conversion to plate tectonics among the applied researchers.

> In the academic world the official Communist Party line is that drift is a "fact." Therefore, in the academic world, which is the Academies of Science . . . you will find a good solid 60–70 percent who agree with it. The Communist Party line is quite clear on this point; it was only published two years ago [in the mid-1970s], when they finally took a stand. Ironically, that is where you find the fewest members of the Party . . . in the Academies. In the Applied Ministries . . . you have to be a member of the Communist Party [but] you find that practically no one believes in continental drift or is even interested in it. They think it is hogwash.

In the Soviet Union it appears that the "state" has been and still is promoting specific theories, but even it cannot easily change "belief systems" that were "constructed" in previous negotiations involving both social (even political) and scientific resources. Once the belief systems are constructed, they become very difficult to overthrow because they become the basis of the new generation's reputation. The same is likely to be true for plate tectonics.

The acceptance of a paradigm and the start of "normal science" does not mean that interests no longer play a role in scientists' decisions about what beliefs should be accepted. Not only did interests help establish the conventional assumptions behind "normal science," but there is still conflict over how to apply the "paradigm" to specific problems, as noted by Patrick Hurley in a 1974 interview.

> If you mean that everything related to plate tectonics is totally accepted by everybody, it hasn't happened yet. What has happened is that there has been a terrific flowering of thought and explanations that now seem to have a more fundamental basis to them . . . [but] each time you put out another explanation that is based on plate

> tectonics, you have a great deal of argument and debate on whether the explanation is correct. So there is just as much division of thought. . . . The unification comes about only through having people spend their time and attention upon relating things to plate tectonics, but not in the outcome.

As noted several times already, plate tectonics cannot definitely specify the "proper" interpretation of geological data. Undoubtedly, there will be increasing consensus on how to relate the general theory to such phenomena, but if the interests proponents are correct, this will only be possible because of the effects of both social and scientific interests.

Even when there is a consensus on assumptions, analogies, research procedures, and other tacit beliefs, conflict is still present, but in the form of competition for scarce resources and recognition. Both of these were present in Cox and Doell's work on developing the magnetic reversal chronology. Glen (1982) documents the troubles that Cox and Doell had in getting access to the USGS lab for establishing K-Ar dates for their samples, and the competition they had with McDougall in identifying new reversal periods.

The desire for recognition is a major theme among functionalist and a source of conflicting interests or a powerful negotiation resource among constructivists. Other common features exist between these perspectives, as well as unique problems. We can conclude this section with a discussion of these communalities and their unique problems.

FUNCTIONALISTS AND CONSTRUCTIVISTS: SIMILARITIES, DIFFERENCES, AND PROBLEMS

Both functionalists and constructivists use an implicit decision-making approach to explain how scientists reach a consensus on their beliefs. Even some of the decision-making factors are similar. For example, the relativists suggest that a scientist's previous recognition provides a powerful negotiation resource in developing persuasive communications, which is called the "Matthews Effect" by the functionalists. Similarly, the functionalists and interests proponents both emphasize scientists' desire for recognition and the resulting conflict and competition. Finally, the relativist's concern with persuasion and influence is closely related to the functionalist's concern with recognition. Recognition is simply acknowledged influence.

Despite these common features there are major differences in these perspectives. Functionalists seek to study what is unique about the social structure of science and propose it can be found in a special normative structure directing how scientists should make decisions (Gieryn, 1982), especially those decisions providing the social context of research. They tend to neglect the research process itself and the implications of recent philosophy for this process. Consequently, they seek quantitative and global or "macro-sociology" statements about the social structure of groups of scientists, such as entire disciplines. In contrast, the constructivists start from recent philosophical and historical studies of science that justify the possibilities of social influences in research decision-making. Science is seen as similar to everyday knowledge production processes and its study is emphasized by constructivists only because these processes are easier to study in science. They emphasize the "micro-sociological" study of interactions among scientists, dislike quantitative techniques, and tend to focus on the research process itself. Typical applications of these different perspectives illustrate some of their problems.

The functionalist studies tend to ignore "intellectual" elements even though this approach specifies the key importance of these elements in the allocation of recognition. For example, "universalism" implies that only the intellectual elements in a scientist's contributions should predict the recognition received by a scientist, while other characteristics, such as gender, religion, the prestige of current affiliation, and previous recognition, should be irrelevant. Yet typical tests of universalism fail to measure adequately these intellectual elements. A typical study would sample *individual scientists* and quantitatively predict the total number of citations received by the scientist—a measure of recognition (or influence)—using several characteristics of the scientists writing the cited papers. The measure of the scientist's "intellectual" contributions is simply the number of articles written by the scientist. There are a number of problems with this simple approach (Stewart, 1983), but two are particularly important. First, publications, even those by the same scientists, will vary in their content and quality (their intellectual elements), so a simple count of papers is a poor measure of these elements. Second, typical studies find that even after controlling for the number of papers, other characteristics of the scientists, such as the prestige of their department, remain significant predictors of recognition. This empirical result has two possible interpretations with respect to the universalism norm. Either scientists in more prestigious departments write better papers, which are cited more by other scientists, *or* other scientists are more likely to cite papers by those in prestigious departments, while ignoring equally good papers by scientists in other departments. Both of these processes would produce a positive correlation between citations and departmental prestige *when the unit of analysis is the individual scientist*, but only the former explanation is consistent with universalism, whereas the latter explanation is an example of the Matthews Effect. Consequently, the studies of the recognition given individual scientists have not provided adequate tests of the universalism norm.

In addition, functionalists have not pursued their goal of describing the *unique* aspects of the social structure of science. Most of their studies focus only on science and fail to make comparative studies to other social organizations. The most comparative studies have been restricted to different disciplines and countries within science itself (Gaston, 1978; S. Cole, 1978). If the unique aspects of science are to be found, then we must develop a general procedure for the analysis of *all social organizations*.

"Anthropological" studies of laboratory life have produced the most persuasive evidence for the constructivists' approach (Latour and Woolgar, 1978; Knorr-Cetina, 1981; Latour, 1987). When sociologists literally live in the lab with the scientists and observe scientific behavior, they find that scientists use reasoning by analogy more than formal logic, compete for organizational resources, cannot relate empirical results to theories without additional assumptions, negotiate the meaning of experimental results, and compete for recognition or influence. As Knorr-Cetina (1981: 5) put it, empirical results are created through "chains of decisions and negotiations." However, these studies and constructivist studies in general tend to focus on detailed studies of the interactions within very specific groups or settings without much effort to generalize to a wider population of interactions. Like the functionalists, the constructivists have not made comparative studies in other social organizations to demonstrate the use of similar knowledge production processes. The closest they have come is the study of parapsychology (Collins, 1985).

More serious is their failure to specify how they determine the relative importance of "social" and "scientific" interests or negotiation resources. For example, geoscientists mentioned both Blackett's paleomagnetics data and his Nobel Prize as some of the reasons they accepted continental drift. A relativist who follows Collins's methodological rule to explain as much as possible by social factors and assume the rest is due to "Nature" should attribute most of Blackett's influence to his Nobel Prize. A better procedure should attempt to assess the *relative* importance of these two types of negotiating resources.

A similar problem occurs with the interests approach. For example, MacKenzie and Barnes (1979) argue that the nineteenth century controversy between Biometricians and Mendelists over statistical theory can be related to their different political interests. The Biometricians tended to be members of a rising professional class favoring eugenic interventions for social change, whereas the Mendelists had ties to more conservative political interests. The problem is that each of these groups undoubtedly had other characteristics that might have influenced its choice of statistical theories. For example, the Biometricians might have been recruited from more quantitative disciplines, as well as being members of the rising professional class. MacKenzie and Barnes subjectively assess one characteristic as more important than another, but a more "objective" procedure might be preferable.

These critiques of the functionalist and constructivist approaches have focused on their methodological problems. The next section proposes a quantitative procedure loosely based upon a constructivist approach, but offering solutions to some of the methodological problems of both sociological approaches. More importantly, the proposed measuring "instrument" can be related to previous studies of social organizations in general, which should aid the comparison of science with other social organizations.

Measuring the "Global" Properties of Decision-Making in a Scientific Discipline

Before developing the proposed measurement procedure, an analogy to the geosciences will clarify what is being attempted and the probable reactions to the suggested procedure. We have seen that plate tectonics theory was based largely on geophysical data providing "global," quantitative characteristics of the earth, even though these results could not be related easily to the concrete data studied by geologists. This provided some of the reasons that geologists resisted the new theory. In a similar fashion, the proposed measurement procedure provides "global," quantitative properties of the decision-making process in the social organization of science. Hopefully, it too will help in the development of a broader theory of how knowledge is produced in social organizations, but it is certain to be seen as too "crude" by those, such as philosophers and historians, concerned with the details of decision-making in science. Furthermore, the results of the procedure can be interpreted from a number of different perspectives on science, just as geophysical data do not constrain geological models without use of other assumptions. Thus the ultimate success of the procedure will depend on whether I can present it in a persuasive manner and whether it can produce results that appeal to the "interests" of others. I will attempt to accomplish this difficult task in the remainder of this book.

Our major concern is the description of how a consensus forms in a scientific community. The relativists suggest that this occurs through a process of interactions and communications among scientists as they negotiate what results, interpretations, and beliefs will be accepted without extensive challenge. Although it is always possible to challenge these, some become accepted by other scientists because their proponents have more negotiation resources and are therefore able to present more persuasive or influential communications. These resources include both social and cognitive elements, such as established scientific eminence or finding previously predicted results.

This constructivist approach is taken as a description of the general decision-making process, but to obtain "global" characteristics of this process we must solve some methodological problems of constructivist studies. Some of these solutions incorporate techniques used in functionalist studies. For example, a key issue is the relative importance of the social and cognitive resources in these conversational negotiations. Rather than emphasizing the social resources, it would be better to quantitatively determine the relative importance of these different resources. Functionalists use regression analysis to determine quantitatively the relative importance of different variables in a population of individuals. Furthermore, their use of citations as a measure of recognition or influence provides a suitable dependent variable to be used in the regression analysis. However, typical functionalist studies use the individual scientist as the unit of analysis and fail to measure adequately the intellectual elements in their interactions with other scientists. A more "micro" unit of analysis is needed that is more consistent with the constructivist's emphasis on the conversational interactions among scientists.

Randall Collins (1975: Ch. 9) has suggested that studies of science might adopt a crude analogy: *take scientific articles as analogues for scientific conversations.* This analogy has a number of advantages. First, articles contain the intellectual elements in science and they are publicly available for inspection and measurement of their content. Not only can we measure important aspects of their contents, but the references in an article provide a means of relating the article to previous research results, which contain the "exemplars"—and hence the cognitive structures—used as the basis for the reported results (DeMey, 1982). Second, citations in science generally refer to specific articles, and less so to their authors. Thus citation counts to articles provide a convenient measure of the influence of these "conversations" on later researchers. Third, the decision as to whose research will provide the foundation of one's own research is a key factor in the formation of a consensus in science. If this research is actually cited, then citations to an article provide a measure of an article's influence on new knowledge.[4] Fourth, the study of articles is much closer to the constructivist's emphasis on conversational interactions among scientists than is the study of individual scientists. Finally, the study of articles allows a convenient separation of the "social" and "cognitive" resources of these "conversations." Cognitive resources would be given by measures of the article contents, whereas characteristics of the author(s) provide a measure of the social resources of the "speakers" in these "conversations." This separation is not complete because "style" is a property of the article, but might be labeled a "social" resource. This is a minor problem compared to those mentioned in the next section, which justifies the measurement procedure using the implications of a simple iconic model of the decisions that generate a citation to a scientific article.

AN ICONIC MODEL OF THE DECISIONS YIELDING A CITATION TO AN ARTICLE

In a "rhetorical" sense, my best strategy would be to engage in what geologists call "arm-waving" and skip directly to the next section and discuss what we might learn from a study predicting citations to articles, leaving implicit the underlying iconic model. Instead, I will propose a simple iconic model of the decisions needed to generate a citation to an article, then use this model to develop the quantitative procedures for studying the "global" aspects of this decision process. This iconic model approach highlights the need for many dubious, *but commonly made*, assumptions. By making them explicit, I will arm the skeptic with many potent criticisms, but perhaps give sympathizers interesting "problems" to work on.

Figure 6.6 is a very simplified representation of some of the decisions that must be made before the scientific literature contains a citation to an article, say article "*j*." First, a scientist must decide if article *j* is worth reading. The probability that the scientist will read the article, $P(R_j)$, is assumed to be a function of the characteristics of article *j* and its author(s).[5] Before this scientist can generate a citation to article *j*, he must be writing or planning to write an article. The probability of writing an article, $P(W)$, is a function of numerous other previous decisions by the scientist and by others who control access to needed resources. Given that the scientist has read article *j* and is also writing an article, then the scientist must decide whether to cite article *j*. The probability of citing article *j*, $P(C_j)$, is assumed to be a function of the characteristics of the article and its authors. Finally, after the scientist submits the article for publication, there is a crucial decision by the journal editors about whether to accept the article for publication. The probability of acceptance, $P(A)$, is based upon the characteristics of the submitted article and its author(s). Only when all of these decisions are made in an affirmative sense do we find a citation to article j in the literature.

Although this is a highly simplified scheme, it does illustrate several points. We can see why constructivists might doubt the utility of citation analysis. There are obviously many important decisions involved in producing citations to articles, and constructivists consider it more productive to examine each of these decisions in detail. For example, Williams and Law (1982) provide a detailed analysis of the actual conversations among a group of coauthors about which articles should be cited in their paper. Their findings support the viewpoint used here: articles do function in the same manner as conversations. Williams and Law show how the coauthors tried to construct an array of previous findings *and* researchers so that other researchers will be persuaded to combine this array with their own cognitive structures and cite the article in their own papers. Many functionalist studies also tend to focus on specific decisions in the citation process. For example, Zuckerman and Merton (1971) examine the journal review process to see if universalism is violated and Garvey, Lin, and Nelson (1970) examine many of the steps involved in the production of articles.

Additional problems are apparent when we use this simple iconic model to

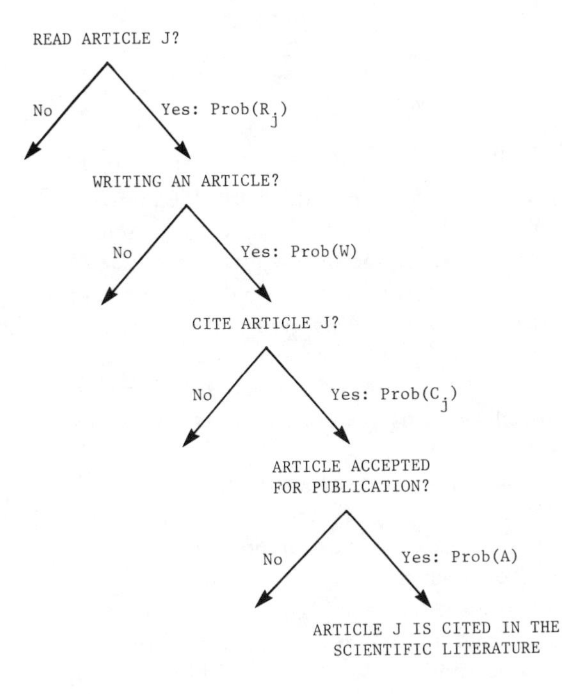

Figure 6.6 A highly simplified model of a scientist's major decision points in the process that produces a citation to article j in the scientific literature. The probabilities with a j subscript are those that may depend on the characteristics of article j or its author(s).

generate a quantitative procedure for studying the number of citations to a set of articles. The iconic model in Figure 6.6 implies that the expected number of citations to article j, $E(C_j)$, is given by the following equation:

$$E(C_j) = N \cdot P(R_j) \cdot P(W) \cdot P(C_j) \cdot P(A), \qquad [1]$$

where N is the number of scientists exposed to article j and the $P(-)$'s are the previously defined probabilities, which are assumed to be the same for all scientists; see note 5. Taking the logarithm of this equation provides an equation more susceptible to traditional regression analysis.

$$\log [E(C_j)] = \log N + \log [P(R_j)] + \log [P(W)] + \log [P(C_j)] + \log [P(A)] \qquad [2]$$

To relate citations to article j's intellectual and social resources, we must specify how $P(R_j)$ and $P(C_j)$ are related to these resources. A multiplicative function provides a suitable specification.[6]

$$P(R_j) = B_0\, I_j^{B_1} \cdot S_j^{B_2} \cdot u_j \qquad [3]$$

$$P(C_j) = D_0\, I_j^{D_1} \cdot S_j^{D_2} \cdot v_j \qquad [4]$$

where I_j and S_j represent, respectively, the general intellectual resources (article characteristics) and social resources (author characteristics) of article j and u_j and v_j represent random error terms. Taking the logarithms of Equations [3] and [4] and inserting them into Equation [2], followed by collecting common terms, gives an equation relating citations to article j to its resources.

$$\log [E(C_j)] = (B_1 + D_1) \cdot \log I_j + (B_2 + D_2) \cdot \log S_j + \log N + \log [P(W)] + \log [P(A)] + (\log B_0 + \log D_0) + (\log u_j + \log v_j) \qquad [5]$$

Before discussing some of the problems and assumptions with this approach, I will first describe what we might learn about the citation process in science when the citations to an article are predicted by variables measuring article (I_j) and author (S_j) characteristics. The estimated (sum of the B and D) coefficients for any I_j or S_j variable in the first two terms of Equation [5] provide information about the importance and relative importance of intellectual and social resources in causing article j to be cited. That is, they tell us something about the decision-making rules and procedures followed by many scientists as they use the decision-tree represented in Figure 6.6.

There are a number of questionable assumptions made in such an analysis. First, Equation [5] indicates that citations to an article are influenced by other factors that would not be included in the regression equation predicting citations to articles. In particular, the number of scientists aware of the article, N, and the probabilities of the scientists writing an article, $P(W)$, and having it accepted, $P(A)$, are not included in the actual estimation process and probably vary across articles and scientists. In examining citations to articles, we would have to assume that these factors were uncorrelated with the characteristics of the articles we were studying. This assumption is often made (implicitly) in most regression analyses. Second, we are assuming that all scientists use the same *B* and *D* coefficients in their decisions to read and cite an article. Although this is problematic, it again is a common assumption in regression analyses. Finally, the choices are dichotomized into affirmative or negative decisions and the decision-tree implies there is only one possible structure of decisions yielding a citation to article *j*. It is likely that scientists may cite articles that they have never read, but this possibility is not included here.

To return to the geoscience analogy used to introduce this measurement approach, it should be apparent that the above quantitative procedure has many analogies with Pearce and Cann's "instrument" for classifying igneous rocks discussed in the first section of this chapter or the empirical data produced by a magnetometer. What we are lacking, of course, is a "theory" that relates the "output" of our "global" measurement procedure to other phenomena. Vine and Matthews had a seafloor spreading theory that implied a connection between the magnetometer output *and* the distance from the mid-ocean ridge. Thus the next subsection discusses some of the possible applications and interpretations of this methodology and the last subsection relates it to studies of "organizational structures," where there are more developed theories. If there are plausible interpretations of possible results and actual applications produce results satisfying to our various interests in understanding decision-making processes in science, then this procedure may find some general appeal and the "dubious" assumptions will be forgotten as they become part of our "tacit" knowledge of conventional procedures.

PREDICTING CITATIONS TO ARTICLES: POSSIBLE RESULTS AND INTERPRETATIONS

Figure 6.7 presents a simplified "path diagram" that would summarize the results obtained when regression analysis is used on a sample of articles to predict how many times each was cited, which is assumed to measure the influence of the article or "conversation" on later researchers. The predictor variables have been grouped into two general categories without listing specific variables. Variables based on the contents of the articles are assumed to measure the intellectual resources of the article, whereas variables based upon the characteristics of the author(s) measure the social resources of the "speakers." The regression of citations on the specific variables in these two groups yields coefficients giving the effects of each variable. The "path coefficients," indicated by the *P*'s in Figure 6.7, measure the *relative* effects of the different predictor variables. (They do not indicate probabilities.) Since each variable in the two groups of predictor variables would have a specific path coefficient, the *P*'s in the figure represent numerous specific coefficients, but these are not shown here.

First, we can consider the functionalist's concern with the norm of universalism. In the case of citations to articles, this norm specifies that scientists should be influenced only by the intellectual elements in the articles and not by any of the characteristics of the authors. In other words, if scientists are

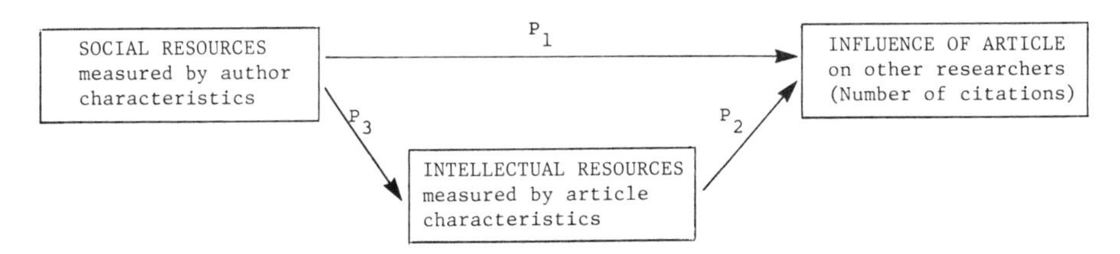

Figure 6.7 A simplified path model showing how an article's influence can be predicted by two general categories of variables: those based on the article's characteristics and those based upon the characteristics of the author(s). The *P*'s in the figure represent standardized regression coefficients, not probabilities.

following the norm of universalism, we should find that only article characteristics should be significant predictors of citations. That is, only the P_3 coefficients should be significant without any significant P_1 coefficients. This test of universalism in the recognition given articles allows better control of intellectual factors than previous studies, which have used individual scientists as the unit of analysis and only used number of articles as a measure of intellectual contributions. If P_1 paths are found significant, then one has evidence for the Matthews Effect (Stewart, 1983).

Second, the comparison of the P_1 and P_3 paths provides a convenient means of assessing the *relative* importance of social and intellectual resources in producing influential conversations in science. We do not have to follow the relativist's questionable procedure of assigning priority to the social factors. Instead, we can quantitatively assess the relative importance of social and intellectual factors in the results from one regression analysis. If communications with a high influence (citations) represent those used by scientists to develop new beliefs, then this causal model shows the relative importance of social and cognitive resources in the construction of *new* knowledge.

Even those constructivists adopting an "interests" approach to the choice of scientific theories may find pertinent results in the regressions represented by the P_2 coefficients. These represent the results obtained when article characteristics are regressed on author characteristics. If one of the measured article characteristics is the degree of support expressed by the article for a specific theory, then the P_2 path coefficients provide quantitative measures of the author characteristics associated with support for the specific theory. Both of these types of analyses will be illustrated in the next chapter.

Finally, it is useful to speculate on the results of comparative applications of this regression procedure. For example, suppose this procedure was applied to the geosciences in the 1920s or 1930s and then again after plate tectonics was accepted. Before the acceptance of plate tectonics it seems likely that geoscientists had fewer shared cognitive resources available for developing persuasive communications and that social resources played a larger role than after the acceptance of plate tectonics. This would be reflected in a *relative* increase in the importance of the P_3 paths compared to the P_1 paths. A comparison of the scientific communities in different countries is also possible. For example, one could use this procedure to assess the "entrenched bureaucracy" (Kerr, 1978*b*) alleged to exist in the Soviet Union. In this case, a comparison of the regression results for the Soviet Union and the United States might show that author characteristics and affiliations were more important predictors of citations in the Soviet Union than in the United States. The same reasoning suggests that it may be profitable to compare disciplines that vary in their level of intellectual development. For example, S. Cole (1978) tested the degree that various social and physical science disciplines follow the norm of universalism and found no difference, but he used the *individual* scientist as the unit of analysis. Quite different results might be found if he had used scientific *articles*.

Some of the most important comparative studies would be between science and other social organizations. These studies could *empirically* test whether the scientific community is unique among social organizations. For example, we might find that in other social organizations the influence of a person or conversation is determined much more by "who one is" than by "what is said." Although this seems plausible, the more important reason to consider studies in other social organizations is that useful analogies for the study of science might be found in theories about organizational properties *and* vice versa. In fact, R. Collins (1975) argues that science studies provide one of the best situations to test and develop theories about social organizations and has proposed a number of propositions about science based on organizational theory. To facilitate this exchange it is helpful to relate the above measurement procedure to procedures used in studies of organizational structures.

THE STRUCTURAL DIMENSIONS OF ORGANIZATIONS

The structural dimensions approach to organizations grew out of attempts to compare the Weberian "ideal type" of bureaucracy to real organizations. Weber (1947) suggested that, among other characteristics, bureaucracies tend to have more formalized relationships, a complex specialization of tasks, and centralized authority systems. Hall (1963) attempted to measure these separate dimensions in a sample of organizations and found that they were not highly correlated with each other. Further research has developed different procedures for measuring such dimensions, examined the relationships among them, and examined their relationships to other aspects of organizations, such as technology, size, and environmental factors (e.g., Hage and Aiken, 1967, 1969; Hickson et al., 1969; Child, 1973; Lawrence and Lorsch, 1967).

These studies generally adopt a macro-sociological approach and treat these structural dimensions as group properties that are causally independent of the actions of individuals and as having their own course of development and interrelationships. However, a more micro-sociological approach will be developed here, in which the structural dimensions of an organization are summary statements about the relative importance of different resources in the numerous negotiations controlling the distribution of influence or power in the organization. A discussion of the structural dimension of "centralization" will illustrate this approach.

A typical definition of centralization is the following: "A highly centralized organization is one in which control is quickly lost as one moves away from the chief executive" (Pugh et al., 1963: 304). This definition emphasizes the *relationship* between two variables: control (or influence) and hierarchical position. It suggests that in an organization where hierarchical position is a powerful resource for achieving the desired goal of influence, then the organization is highly centralized. Among the many procedures for measuring centralization, Tannenbaum's (1968) "control graph" procedure provides a basis that can include other structural dimensions within a general framework that is compatible with the citation model for science.

A "control graph" is constructed by asking organizational members to indicate how much "say or influence" is present at their level in the organization, and the average influence for each level is graphed against the level's position in the organizational hierarchy. More centralized organizations have steeper control graphs. For example, Figure 6.8 provides the control graphs for two hypothetical organizations, where organization "A" is more centralized than organization "B." Clearly Tannenbaum's procedure is equivalent to the regression coefficient obtained when a person's level in the hierarchy is used to predict the person's influence in the organization.

In other words, Tannenbaum has used regression analysis to measure the importance of *one* social resource (hierarchy level) for attaining influence and

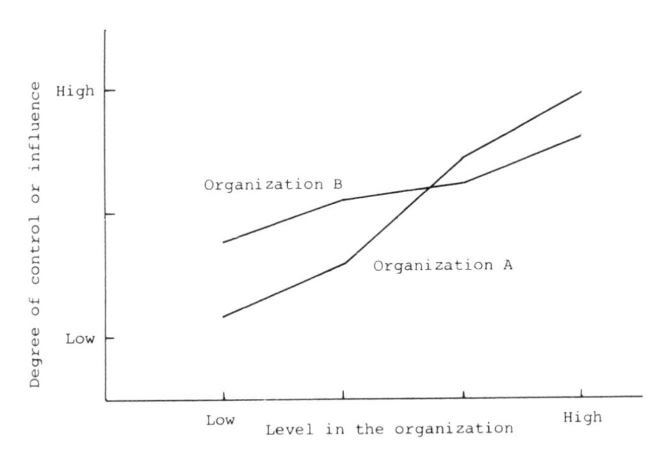

Figure 6.8 Simple "control graphs" for two organizations. Control graphs are constructed by plotting the average degree of reported influence at each level against the level in the organization. Organization A, which has a more rapid loss of control as one moves away from the higher levels in the organization, is more centralized than organization B.

the resulting regression coefficient represents one structural dimension (centralization) of the organization. This approach needs to be generalized to include other resources that help individuals in organizations attain influence, where the *relative effects* of *several* resources are determined by *multiple* regression analysis.

To illustrate this basic approach, we will assume that only three resources determine the influence of an organizational member: hierarchy level, amount of educational training, and number of written rules constraining the person's behavior. Figure 6.9 represents the regression of influence on these three resources. The regression coefficient for the effect of hierarchy level on influence is still taken as a measure of centralization, *but* now the effects of the other resources are controlled. If the regression coefficient for the effect of hierarchy level represents the structural dimension of centralization, do the other regression coefficients correspond to structural dimensions studied by organizational researchers? A tentative correspondence is proposed in Figure 6.9. The structural dimension of "professionalization" is measured by the size of the regression coefficient for the effect of education on influence in the organization. That is, an organization is highly professionalized if the amount of education is a strong predictor of a person's influence in the organization. Similarly, if the number of rules pertaining to a person's status in the organization is a strong (negative) predictor of influence, then the organization has a high score on the structural dimension of "formalization."

This general approach to the measurement of some of the structural dimensions of organizations has several plausible characteristics. First, it is a logical extension of Tannenbaum's control graph that is consistent with organizational theory. For example, Hall (1972: 190) suggests that "formalization and profes-

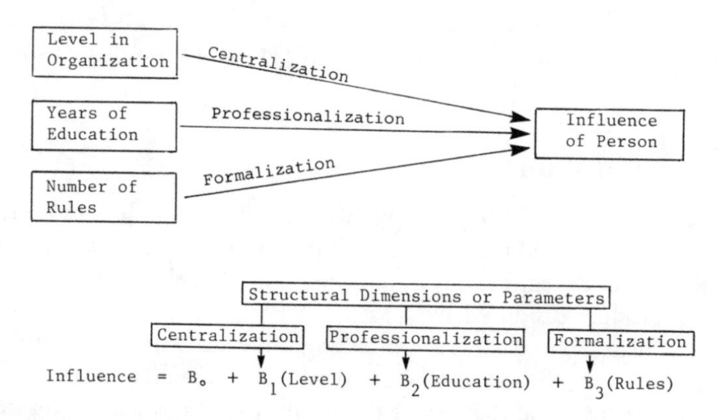

Figure 6.9 A path model and the associated regression equation for measuring the traditional structural dimensions of "centralization," "professionalization," and "formalization" as regression coefficients or structural parameters. This equation would be applied to each individual organization to calculate its structural dimensions.

sionalization are actually designed to do the same thing—organize and regularize the behavior of the members of the organization." Perrow (1972: 27) argues in a similar fashion that professionals are people with rules built in. There is a similar trade-off between emphasizing formalization or centralization (Hall et al., 1967). Each of these arguments has the same basic premise: rules, education, and level are basic resources that affect the distribution of autonomy or influence, and organizations can vary in their emphasis on these resources. Thus the multiple regression equation in Figure 6.9 provides a means of measuring the relative importance of these resources in an organization.[7]

A second advantage of this approach is its treatment of the organization as more than the "sum of its parts." Instead of measuring the degree an organization is formalized by counting the total number of rules or measuring professionalization by the average amount of education, the regression approach extracts information from the *multivariate* relationships among the variables instead of using measures based upon simple sums or averages. A third advantage is that the regression coefficients, which are often called "structural parameters," provide the best summary measure for the comparison of different groups or organizations (Duncan, 1975).[8]

Finally, this approach is similar to the regression model for examining the important resources determining the influence of scientific articles. In other words, the regression procedure outlined in the previous section can be seen as studying the structural dimensions of the social organization of science. There is one important difference between these two approaches. The study in science uses articles or *"conversations"* as the unit of analysis, whereas the regression procedure for organizations uses *individuals* as the unit of analysis. One could study conversations in organizations and this would allow better measurement of the cognitive resources in organizational negotiations for influence, but this would be much more difficult than the study of "conversations" in science. Despite this difference, it is clear that the proposed article analysis is a specific application of a procedure that can be used for studying social organizations in general.

Summary and Conclusions

The major emphasis of this chapter is that scientific research is more than the product of well-socialized individuals testing theories about natural phenomena through the means of empirical research and logic. Rather, scientific knowledge is a cultural product produced by a particular type of social organization. Intellectually, "plate tectonics" is a loose "cognitive structure" consisting of interrelated models, preferred analogies, symbolic generalizations, selected instrumental procedures and their empirical results, conventional assumptions and interpretative procedures, values, and exemplars illustrating how these elements can be related to each other. Logic or other "rules" of inference only tie together a very small fraction of the elements in this network. Most are tied together by "social conventions" about what seems "reasonable" or "rational" to the community of researchers as they try to understand their research results. Thus plate tectonics research consists of producing an "adequate explanation" for some empirical phenomena by the creative rearrangement of these elements within the constraints imposed by conventions, rather than logic. Hence scientific explanations ultimately are "social constructions" and can always be "deconstructed" by opponents willing to contest the accepted conventions, but doing so risks being labeled "irrational." These widely accepted conventions become "scientific" resources in future negotiations.

To understand which elements a scientist is likely to use in constructing her explanation, we must know her "interests" in the various elements. These interests result from her career trajectory through numerous groups which impart special research skills and styles and the other elements in a "paradigm." However these interests are complemented by an interest in recognition from her scientific peers and this must be gained by producing persuasive communications that appeal to the interests of others in the community. These provide "scientific" resources in persuading others, but she also may have "social" resources, such as an established reputation or an appointment in a prestigious department. The combination of these resources determine the degree her communication will influence the construction of future knowledge.

This orientation toward recognition from one's scientific colleagues and the allocation of other social rewards, such as income and job security, on the basis of this recognition are distinctive features of the scientific community. This is a "sociological" reason for special trust in the knowledge produced by the community because it deals with the reward structure and autonomy of the scientific community. Another possible reason is closer to philosophers' concern with the "method" of reaching a consensus within the community. Scientific knowledge is supposed to be "objective" and "rational," but when sociologists examine closely how scientists actually use these concepts, both of these concepts are themselves largely "social constructions" (Mulkay and Gilbert, 1981).

In this chapter, this concern with "method" was given a sociological translation into the relative importance of "social" and "scientific" resources in producing persuasive communications in science. Furthermore, it was considered an *empirical* issue, rather than a normative *prescription* as to which should be more important. Thus the final sections of this chapter proposed a procedure for measuring the relative importance of social and scientific resources and noted how it was appropriate for the study of social organizations in general. Perhaps future empirical study will establish that the "method" of reasoning in science is indeed "less biased" by social factors than the reasoning in other social organizations. The analyses in the next chapter illustrate how these future studies might proceed.

CHAPTER SEVEN

Quantitative Studies

The previous chapter suggested that science is just one of many types of social organizations, all of which have belief systems about the nature of their external environments. These belief systems result from the interactions among the organization's members as they study their external environment and discuss how to interpret the results of their efforts. Some constructivists—the relativists—argue that the consensus in belief systems results from conversational negotiations among members who have varying resources for presenting persuasive communications, but in science the key issue is the *relative* importance of "social" and "scientific" resources. To measure this aspect, we used scientific articles as analogues for "conversations," proposed a simple "iconic" model of the decisions producing citations to an article, and "derived" a statistical procedure or "instrument" measuring the importance of social and scientific resources. This required acceptance of many questionable assumptions, but the ultimate basis for evaluating the instrument is the same as in science in general: Does it produce "useful" results? That is, does it appeal to the "interests" of other scientists by using their special skills, producing results fitting their favored theories, creating interesting "puzzles," or satisfying their values?

Consequently, we cannot determine the "value" of this instrument solely on the basis of the "rationality" of its foundations. It must be applied and the results interpreted. The first section of this chapter uses the statistical model to examine what caused some geoscience articles to be cited more than others. It also considers and proposes solutions to some of the methodological problems that arise in such a study.

The second and third sections examine a related issue that is ignored in the first section: Conversations involve "speakers" *and* "listeners." The first section's analysis ignores the characteristics of the people who actually cite the article, the "listeners," but proponents of the "interests" perspective would suggest that both speakers and listeners have interests that determine their theory choice. The second section approaches this issue by using the characteristics of an article's authors to predict the degree an article expresses support for seafloor spreading. The third section contains a similar analysis, but the articles studied were all published before 1950 and dealt with the subject of continental drift. Both of these studies find that some author characteristics predict opinions on drift related theories, but classifying the predictors as "scientific" or "social" requires additional assumptions. The concluding section suggests what we might learn from additional research using the procedures in this chapter to study other scientific and non-scientific organizations.

Predicting the Influence of Scientific "Conversations"

This section contains two different analyses. The first is a "cocitation analysis" that identifies the "exemplars" for two major research areas in the 1970 geoscience literature, one of which was plate tectonics. The second analysis uses this information to help predict the number of citations to a sample of articles published in 1968.

IDENTIFYING "EXEMPLARS": A COCITATION ANALYSIS

The basic assumption of cocitation analysis is that when two references are frequently cited together ("cocited") in the same articles, they are likely to be on similar topics (Small, 1973). Since this procedure restricts analysis to only highly cited references, the results may identify the major "exemplars" in different research areas. Various studies support this interpretation. Cocitation analysis has been used to map the specialty structure of science (Small and Griffith, 1974; Griffith and Small, 1974) and to study national differences in research emphases (Sullivan, Barboni, and White, 1981) and theory development in particle physics (Koester, Sullivan, and White, 1982). Small (1977) has surveyed scientists to validate the ability of cocitation analysis to identify major intellectual developments in the area of collagen research.

Methodological Procedures

Henry Small of the Institute for Scientific Information (ISI) generously provided the cocitation data analyzed in this chapter. He produced these data by the following steps.[1] First, all the publications (items) cited fifteen or more times in the 1970 *Science Citation Index* were separated from the other publications. Second, all possible pairs of these highly cited items were formed and the number of cocitations to each pair were counted. Third, the cocitation counts were "normalized" to correct for different citation rates. If C_A and C_B represent the number of citations to articles A and B, respectively, and C_{AB} is their cocitation count, then the normalized score is given by $C_{AB}/(C_A+C_B\text{-}C_{AB})$. Finally, the matrix of normalized scores among the pairs of articles was analyzed by a single-link clustering program that formed clusters of articles directly or indirectly linked together with normalized scores of 0.16 or above.

These steps produce two types of lists: a list of the major clusters and lists of the items in each cluster with their pair-wise normalized scores. The information contained in the list of items in a cluster can be represented visually by using smallest space analysis (SSA) to "map" each item into a two-dimensional space. SSA uses the normalized score as a measure of the "strength" between any pair of items and plots each item in a two-dimensional space so as to minimize the total "stress" created by forcing some items closer or further apart than indicated by their original *pair-wise* strengths. In such an analysis each item would be represented by a "point" in the two-dimensional space with shorter "distances" between two items indicating they were on more similar topics.

However, Small's (1974) "hill" model provides a more aesthetic means of portraying the relationships between the items within a cluster. He suggests that each item be represented by a hill equal to a bivariate normal density curve with a total volume equal to the number of citations to the item. The "distance" between two articles is found by overlapping their respective hills until the volume of overlap is equal to the number of cocitations between the two items; the distance measure is given by the distance between the peaks of the two hills. This procedure was modified by allowing the bivariate "hills" to have different standard deviations, so that "the hill representation of a cluster of documents would then resemble a landscape of the most varied sort, having slender peaks and rolling hills. If successive annual cumulations of citation data were used to study the cluster, a shifting landscape would be observed, with the hills and peaks changing in size and shape and shifting relative to one another" (Small, 1974: 400). Obviously, this technique provides a pleasing way to picture the relationships among a set of geology articles related to continental drift. Each cluster of items would represent a "conceptual continent" and the items in the cluster are "hills" that may "drift" away from the continent and over time join other hills to create new continents.

This procedure was used to determine the "distances" between the pairs of items in a cluster, then smallest space analysis (Guttman, 1968: Lingoes, 1965) determined their relative locations in a two-dimensional space. Finally, a computer graphics program plotted the upper surfaces of the hills for the items within a cluster, but to make the hills more distinct, the standard deviations for all the hills in any figure were reduced proportionally. Thus within a pictured cluster the volume of a hill reflects the total number of citations to the item and the distance between the hills reflects the two-dimensional distances given by the smallest space analysis, but the apparent overlap between hills does not reflect the true extent of cocitation between the pairs of items. Finally, the portrayed hills do not include all of the items in the original clusters because the ISI data did not list a normalized score between a pair of items if it was below 0.10. If an item was missing too many of these scores with other items in a cluster, it was not included in the hill figure.

Results

One of the clusters that appeared in 1970 and later years included the major plate tectonics articles. Figure 7.1 is the hill model for this cluster of articles in 1970. It shows the progressive development of a seafloor spreading "ridge" that starts with Hess's (1962) and Dietz's (1961) initial proposal of the seafloor spreading hypothesis, followed by Vine and Matthews's (1963) hypothesis connecting the linear magnetic anomalies to the seafloor spreading model, Vine's (1966) verification of this connection, and culminating in the Heirtzler et al. (1968) application of the model to many of the major oceans of the world. The central and major peaks are three articles developing and applying plate tectonics. These are the LePichon (1968), Morgan (1968), and Isacks, Oliver, and Sykes (1968) articles. The hill for the Morgan (1968) article is partially covered by the

LePichon hill. McKenzie and Parker (1967) is also associated with this central group.

The Wilson (1965*a*) article proposed the plate concept as an implication of his conception of transform faults, which was supported by the Sykes (1967) article. These two articles form another ridge connecting seafloor spreading to the central plate tectonics peaks. Among the remaining articles in this cluster, only the Sykes (1966) article on Benioff zones had a sufficient number of co-citations to these major articles to allow inclusion in this picture of the "conceptual continent" of plate tectonics. The other articles include the Pitman and Heirtzler's (1966) analysis indicating the acceptance of seafloor spreading at Lamont, Oliver and Isacks (1967) study of the Tonga Trench, Barazangi and Dorman's (1969) analysis of earthquake epicenters, and several other articles directly related to plate tectonics (McKenzie and Morgan, 1969; McKenzie, 1969; Heirtzler, 1968; Isacks and Molnar, 1969).

The cocitation analysis for 1970 indicated only one other major cluster of geoscience articles: an experimental petrology/geochemistry group of articles in Figure 7.2. These articles emphasize experimental studies on the distribution of different elements in the solid and liquid phases of natural and synthetic rocks subjected to high temperatures and pressures in the laboratory. The results constrain the possible models for the origin of different oceanic basalts. The

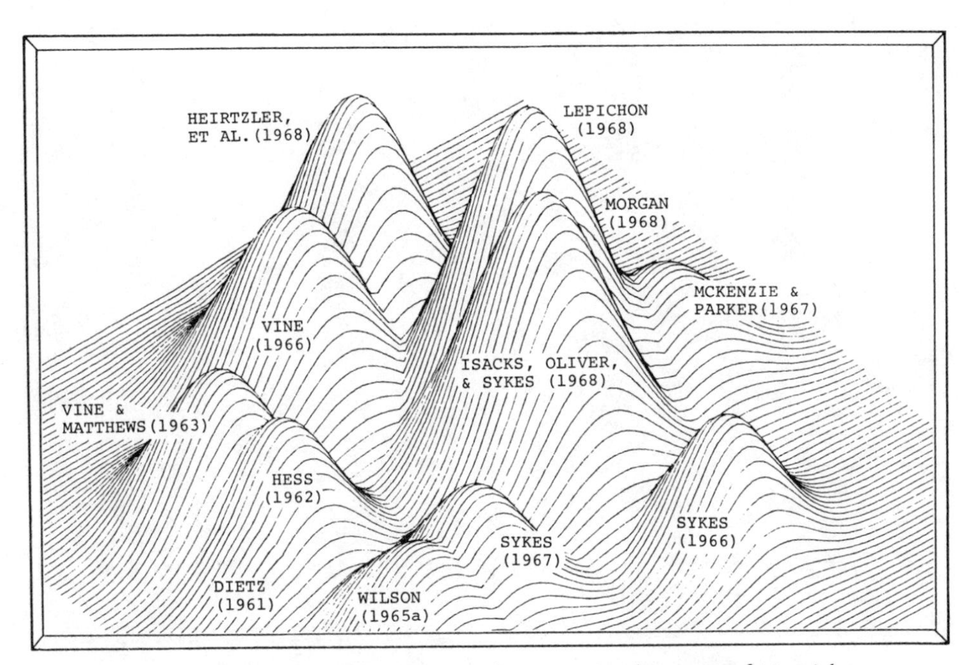

Figure 7.1 Hill model for the 1970 plate tectonics cluster: 12 of 19 articles. (From Stewart, Hagstrom, and Small [1981], reproduced with permission from *Geoscience Canada*.)

Yoder and Tilley (1962) article is one of the classics in this area, and received fifty-five citations in 1970. The article by Gast (1968) is distinctive because it focused on the distribution of trace elements in igneous rocks and related the results to the seafloor spreading model. Other articles associated with this cluster, but not shown in Figure 7.2, were by Melson et al. (1968), van Andel and Bowen (1968), and O'Hara (1965).

Discussion

It is apparent that cocitation analysis has identified the core articles in the plate tectonics revolution, which provide the "exemplars" of the plate tectonics "paradigm." Similarly, the petrology/geochemistry cluster of article represents the exemplars in another key area of geoscience research in 1970. Stewart, Hagstrom, and Small (1981) describe the resulting "drift" and "collision" of these and other new "continents" from 1970 to 1975, but we only need the information on these two clusters in the next analysis. Since these exemplars contain the "cognitive structures" used by the researchers in these two areas, the next analysis uses the "connections" between an article and these two clusters as a measure of the article's use of these different cognitive structures.

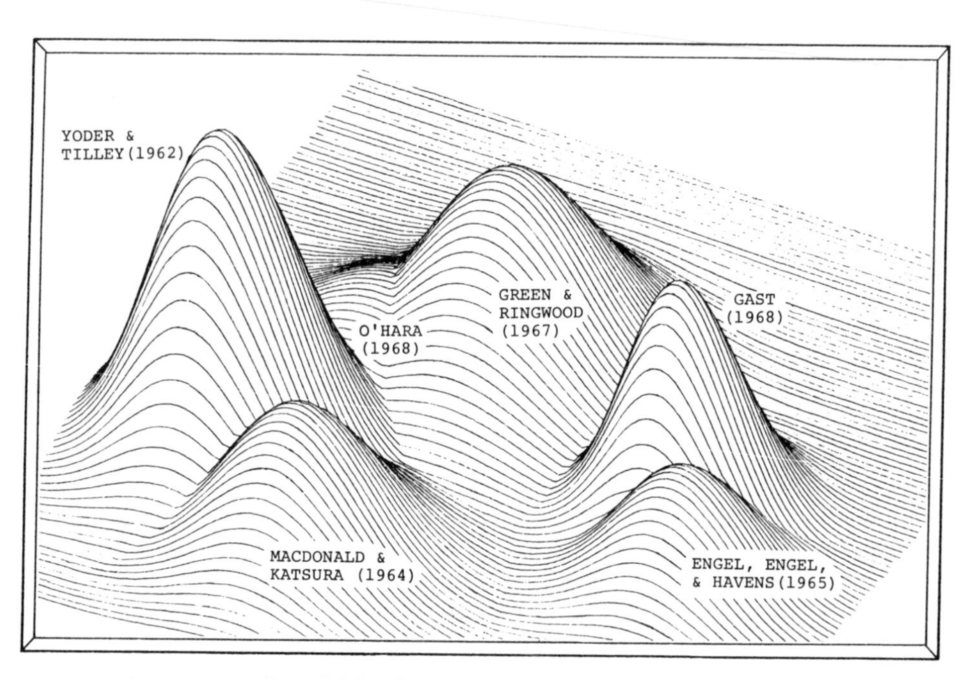

Figure 7.2 Hill model for the 1970 petrology/geochemistry cluster: 6 of 9 articles. (From Stewart, Hagstrom, and Small [1981], reproduced with permission from *Geoscience Canada*.)

PREDICTING CITATIONS TO GEOSCIENCE ARTICLES

The 139 articles studied in this subsection are a subsample of articles studied by the Center for Research on Scientific Communications at Johns Hopkins University (Garvey, 1973). The JHU study focused on publications associated with a number of scientific organizations, including the American Geophysical Union (AGU). All of the AGU related articles were published in 1968 and only those related to topics in geochemistry, geology, or solid-earth geophysics were included in the present analyses. Most of the excluded articles were on the atmosphere, other planets, or the oceans' waters. The major publishing journals were *Nature, Science, Journal of Geophysical Research*, and *Geochimica et Cosmochimica Acta*. Subjective assessments indicated about one third of these articles were related to plate tectonics and its development.

William Garvey of Johns Hopkins University kindly provided their information on these articles, but this was supplemented by reading each article and coding information about its contents. The influence of the articles was measured by the citations to the articles appearing in the *Science Citation Index* (Institute for Scientific Information, 1972 and 1975) for the first six years following publication. Citations from later articles by any of the authors were not counted since our measure of influence should reflect only the impact of the article on other researchers.

Measuring Article and Author Characteristics

The measurement of the article characteristics presents the greatest problems, especially if quantitative measurements are desired. For example, "logically coherent" articles might be cited more frequently, but it would be difficult to measure this feature. However, we can measure easily some of the important characteristics of an article. For example, the constructivist approach suggests that the degree an article builds upon previous literature is an important negotiating resource in building a persuasive argument (Gilbert, 1977; Williams and Law, 1982). The total number of references in the article provides a rough measure of this characteristic of articles, but scientific knowledge ages rapidly, so the number of recent references was also used.

Neither of these measures considers the specific body of previous research that an article builds upon, but we can use several methods to relate the references in an article to the two major cocitation clusters identified above. One plausible linking procedure would be the percentage of our article's references that are also items in the cluster. Other linkages between a specific article and clusters are possible: the 1968 article itself might be a member of a cluster defined by the 1970 publications, or the items in a 1970 cluster may cite the 1968 article.

Since most of the 1968 articles will not have such direct links to specific clusters, we need a more general procedure. If the references in all the items in a cocitation cluster are combined to form an *aggregated* set of references for a "cluster-article" representing a cocitation cluster, then two additional procedures can be used to link the 1968 articles to the "cluster-article." First, one can calculate the percentage of an article's references matching those in the "cluster-article"—a "bibliometric coupling" score (Kessler, 1963). Second, some of the references in the "cluster-article" will be more important than others in that they will be used by more than one of the cocitation cluster items. The more a 1968 article includes these key references in the "cluster-article," the more it should be related to the cognitive structure represented by the "cluster-article."

Rather than accepting any of these linkage measures as the "best" one, a factor analysis was used to combine the following linkage measures between the 1968 articles and the 1970 cocitation clusters.

1. Whether the 1968 article was a member of the cocitation cluster (MEMBER).
2. The percentage of the cocitation items that the 1968 article was directly linked to by either referencing a cluster item or by being referenced by one of the items (% LINK).
3. The bibliometric coupling between the 1968 article and the aggregated references in the cocitation "cluster-article" (COUPLING).
4. The average "importance" of the references shared between the 1968 article and the "cluster-article," where importance was measured by the average number of citations to the reference from the cocitation items (IMPORTANCE).
5. A subjective assessment on a one to three scale for increasing relevance of the article to plate tectonics or continental drift (CD/PT).

Each of the articles in the 1968 sample was linked to *both* the plate tectonics and petrology/geochemistry clusters by *each* of the first four measures: MEMBER, % LINK, COUPLING, and IMPORTANCE. With the subjective assessment measure, CD/PT, there were a total of nine indicators of the links between an article and the two major cocitation clusters identified for 1970.

Table 7.1 summarizes the results of a maximum likelihood, oblique factor analysis on these nine measures for the articles in the sample. Two factors were extracted and accounted for 64 and 33 percent of the variance among the nine variables, or a cumulative proportion of 97 percent of the variance. These two factors were correlated at the .24 level. Table 7.1 presents the PROMAX rotated factor pattern matrix, where the squared scores show the direct contribution of each factor to the variance of the different measures. The first factor clearly represents the degree an article is related to the topic of plate tectonics, and the second factor represents the geochemistry/petrology topic. The two factor scores from this analysis are assumed to measure how much these articles use the "cognitive structures" in the exemplars for plate tectonics and geochemistry/petrology.

The measurement of the cognitive structures used in an article is important, but we can measure numerous other article characteristics. For example, the total amount of information in an article should be related to the total length of the article. Counting the number of tables or figures presenting empirical

TABLE 7.1 Results of the Maximum Likelihood Factor Analysis with Oblique (PROMAX) Rotation for Nine Indicators of an Article's Relationship to the Plate Tectonics and Petrology/Geochemistry Cocitation Clusters Determined by the 1970 ISI Cocitation Results

	ROTATED FACTOR PATTERN MATRIX	
Assigned Factor Name =	Plate Tectonics	Petrology/Geochemistry
1. Indicators related to Plate Tectonics		
a. CD/PT	.769	.077
b. MEMBER	.415	−.093
c. % LINK	.830	−.027
d. COUPLING	.967	−.086
e. IMPORTANCE	.818	.122
2. Indicators related to Petrology/Geochemistry		
a. MEMBER	−.022	.599
b. % LINK	−.164	.752
c. COUPLING	.021	.953
d. IMPORTANCE	.202	.568
3. Factor eigenvalues	3.457	1.772
4. Proportion of variance	.642	.329
5. Cumulative proportion	.642	.972
6. Squared multiple correlation of variables with each factor	.941	.933

NOTE: The results were obtained with SAS 76. The text describes the different indicators and their construction. All eigenvalues beyond the first two were less than .40.

results, which Latour (1987) has called "inscriptions," gives more detailed information about the type of information contained in an article. Even a simple count of the number of equations in the article may indicate the degree the article uses formalized or mathematical reasoning. Finally, a set of "dummy" or zero-one variables can indicate other aspects of articles, such as the publishing journal, the type of data used in the article, the general subject matter, or the instrumental techniques employed.

In addition to measuring the article content we need information on the "social resources" of its author(s). A variety of biographical sources provided information on the training, professional age, rank, and specialization of the authors. The 1965–1969 *Science Citation Index* (Institute for Scientific Information, 1972) and the corresponding *Source Index* provided a number of measures of the authors' previous productivity and its influence on other researchers. These counts excluded self-citations and the articles in the present sample. Since departmental prestige scores were not available for foreign universities and private or governmental organizations, the logarithm of the number of departmental publications listed in the 1965–1969 *SCI Corporate Index* was used as a substitute. This was a reasonable proxy since it correlated at the .82 level with the ACE prestige scores (Roose and Anderson, 1970) in the subsample of departments with ACE scores. Other measures included professional age in 1968 and dummy variables for type of institutional affiliation, country, and discipline and specialty interests.

When an article has more than one author, we must combine the information on the separate authors. For quantitative variables, such as measures based on previous productivity, the aggregate sum over all the authors provides a better predictor of citations than a simple average of the authors' scores (Stewart, 1983). For nominal characteristics, such as type of institutional affiliation or discipline of degree, we can use the percentage of authors with different values. Table 7.2 lists and describes the variables found significant during the analyses in this chapter.[2] Table 7.2 lists the variables according to an assumed causal order.[3]

Methodological Issues

Several methodological issues must be considered before a quantitative analysis is possible.[4] First, we must consider whether the obtained citation counts to these articles adequately measures their actual "propensity" to be cited. Since citations occur only as discrete values, if the length of time allowed for citations to occur is too short, then the observed counts may not show as much variation as would occur over a longer counting period. In other words, the observed citation counts might be *unreliable* measures of the actual tendency for these articles to be cited. Allison (1978) has shown how to estimate the reliability of such discrete variables. For the present article data the observed citation counts for the six years after publication have a 0.97 estimated reliability.

The second methodological issue is important, but often neglected: the functional relationship between citations and the predictors of citations. Previous studies using individuals as the unit of analysis generally assume a simple linear relationship between citations and the characteristics of the scientists, but these studies have not tested alternative functional forms or examined the nature of the residuals, which can influence the significance tests if the residuals are not normally distributed or "homoskedastic" (Maddala, 1977). In the previous chapter we used a logarithmic transformation of the citation counts because it "linearized" the relationship between logged citations and the characteristics of the articles and their authors. Additional support for this procedure is provided by Stewart's (1983) test of several functional forms. These results indicated that the logarithmic transformation was best on the basis of several criteria, including the production of residuals with desirable properties.

The final methodological issue concerns the selection of variables used to predict citations. The selection procedure was basically exploratory. After choosing an initial set of plausible predictor variables, "stepwise" regression added any other variables that were significant at the .02 level.

TABLE 7.2 Names and Descriptions of Selected Variables: Listed by Their Assumed Causal Order

Variable Name	Description
INFLUENCE	Log of the total number of citations to article in 1969 and 1970–74 *SCI*, excluding self-citations.
	ARTICLE VARIABLES
RECENT REFS	Log of the number* of references within 3 years.
PUBDELAY	Number of months between acceptance and publication dates.
NUM REFS	Log of the number of references.
LENGTH	Log of the number of pages in the article.
ACCEPTS DRIFT	Subjective coding from 1 to 5 for increasing acceptance of evidence for continental drift; coded only for articles relevant to continental drift.
PLATE TECTONICS PETROL/GEOCHEM	Factor scores measuring degree article's references are tied to articles in 1970 co-citation clusters for plate tectonics or petrology/geochemistry; see Table 7.1.
EMPIRICAL	Subjective coding from 1 to 3 for increasing article use of empirical data.
FIELD STUDY	Dummy variable indicating article emphasizes descriptive field data.
GEOPHY DATA PALEOMAG DATA	Dummy variables for articles using general geophysical or specifically paleomagnetic data.
OCEANIC TOPICS	Dummy variable for studies of oceanic crust.
	AUTHOR VARIABLES
DEPT QUALITY DEGREE QUALITY	Quality of present department or degree department measured by log of sum* over authors for their departments' productivity according to the 1965–1969 *SCI Corporate Index*.
AVE INFLU PRE-1967	Log of the total number of citations to pre-1967 articles cited in the 1965–69 *SCI* divided by the number of cited articles, after each variable is summed* over authors.
CITES 1967–68 PUBS	Log of the sum* of the citations over authors to first-authored, 1967–68 articles cited in the 1965–69 *SCI*; excluding counts to the present article.
NUM CD/PT PUBS	Log of the sum* over authors of articles related to continental drift, seafloor spreading, or plate tectonics published between 1960–68; excluding the present article; based upon a personally compiled bibliography.
PETROL SPECIALTY	Percentage of authors with a speciality in petrology.
JAPANESE	Percentage of authors located in Japan.
UNIVERSITY	Percentage of authors located in a university.
MARINE SPECIALITY	Percentage of authors with a marine specialty.

NOTE: This list is restricted to variables that were dependent variables or stepwise selected variables for the analyses reported in this chapter. Note 2 contains a list of the above variables and the other variables not selected in any of the analyses. The author variables are not listed according to any causal order, since they are treated as exogenous variables. An asterisk (*) indicates that .5 was added to these variables before logging their scores.

Results: Intellectual Resources Are Most Important

The initial predictors of citations were length of article and those measures reflecting the amount and type of connections to previous literature: the number of references, the number of recent references, and the two factor scores reflecting the relationship of the article to the specific literature represented by the cocitation clusters for plate tectonics and petrology/geochemistry. The stepwise procedure added five additional variables. After this addition the total number of references was not significant at the .15 level, so it was dropped from the equation, which did not alter the best set of predictor variables. The variables added by the stepwise procedure included *both* article and author variables. The combined set of variables could explain over seventy percent of the variation in the logged citation counts. Table 7.3 presents the final results.

The absolute values of the standardized or "Beta" coefficients indicate the relative importance of the different predictor variables.[5] These show that the three most important variables are the article length, its relationship to the plate tectonics cluster, and the number of recent references. These all have a positive effect on citations, as did the degree the article was related to the petrology/geochemistry cluster. The stepwise procedure added two other article variables

TABLE 7.3 Regression Analysis Predicting an Article's Influence on Subsequent Research

Dependent variable = INFLUENCE

Independent Variables	Coefficients Raw	Beta	t-value
I. Constant	.097	—	.35
II. Article Variables			
1. LENGTH	.533**	.395	6.69
2. PLATE TECTONICS	.385**	.315	5.45
3. RECENT REFS	.261**	.210	3.65
4. PETROL/GEOCHEM	.224**	.182	3.47
5. PUBDELAY	−.052*	−.159	2.72
6. EMPIRICAL	.260	.115	2.39
III. Author Variables			
1. AVE INFLU PRE-1967	.306**	.203	3.97
2. UNIVERSITY	.404*	.150	3.09
3. MARINE SPECIALITY	.524	.125	2.37
R-square		.71	
Adjusted R-square		.69	

NOTE: Last five variables were added by stepwise regression procedure using .02 significance level. Note 2 describes all the variables that were possible, but not selected, predictors. Table 7.2 describes the above variables.

* Significant at the .01 level.

** Significant at the .005 level.

and these indicate that articles with longer publication delays were less likely to be cited, whereas articles emphasizing empirical results were more likely to be cited. These article content variables are assumed to be "intellectual" characteristics of these articles or "conversations."

However, the stepwise procedure also added some author variables to the equation. The most important of these, and the fourth most important variable overall, was the average influence of the authors' pre-1967 publications. This variable had a positive effect on citations, as did the other author variables: the proportion of authors in university settings and the proportion in a marine specialty. Thus both social (author) and intellectual (article) variables determine how much influence these articles had on later research.

The standardized coefficients reflect the relative importance of the *individual* variables, but not the relative importance of the two general *groups* of variables, i.e., those representing social and intellectual resources. To ascertain this, we can assume a causal order among the two groups of variables and then examine the changes in explained variance when the groups of variables are added to the equation separately and then together (Duncan, 1970). A plausible order can be developed by imagining three different situations: (a) articles are published anonymously without any information on the authors, (b) only information on the authors is given, and (c) both article and author information are given. The most plausible causal order of examining the changes in the explained variance is to contrast the case of anonymous publication with the full information situation.

Before conducting this test it is necessary to select which variables will represent each group. If all the variables in each group are used, the group with more variables will tend to explain more of the total variance. The best five variables from each group were used, where "best" means the chosen variables were the most significant predictors when used alone or when the effects of the other group's variables were included as predictors. The article group of predictors were article length, recent references, publication delay, and the two cluster variables. The author variables were the proportion in universities and marine specialties, average influence for pre-1967 publications, total citations to 1967–1968 publications, and the aggregate number of publications from the authors' present departmental affiliations.

The combined set of variables explained 70.7 percent of the variance in logged citations. When only article characteristics were used, the explained variance was reduced to 61.5 percent—a reduction of only 9.2 percentage points. This means that if articles were published anonymously, we can still explain about 87 percent of the variance obtained when we also include information on the authors. With just the five author variables, the explained variance drops to 32.9 percent. Thus the addition of article variables to author variables alone more than doubles the explained variance.

An alternative procedure for determining the relative importance of each group of variables uses a weighted sum for each group, where the weights are the raw regression coefficients obtained from predicting citations with all ten selected variables (Alwin, 1988). When citations are predicted with these two "composite" scores, the article group has a standardized coefficient of .663, which is more than twice the .327 coefficient for the author group. The general conclusion is that both individually and as a group the intellectual characteristics of these "conversations" are the most important predictors of their influence on other scientists.

Discussion

These results should interest both functionalists and constructivists, but they will have different interpretations of the empirical "facts" produced by this analytical procedure or "instrument." The functionalists might argue that the results indicate "universalism" is much more important than "Matthews Effect" processes in the allocation of recognition in the geoscience community. The constructivists might argue that we have shown only that basing one's arguments on the previous literature is the most effective way to create persuasive communications. This analysis cannot determine which of these two interpretations is "correct" because the "instrument" only measures very "global" characteristics of why some communications are more influential. Other studies, such as those observing actual laboratory behavior and interactions, will be needed to determine if scientists are "negotiating results" or "following norms." As previously indicated, these studies tend to support the constructivist perspectives. The procedures used here, however, have some advantages compared to those used by previous functionalist and constructivist studies. The regression analysis provides a *quantitative* assessment of the *relative* importance of the different resources that create persuasive scientific communications instead of the subjective assessments used by previous constructivist studies. Compared to previous functionalist studies, these procedures give a better test of the universalism norm because they provide much better measurement of the intellectual contributions made by scientists.

The constructivist approach does have some advantages over the functionalist approach because it raises additional questions not addressed by the functionalists. Some of these questions concern the social origins of scientific ideas and can be approached by further analysis of these data on geoscience articles. For example, some constructivists argue that scientists will resist or adopt a theory on the basis of their "interests" in it. The next two sections examine this idea.

Assessing the Importance of Social and Scientific Interests

Relativists argue that scientists structure their articles in a manner aimed at persuading other scientists to accept the reported results as the basis of their own research. The previous analysis provides information about how other scientists responded to the articles, but it does not directly address the "interests" proponents belief that the citers of these articles are following their particular scientific and social "interests." The examination of what author and article characteristics predict the positions stated in these articles is relevant to this approach. In particular, we will examine what predicts how much an article is related to the plate tectonics cluster of exemplars and how much an article expresses support for plate tectonics. The major advantage of this procedure is that it avoids the use of subjective procedures to assess the relative importance of different interests, but we will find that labeling the predictors as "scientific" or "social" interests is often problematic when examining *current* science. However, when we study the opinions of *past* scientists, which is done in the next section, we can more easily classify predictors into "social" and "scientific" categories because we *now* "know" what theories should have been acceptable "working hypotheses."

PREDICTING OPINIONS ON PLATE TECTONICS

The articles studied in the previous section have two measures of an article's relationship to the theory of plate tectonics. The first one, PLATE TECTONICS, is the article's factor score for its relationship to the cluster of plate tectonics articles. All the articles have a meaningful score on this variable, although the majority have the same lowest score. The second variable, ACCEPTS DRIFT, is a subjective assessment of the degree the article claims support for plate tectonics, seafloor spreading, or continental drift. Only articles on these or closely related subjects were assessed on a one to five scale for increasing acceptance. Although this variable is most relevant to the interests approach, predictors of both variables are given in Table 7.4. Considering first the PLATE TECTONICS variable, a simple stepwise regression with a .02 significance level gave results consistent with the earlier history chapters. The most important variable is a simple "dummy" variable indicating that the article reports information about the ocean basins. The next most important predictor is also expected; the more the authors have written on this topic in the past, the more likely the present article will be related to this topic. Finally, articles emphasizing simple field reports of geological or geophysical measurements are less likely to relate their results to the plate tectonics literature. These results only tell us which articles had more ties to the plate tectonics cocitation cluster of articles. They do not tell us the position taken with respect to this theory.

This can be approached by predicting the ACCEPTS DRIFT variable, but the analysis has a number of problems. First, most of the studied articles were not relevant to this theory. Subjective judgments of this relevance left only 46 articles sufficiently relevant to be included in the subsample. Second, one often had to "read between the lines" to determine the position taken on drift-type theories. Third, only two of the articles were coded as "somewhat" negative and none were assessed as "strongly" negative. Thus this analysis really reflects differences between the nine articles coded as negative or neutral and the remaining 37 coded with two levels of positive assessment: 10 were "somewhat" and 27 were "strongly" positive. Finally, the analysis makes some causal order assumptions about which of the content variables will be used to predict ACCEPTS DRIFT: see note 3.

TABLE 7.4 Regression Results for Predicting Characteristics of Articles Related to Continental Drift, Seafloor Spreading, and Plate Tectonics

Dependent Variables					
PLATE TECTONICS			ACCEPTS DRIFT		
Independent Variables	Raw	Beta	Independent Variables	Raw	Beta
Constant	−.152	—	Constant	4.17**	—
OCEANIC TOPICS	.970**	.461	PLATE TECTONICS	.31**	.392
FIELD STUDY	−.411**	−.203	PALEOMAG DATA	.78**	.348
NUM CD/PT PUBS	.320**	.322	GEOPHY DATA	−.51**	−.243
			PETROL/ GEOCHEM	.16*	.240
			PETROL SPECIALTY	−2.30**	−.562
			NUM CD/PT PUBS	.29**	.305
			JAPANESE	−1.48**	−.242
Number of Observations		139			46
R-square		.53			.83
Adjusted R-square		.52			.79

NOTE: Table 7.2 describes the dependent and independent variables. Note 2 describes all the variables examined as possible predictor variables. Stepwise selection with a .02 significance level was used to select the predictors of PLATE TECTONICS. PALEOMAG DATA was forced into the ACCEPTS DRIFT equation and the other variables were chosen by stepwise selection with a .05 significance level. For ACCEPTS DRIFT the articles studied were restricted to those subjectively assessed as relevant to the issue of continental drift and coded from 1 to 5 for increasing acceptance of a belief in continental drift. The text gives the distribution of scores on this variable.
* Significant at the .01 level.
** Significant at the .005 level.

The initial predictors were PALEOMAG DATA and PLATE TECTONICS, but the stepwise procedure added some other variables that were significant at the .05 level. Table 7.4 shows that about eighty percent of the variance can be explained by the predictors. Articles expressing stronger acceptance tended to be on paleomagnetic subjects, related to the plate tectonics literature, or by authors with more previous publications on plate tectonics or continental drift. Articles related to the specific literature represented by the petrology/geochemistry cocitation cluster of articles were more likely to express acceptance of drift theories. The remaining variables had negative effects. Articles with a larger proportion of authors located in Japan or with a petrology specialty were less likely to express support for plate tectonics, as were articles on general geophysical topics.

The variables with positive effects are not surprising. Even the negative effect of the JAPANESE variable is consistent with Allègre's (1988) suggestion that Japanese geologists were much slower to accept plate tectonics. Perhaps petrologists *in general* find it less relevant, but those citing the literature on the origins of basaltic magmas (PETROL/GEOCHEM) find it more relevant. However, the negative effect of geophysical data sources is somewhat puzzling, given the importance of this type of evidence for seafloor spreading. Perhaps after controlling for *paleomagnetism*, prior publications on plate tectonics, and the geophysical topics represented in the PLATE TECTONICS factor score, the articles using *other* geophysical data sources tended to be more neutral toward plate tectonics.

DISCUSSION

This section illustrated some of the other possible analyses in the study of the characteristics of scientific articles. In particular, we have *quantitative* evidence that scientists with different "interests" differ in opinions on "drift" theories. For example, articles by scientists with more previous publications related to continental drift (or seafloor spreading and plate tectonics) are more likely to express support for such theories. However, the results are more dubious than those given when the dependent variable is the article's influence. There is the problem of valid and reliable measurement of both the dependent and independent variables, e.g., assessing the degree of acceptance requires a subjective judgement. Note, however, that subjectively assessing opinions on a theory is less difficult than subjectively determining the most important predictors of these opinions when these predictors are numerous and intercorrelated with each other (Faust, 1984). The regression analysis provided the latter information.

Furthermore, we had to make some assumptions about causal ordering among some of the article variables. The results could depend strongly upon these assumptions. Finally, it is difficult to classify the predictor variables into "scientific" and "social" categories. Those variables related to characteristics of the article, such as PALEOMAG DATA, might reasonably be called "scientific" interests, but author variables might be classified either way. For example, articles by Japanese scientists seem to express less support for drift theories. This could occur because Japanese scientists have some "interest" in not supporting plate tectonics or in 1968 they may have been less informed about the recent evidence for plate tectonics. We cannot determine which is the case from this analysis, but if the reason is lack of information, we might be reluctant to label this a "social" interest.

Only when we have fairly strong *prior expectations* can we accomplish the most interesting aspects of the interests approach: classifying the reasons for supporting or rejecting a theory into "scientific" and "social" categories. For example, if we had good reason to believe that political beliefs should have no effect on choice of theories, but found quantitative support indicating that articles by more liberal geologists were more likely to support drift, then we might suggest that social factors were important. The analysis in the last section suggests that accepting a specific scientific theory as "correct" provides a useful prior expectation.

Predicting Published Opinions on Continental Drift: 1900–1950

Most of the histories of plate tectonics emphasize the recent quantitative evidence given by magnetic and seismological studies, but they also mention the evidence used to argue for continental drift in the first half of this century. Although earlier proponents of drift gave similar evidence, they were unable to persuade most geoscientists to even consider drift as a plausible theory or a "working hypothesis." Consequently, we must conclude *in light of current beliefs* that this non-quantitative evidence was either so ambiguous that conflicting interpretations were possible or that most geoscientists in the first half of this century resisted drift theory for some non-cognitive reasons. *In either case*, one must consider why many geoscientists ignored or rejected the evidence *or* emphasized only one of several possible interpretations.

As noted earlier, a number of geoscientists explain this early rejection of drift theory in terms of the "investment in orthodoxy" of the opponents (Bullard, 1975*a*) or the idea that continental drift implied that the previous "geological synthesis might need rebuilding" (Blackett, 1965: 2). This is similar to the interests perspective about science in general. This section tests this hypothesis by examining the author characteristics predicting the opinions on continental drift expressed in a sample of articles that dealt with drift theory and were published between 1900 and 1950.

DATA DESCRIPTION

Few scientists fully adopted continental drift theory between 1900 and 1950, but a number of them discussed it in some of their publications. The sample of these publications was developed by first compiling a comprehensive bibliography of articles on continental drift, based upon articles listed in Kasbeer (1973), discussed by Wood (1985), Wegener (1924, 1966), or du Toit (1937), or listed in the *Bibliography of North American Geology* (*BNAG*) or in the *Bibliography and Index of Geology, Exclusive of North America* (*BIGENA*) under "continental drift," "continental displacement," or the "earth's crust, continental drift." The English language articles available at the University of Wisconsin libraries were read and coded subjectively on a one to five scale for increasing acceptance of drift theory. Some articles in a foreign language or not available at the University of Wisconsin were included if their stated position on drift was discussed explicitly by Wood (1985), Wegener (1924, 1966) or du Toit (1937), all of whom devoted substantial sections of their books to discussing previous articles on drift.[6]

Although a few scientists wrote more than one article related to drift theory, only a couple of scientists appeared to change their mind between earlier and

TABLE 7.5 Description of Variables Used in the Study of Opinions Expressed toward Continental Drift Theory in Published Articles: 1900–1950

Variable Name	Variable Description
OPINION	Subjective coding from 1 (very negative) to 5 (very positive) of the scientist's first published opinion on drift theory.
YEAR	Year of publication for the stated opinion.
LOG PUBS	Log of the sum of the number of publications listed in *Bibliography of North American Geology* and *Bibliography and Index of Geology, Exclusive of North America* for ten years prior to the year of the stated opinion; see Note 8.
AFRICA & INDIA, EUROPE, DUTCH, BRITISH, SWISS, GERMAN, N-AMER	Non-exclusive dummy variables for whether the scientist had educational, training, or expedition experience in South Africa or India, continental Europe, Netherlands, Britain, Switzerland, or Germany, or was born in North America.
AGE	Age at time of stated opinion.
PHD	Dummy variable for whether the scientist had a PhD degree.
UNIV, GOVT, AND SURVEY	Mutually exclusive dummy variables for type of institutional affiliation of the scientist at time of stated opinion.
RANK	Ordinal coding of the rank of the scientist at the time of the stated opinion: 0 = student, 1 = assist prof or equivalent in government or survey, 2 = associate prof or equivalent, 3 = full professor or equivalent, 4 = directors, administrators, or "professors" in European universities.
PALEO, STRAT, PETROL, MINE-ECON, GEOPHY, GEODESY, GEN GEOL, BIOSCI AND OUTSIDER	Non-exclusive dummy variables for the scientist's specialties: paleontology, stratigraphy, petrology, mining or economic geology (including mineralogy), geophysics, geodesy, general geology, biological sciences (botany or zoology), or some other discipline outside of geology (botany, zoology, math, physics, astronomy, geography, engineering, chemistry, or meteorology).
NOINFO, NORANK, NOSPEC, AND NOINST	Missing data dummy variables for whether no biographical data, no rank data, no speciality data, or no institutional data were available for the scientist.

later publications during this period. The analysis here uses only the articles containing *the first stated opinion* on continental drift.[7] The predictor variables are various characteristics of the scientist measured at the time of the stated opinion. This information was collected from various biographical sources, which included the different editions of *American Men of Science*, *Directory of British Science*, *Who's Who of British Science*, *Prominent Scientists of Continental Europe*, *Who's Who in Science in Europe*, and *Poggendorff's Biographisch-Literarisches*

Handworterbuch. Sarjeant's (1980) *Geologists and the History of Geology* provided general information for many of the geoscientists and gave references for biographical sketches in the scientific literature. About twenty of the 108 scientists in the final sample had incomplete biographical data. The sum of the publications listed in *BNAG* and *BIGENA*[8] for the ten years prior to the time of the stated opinion provided our measure of the scientist's "investment" in non-drift perspectives. This sum (plus one-half) was logged to reduce its highly skewed nature. Missing data on specific variables were replaced with the mean scores for each variables *within* one of two subgroups: those born in North America and the remaining geoscientists. Dummy variables were used to indicate when data were missing on selected variables. The variables used in this analysis are listed in Table 7.5.[9]

RESULTS

Table 7.6 shows the general distribution of opinions on continental drift from 1900 to 1950 for scientists born in North America and the "other" scientists, most of whom were British or Europeans. In both groups, the positive positions expressed by the original proponents in the period prior to 1915 are followed by more negative reactions in the 1920s, especially among the North Americans. The number of *new* scientists entering the debate peaks in the 1920s and drops off rapidly. Finally, the "Total" column clearly indicates the polarization of opinions of drift. The "average" opinion is about "neutral," right where historians of this debate might argue it should have been (Laudan, 1987). However, only a few individuals adopted this "correct" position, so we want to explain the variation of the individuals about this average.

Table 7.7 gives the stepwise-selected predictors of the individuals' opinions on drift for two different groups: those born in North America and "other"

TABLE 7.6 Distribution of First Published Opinions on Continental Drift by Year, for Those Born in North America and All Others

N. Amer. Opinions	1900–1914	1915–1919	1920–1924	1925–1929	1930–1934	1935–1939	1940–1950	TOTAL
Positive	100.0	–	0.0	37.5	50.0	50.0	42.9	40.7
Neutral	0.0	–	0.0	12.5	0.0	50.0	14.3	11.1
Negative	0.0	–	100.0	50.0	50.0	0.0	42.9	48.2
(N)	3	0	5	8	2	2	7	27
Other's Opinions								
Positive	100.0	50.0	58.3	50.0	64.3	41.7	57.1	56.8
Neutral	0.0	0.0	4.2	18.8	7.1	16.7	28.6	11.1
Negative	0.0	50.0	37.5	31.3	28.6	41.7	14.3	32.1
(N)	4	4	24	16	14	12	7	81

TABLE 7.7 Regression Analyses Predicting a Scientist's First Published Opinion on Continental Drift Theory, for Those Born in North America and All Others

NORTH AMERICANS			OTHERS		
Independent Variables	Raw	Beta	Independent Variables	Raw	Beta
1. Constant	5.34**	—	1. Constant	4.89**	—
2. LOG PUBS	−0.46**	−0.479	2. LOG PUBS	−0.54**	−0.464
3. NOINS	1.39	0.283	3. NOINS	−0.90	−0.257
4. RANK	−0.54	−0.268	4. AFRICA&INDIA	1.06*	0.331
5. OUTSIDER	1.02	0.256	5. SWISS	1.74	0.236
			6. EUROPE	−0.67	−0.203
R-square		0.791			0.303
Adjusted R-square		0.753			0.256
Number of Observations		27			81

NOTE: In all equations the predictor variables were selected by stepwise regression using a .05 level of significance. Only the selected variables are given in this table. All variables are described in Table 7.5.
* Coefficient is significant at the .01 level.
** Coefficient is significant at the .001 level.

scientists. The most striking result is that in *both* groups the number of previous publications is the strongest predictor *and* its effect is negative. Among the North Americans the negative effect of RANK also indicates that "established" scientists were more likely to resist drift theory. Furthermore, the "positive" effect of the NOINS "dummy" variable indicates that in cases where information is lacking on a scientist's institutional affiliation, that scientist was more likely to accept drift. If such scientists are less "established," hence omitted from biographical sources, then this variable also indicates that more established North American scientists tended to resist drift theory. Finally, North American scientists "outside" the traditional geoscience fields were more likely to accept drift. Together these variables can explain 75 percent of the variation in the opinions on drift among the North Americans.

For the "other" scientists, we can explain only 25 percent of the variation in OPINION, even though OPINION's variance is similar in both groups. The number of publications has a strong negative effect, but the next most important predictor was education or research experience in South Africa or India (AFRICA&INDIA). This has the expected, positive effect on opinion toward drift. The positive effect of the SWISS variable and the negative effect of the EUROPE variable support Carozzi's (1985) historical analysis indicating that Swiss geologists supported drift, but most Europeans did not. The NOINS variables has a negative effect, which is the opposite of its effect among the North Americans. This is puzzling because, if interpreted as we did above, it means more "obscure" scientists tended to *oppose* drift. Perhaps it simply indicates the relative difficulty of finding institutional data on Europeans: It has a positive correlation with EUROPE and a negative correlation with BRITISH.

Although the analyses are not presented here, it is interesting to note some of the variables that predict the LOG PUBS and AFRICA&INDIA variables. In both groups, paleontologists tend to have more publications, but only among the "other" scientists is AGE one of the predictors. Thomas Kuhn's suggestion that younger geologists are more likely to accept novel theories is not supported here. Publication "investments" in the older viewpoint has a much stronger effect and "mediates" the effect of AGE among the "other" scientists. However, the direct effect of the OUTSIDER variable among North Americans does support his suggestion that "outsiders" may be more willing to adopt a new theory or paradigm. The strongest predictor of AFRICA&INDIA among the "other" scientists is specialization in mining or economic geology. This may indicate economic interests in the mineral resources of the colonies. If so, then we have a clear example of how a "social" interest has indirect effects upon opinions about a scientific theory

DISCUSSION

These results illustrate how interests proponents might provide quantitative evidence for their perspective. In this specific case, it appears that scientists with larger publication "investments" opposed the new theory of continental drift. However, this conclusion is based upon only one possible interpretation of this quantitative evidence. If this study had been done in 1950, an alternative interpretation would have been more "plausible." In 1950 most geoscientists did not believe in continental drift, so one could argue that this study only shows that those scientists who know the most geology—the high producers—have rejected drift. Similarly, the acceptance of drift by outsiders would only show their ignorance of the geological evidence against drift. This alternative interpretation is less plausible *if we accept the present beliefs of geoscientists* that the earlier evidence should have been sufficient to keep drift theory at least among the possible "working hypotheses." Thus the acceptance of a current theory provides some of the prior expectations that support the interpretation of LOG PUBS as a "social" instead of a "scientific" interest.

Conclusions

The starting point for the analyses in this chapter was the processes yielding a citation to a scientific paper. From this model, we "derived" a statistical model for predicting citations to scientific articles and indicated how it was related to studies of other "social organizations." The first section illustrated how one might use this "instrument" for measuring "global" properties of a scientific community.

There are several ways to judge whether this "instrument" has given "useful" measurements. First, the ability of the "instrument" to produce quantitative results with fewer subjective judgments were some of the reasons for developing it, but this criterion will appeal only to those who like quantitative arguments. Second, we might consider whether the results are relevant to the various theories about the behavior of scientists. Certainly the results in the first section of this chapter are relevant for the functionalist and relativist perspectives on science, even though we cannot use the results to choose which of these is the "better" perspective. The results in the last two sections of this chapter are relevant to the interests perspective, but other assumptions were needed to classify "social" and "scientific" interests.

Another criterion is purely statistical: How well do the procedures predict citations or expressed opinions? Again the "instrument" worked quite well. We were able to explain more than 70 percent of the variation in citations to articles. Similar studies of citations to individual scientists rarely do *half* as well. We did an even better job explaining the variation in opinions on plate tectonics in the second section and the opinions on continental drift among North Americans in the third section. The models explained 80 percent of the variation in opinions in both of these analyses. Among the "other" scientists we only achieved a modest 25 percent, but this is a more diverse group with a tendency to have more missing biographical data.

Perhaps the most important criterion is the degree the results fit our "theories" about the behavior of scientists. Here we encounter two problems. First, our theories are quite incomplete. For example, we must specify *when* and *where* "social" and "scientific" resources will be more or less important predictors of influential communications in science, or when scientists are more or less likely to follow the "universalism" norm. This has not been done by either functionalists or constructivists. At the moment we only have intuitive speculations, e.g., that the "universalism" norm is less likely to be followed in the "softer" social sciences. The second problem pertains directly to the analyses in this chapter. The citation analysis is only for one discipline at one point in time. Even though the results indicate that "intellectual" resources were much more important than "social" resources in constructing influential articles, they only apply to the geosciences in 1968. Many additional studies are needed to see if the *relative* importance of social and intellectual "resources" varies in some systematic manner with other characteristics of scientific fields.

An analogy with the development of seafloor spreading theory seems particularly appropriate. The citation analysis results are like a *single* magnetic field measurement at one location in an ocean basin. Before seafloor spreading theory was considered plausible, Vine and Matthews had to indicate its relevance to the quantitative magnetic data and these data had to be extensive enough to show the variations called "anomalies." This analogy indicates the immense task ahead for sociologists of science. We need both better theory and more measurements in other "ocean basins." Whether these measurements should include those produced by the "instrument" used in this chapter is certainly debatable. Perhaps its results are more like a single water temperature measurement, which has almost no relationship to seafloor spreading. However in its defense, we might note that even after Vine and Matthews proposed their explanation of the linear magnetic anomalies, some oceanographers were so tired of seeing the magnetometer readings "go up and down" without apparent cause that they didn't want to bother with it.

The previous criteria mentioned for evaluating the usefulness of this "instrument" suggest that it might provide a valuable addition to the other "tools" used in science studies. Another reason to include it is that it is similar to the methods used to study other social organizations, especially bureaucracies. The citation analysis model can be generalized to examine the relative importance of "speaker" and "content" variables in communications in all types of "social organizations." These other types of organizations may provide the best "ocean basins" for future studies so that we can determine the variations in the measurements given by this instrument and the degree that scientists' "method" of reasoning differs from "everyday" reasoning.

CHAPTER EIGHT

Reflections

This book has examined several levels of changing "cognitive structures." At "level 1," we have geologists studying *physical* phenomena and interacting with each other as they develop their "cognitive structures" or theories about natural phenomena. However our primary interest is comparing the cognitive structures used at "level 2" by observers of science—philosophers, historians, and sociologists—to explain how the cognitive structures of scientists change over time. At the end of Chapter 5, we even considered briefly a "level 3" analysis, when we summarized the changes in the cognitive structures of the observers of science as new "paradigms" in epistemology. In the latter case we used a "level 2" cognitive structure to explain reflexively its own level's development. In addition, I occasionally suggested that the development of plate tectonics theory (level 1) may provide a good model of what is needed for the development and acceptance of a "level 2" paradigm. Given these confusing levels of analyses and the use of analogies from different levels to analyze other levels, it is helpful to recapitulate the major points of the various chapters in light of the conclusions in later chapters. The first part of this chapter provides this recapitulation. The second part discusses some of the implications of this book for our understanding of science and its further study. The concluding section is more reflexive and illustrates how the perspectives discussed in this book might apply to its own development and its reception by others.

A Recapitulation

Chapter 2 provided a brief history of continental drift theory before World War II. We saw how drift theory provided one of many possible explanations of the available evidence, but it was treated rather harshly, especially in North America. Drift theory did not gather many followers even though Holmes and du Toit removed many of its empirical and conceptual problems before World War II. In light of later geological evidence, historical studies, comments from interviews, and the quantitative analyses presented in later chapters, there seem to be several reasons why drift theory failed to win many followers.

First, the "localism" (Le Grand, 1986*a*, 1986*b*) or "provincialism" (Frankel, 1984) of the geosciences was important in several ways. Even before Wegener's drift theory, geologists had lost interest in such global theories and increasingly focused on detailed specialty studies (Greene, 1982). Within these smaller communities a sense of "progress" seemed possible because there was more agreement on a "disciplinary matrix" of metaphysical assumptions and models, instruments, interpretative procedures, values, and exemplars. The contracting earth "theory" became more of a vague assumption providing an implicit force in the geosyncline model, which represented the largest scale of analysis with much continuing interest. Consequently, specialists did not recognize the integrating ability of continental drift theory, but only saw its defects (or advantages) within their own areas of expertise.

Preferred regions of geological study also indicate "localism." North Americans tended to focus on their own continent, whereas the geological evidence for drift from the southern continents was more easily related to continental drift theory, but was either unknown to most geologists in the northern continents or reinterpreted in ways consistent with the assumption of stable continents. Such reinterpretation of evidence is always possible in science, but it is not labeled "irrational" until there is widespread, tacit agreement on such things as the proper assumptions, models, analogies, and data sources. This agreement was lacking so scientists were "free" to choose their preferred interpretations.

Second, given the specialty structure and the interpretive flexibility of geological evidence, a "satisficing" type of decision-making naturally led to acceptance of the assumption of fixed continents. Satisficing means taking the first theory that meets some minimum criteria. By the time that drift theory developed, most geologists had interpreted the evidence in the northern continents within a fixist framework, so it was the "natural" choice. Or, as Raup (1986) put it, new theories are guilty until proven innocent and old theories are innocent until proven guilty.

Finally, the last analysis in Chapter 7 indicated that those opposed to drift theory tended to be more productive and were probably the "elites." A reasonable inference is that their opinions influenced others to reject or neglect drift theory, but empirical evidence for this would require a citation analysis among pre–World War II articles. Perhaps the elites exercise more direct control because they would have higher rank and tend to be journal editors and referees, which would let them control access to academic positions and journal space. Some of the interview comments indicate that this may have occurred. The *same type* of explanation can be given for why drift *was accepted* in some localities. That is, it tended to be accepted where the evidence was better, related to particular specialties, and the elites tended to endorse it. In particular, Argand among Swiss Alpine geologists had world-wide geological experience and du Toit among South African geologists had studied in South America.

In Chapter 3 we examined some of the postwar developments in selected specialties. A large influx of defense spending for marine research produced many "surprising" results, some of which depended upon the development or improvement of geophysical instruments. Instrumental improvements were also important for the developments described for seismology, geochronology, and paleomagnetism. However, these results did not produce a widespread sense of a "crisis" among geoscientists or much effort to synthesize the new results into a more global theory. Chapter 5 suggested that the isolated specialty structure or the "multiparadigm" state of the earth sciences could explain these two features. Proponents of drift theory were mainly a few southern hemisphere geologists and the few geophysicists studying the changing locations of the earth's paleomagnetic poles. As before, opponents of drift ignored, dismissed, or reinterpreted this evidence for drift.

By the end of the 1950s several important trends had developed. First, geophysicists were producing more quantitative and global data of increasing relevant to geological subjects. This was aided by a second trend toward interdisciplinary cooperation on the ships engaged in marine research. Finally, Wood (1965) suggested that the increased prestige of the geoscientists in the "marginal" areas of geophysics and geochemistry helped prompt a movement toward more encompassing "Earth Science" departments, but this movement lacked a justifying global "ideology" or theory. This helped motivate the development of more global theories and explains some of the features of the theories offered by Carey, Wilson, Dietz, and Hess, which were described at the end of Chapter 3 and the beginning of Chapter 4. All of these theories were more global and tried to incorporate many of the diverse results from the specialized studies, especially the marine studies. Chapter 4 described Hess's "geopoetry" of seafloor spreading, how it was tested and elaborated in the 1960s, and how it eventually became part of the global theory of plate tectonics.

Plate tectonics provided an integrating framework for *some* of the "facts" accumulated in earlier geoscience research. It did not replace the specialty "paradigms" providing these facts, but provided a broader model that had many, possibly useful analogies for different specialty data. The initial version of the theory provided analogies or "explanations" for many marine magnetics and seismological data, *provided* one was willing to accept certain assumptions and select the "best" data for testing the theory.

Consequently, the new "paradigm" of plate tectonics was accepted rapidly by those with the most "interests" in it. These included geophysicists and oceanographers seeking a "global" theory, especially one based upon their particular instrumental skills, knowledge, and desire for quantitative predictions. Very simple analogies from the initial model satisfied their interests, but other geoscientists were not convinced because they had different interests. Most notable among these were continental geologists or others who had taken a strong stand against continental drift theory or developed alternative theories. The subsequent "negotiations" between proponents and opponents illustrated their conflicting interests. Continental geologists wanted a theory with specific implications for their detailed field studies and preferred theories "built up from the(ir) facts," instead of "down from abstract, geophysical models." Consequently, opponents and proponents tended to have "communication problems" because they had different values and data interests. Furthermore, proponents were busy reinterpreting many traditional geological concepts (e.g., uniformitarianism or geosynclines) into their new model of the earth and creating new concepts (e.g., plates, subduction zones, or triple-junctions). This added to their communication problems even when each group tried to explain the same "facts," as indicated by the discussion of the opponents' views in Chapter 4.

The same basic processes had occurred in the earlier debate over continental drift theory, but this time the proponents of plate tectonics eventually persuaded most of the opponents to work within the new "paradigm." This required use of both "social" and "scientific" resources. The simple plate tectonics model had to be modified to make it more "realistic," that is, more pertinent to geologists' interests. This was done by changing some of the basic assumptions, such as allowing internal deformation of plates, and proposing more complex analogies from the basic model that were appropriate for continental geology, such as "exotic terranes," whose identification required use of the traditional skills of the field geologist. Yet we have seen that even with these modifications, numerous analogies and assumptions are needed to "construct" a relationship between the abstract plate tectonics model and continental geology.

Social resources probably influence this process and may have been important in the acceptance of plate tectonics. Some of the early proponents of plate tectonics were established scientists at major institutions and had effective rhetorical skills. Special conferences were held to recruit other elites. As journal editors, article and grant referees, and teachers in general began to accept the new paradigm, they could "encourage" others to follow, even though it was and still is possible to "deconstruct" much of this evidence, if one is willing to challenge widely held assumptions, analogies, and models. However, the analysis in the first section of Chapter 7 suggests that the "social" resources of authors were not nearly as important as their "cognitive" resources in persuading others to cite their articles.

In Chapter 5 we found ourselves in the middle of another "revolution," but it was about the basic nature of scientific reasoning and it was still in a "revolutionary science" stage, where there were two basic positions. On the one side were those starting from philosophical approaches suggesting that scientific knowledge was special because it was based upon objective tests of the rational deductions of theories or rational appraisal of problem-solving capacity of research traditions. On the other side was Thomas Kuhn (1962), who suggested there were two types of science—normal science and revolutionary science—and neither type was described adequately by the philosophical approaches. In essence, he argued that the special character of science did not reside in a special methodology, but in its use of cultural traditions and its social structure.

Each of the alternative perspectives in the philosophy and history of science described different aspects of the history of plate tectonics, but each had some problems. Those emphasizing the more logical or formal aspects of a scientific theory neglected how scientists made the key decisions and assumptions necessary to apply the logical procedures. In particular, they did not specify how scientists would chose the basic imagery (the "iconic model") and related assumptions that provide the basis for the formal deductions, how scientists decided to accept some results as "facts" when it is always possible to challenge their underlying theories and assumptions, and how scientists decided when the observed results matched the predicted results.

Kuhn emphasized the crucial nature of these decisions and the inability of logical rules to specify how they should be made. He proposed that persuasion on the basis of shared values and the tacit knowledge contained in the cultural beliefs determined these crucial choices. However, he did not specify how one would study this persuasion process or the operation of this tacit knowledge. Despite these problems many of his descriptions of scientific revolutions seemed to occur in the development of plate tectonics, especially if this revolution was seen as a shift from a multiparadigm science. Consequently, we adopted the paradigm perspective and even used it in a reflexive sense to describe the changes occurring in epistemology itself.

Despite the usefulness of the paradigm approach, there were a number of problems needing solutions. One of these problems was a better understanding of the relationships between the metaphysical model, the symbolic generalizations, and the empirical data. This issue was covered in the last substantive section of Chapter 5, which described the models and analogies perspective in the philosophy of science. This helped us understand how the basic imagery—the iconic model—was developed by analogy from source models and how it was related by analogy to the empirical data. Several examples from the geosciences, including the seafloor spreading model, illustrated this general approach. Two important conclusions were reached. First, this approach still required that scientists make some important decisions, but did not specify how they were made. Second, the proponents of models and analogies suggested that the social sciences used the same general approach in their research.

Consequently, Chapter 6 examined a diversity of social science models of the decision-making process and suggested how they could be applied to the behavior of scientists. The description of a simple research project identified the complex nature of the decisions made by the researcher and others before the results could influence the consensus that develops in a scientific community. We considered DeMey's (1982) analysis of how scientific reasoning involved the use of "cognitive structures" and illustrated this process with several geoscience examples to show the cognitive changes occurring in the geosciences with the acceptance of plate tectonics. We even found explicit examples of these cognitive structures in the Pearce and Cann (1973) procedure for classifying basaltic rocks according to their origins in different plate tectonic environments. Our analysis of this example indicated that "Nature" influences the results of empirical study, but must "speak through" a socially constructed "instrument."

Thus, previous decisions influence the "facts" obtained in empirical research and other decisions are needed before these "facts" determine the consensus on a theory. These research decisions are not made by "isolated" individuals, but by *members* of a scientific community or organization who share many norms, values, and beliefs—a "cultural tradition." Thus the studies of decision-making in social organizations ("bureaucracies") were particularly relevant and a brief review indicated that the basic processes identified in science occur in these organizations as well. However the decision-making approaches in science and in organizations required further elaboration in terms of the motivations, "interests," or goals influencing the decisions of individual members.

We examined some of the possible motivations of scientists in Chapter 6 and found that a desire for recognition from colleagues was particularly important. This motivation provided a means of social control in science because scientists could give or withhold recognition to promote or retard research in specific areas. Several examples from the history of plate tectonics illustrated this use of recognition for social control. In addition, peer recognition was the "currency" within the scientific community that could be "exchanged" for non-scientific rewards, such as money or job security, helping to isolate the scientific community from external pressures.

This view was one of the major contributions of the "functionalist" perspective on science, but we found little support for the "norms" that it suggested would direct the interactions among scientists. Only the norm to acknowledge (or recognize) the influence of others on one's own work seemed to be supported by the available data. In contrast, the "constructivist" sociology of science seemed better supported by these data, but proponents of both the "relativist" and "interests" perspectives seemed to have a similar problem with their empirical studies. Both of these groups often recognized a distinction between "scientific" and "social" resources or interests, but they never specified how they subjectively assessed the relative importance of these resources or interests in their case studies.

To solve these problems, we considered a "crude" iconic model for the decisions yielding a citation to a scientific article and derived a statistical "instrument" for measuring the relative importance of social and scientific resources in the construction of influential articles. This required several dubious assumptions, but this often occurs in scientific research and the procedure's plausibility ultimately depends on whether it can produce results that satisfy the "interests" of other researchers, which might include a desire for quantitative results or compatibility with other theories. Consequently, the last section of Chapter 6 briefly illustrated how the study of the influence of scientific articles or conversations was similar to studies of the structural dimensions of social organizations.

Chapter 7 applied the statistical model to several sets of data relevant to the history of plate tectonics and illustrated possible solutions to certain methodological problems in the application of this abstract model. The first section indicated that cocitation analysis could identify "exemplars" and found that in 1968 the scientific resources of articles were much more important than their social resources. Equally important, the procedure could explain more than 70 percent of the variance in the influence of these *articles*, which suggests that this procedure may be a valuable supplement to traditional studies of *individual* scientists.

The other two analyses in Chapter 7 assessed the role of scientific and social interests in the theory choices of scientists in 1968 and before 1950. We quantitatively assessed what caused some scientists to support plate tectonics in their

1968 articles, but we could not easily assign these causes into "social" and "scientific" categories because we lacked some necessary information. However, "accepting" plate tectonics and the evidence for it, allowed us to present plausible arguments that social interests were a major reason for resistance to continental drift before 1950.

The preceding discussion suggests that the processes of social science research are similar to those occurring in the natural sciences. The differences that do exist are matters of degree rather than qualitative differences. The next section suggests some of the relative differences that seem to exist between the sciences and other social organizations and proposes possible research areas that could help to empirically describe these differences.

The Nature of Scientific Knowledge and Suggestions for Future Studies

One of the major themes in the second half of this book concerned why we should trust scientific knowledge more than other belief systems about the natural world. Some of the common justifications were rejected easily. For example, we could not argue that scientific knowledge produces a "true" description of the real world because we lack an independent and accurate description of the real world. Even the more sophisticated justifications of philosophers that scientific beliefs are based on deductive—"rational"—reasoning about the relationships between theories and objective data ignored the importance of analogical reasoning and numerous decisions required by researchers and others. Furthermore, the arguments that scientific theories explain more "facts" or have greater "problem-solving" ability presume a shared cultural tradition, but do not explain how it works to structure the decision-making activity of scientists. Kuhn's appeal for increased "value-satisfaction" has similar problems.

In contrast, a major thesis of this book is that scientific research should be examined from a general social organization perspective. This suggests that reasons for trusting theories produced by scientists can be found in the "organizational structure" of the scientific community as well as its decision-making procedures. From this viewpoint several aspects of the social organization of science seem particularly important. First, its members are strongly socialized into a particular cultural tradition or "disciplinary matrix." Second, scientific disciplines are relatively autonomous from other social organizations. For example, the members of the organization control the socialization and admittance of new members, as well as other members' access to such valued rewards as academic positions, research funds, and journal publication. Furthermore, in universities the access to essential nonscientific rewards, such as salary and job security, is influenced strongly by the evaluation of scientific peers rather than by outsiders. The control exercised by one's scientific peers is through the control of recognition rather than more coercive processes. Consequently, individual scientists have considerable personal autonomy in their choice of research projects and theories. Finally, there is the open nature of the communication system. Scientific research is a process of persuading others that certain interpretations of empirical research are more plausible than others and much of this persuasion occurs in publicly available communications or scientific articles. Ziman (1968), in particular, emphasizes the importance of these factors.

These organizational features imply several important characteristics of the process of scientific research. For example, they orient scientists toward recognition from their peers. Since this recognition flows more to those producing original results, the recognition system encourages innovation and competition among scientists. Although competition and innovation occur within the cultural tradition of the organization, that tradition also requires empirical testing of some of its elements and legitimates the questioning of the assumptions underlying others' results, especially when research fails to produce results satisfying the criteria established by past research. Consequently, the recognition or reward system promotes change in the cultural tradition as new empirical results are reported. Compared to other social organizations, fundamental change in cultural traditions or beliefs occurs faster in scientific organizations.

At least this is the "ideal" version of what should happen in science, but in all social organizations there are competing "interests" and unequal distribution of "resources" for persuading others to make "decisions" compatible with one's own "interests." Ideally, the "method" of persuasion in science relies only on the use of "scientific" resources using appeals to "scientific" interests, but in reality this probably varies considerably by discipline, specialty, country, and time period. The proposed "instrument" might be one way of measuring this aspect of scientific reasoning.

Since this method is similar to those used to study social organizations in general, future research might compare scientific and other social organizations with this measurement procedure. Intuitively, one would expect "social" resources, such as hierarchical position, to be more important in non-scientific organizations. However such studies would be difficult because "communications" in non-science organizations are not "published" and it would be difficult to measure their "influence" on others (cf. LaRoche and Pearson, 1985). Consequently, comparative studies of different disciplines, specialties, or countries would seem to be the best way of testing whether the "instrument" yields interesting differences between scientific organizations. These studies might focus on different "research areas" because they are closer to the level of cognitive organization for most scientific research. Such studies would also make some of the assumptions underlying the "instrument" less dubious, such as assuming equal tendencies to publish an article.

If we used this instrument to study several different research areas and its "output" recorded interesting variations in the relative importance of social and scientific resources, then we could correlate the output with other characteristics of the research areas, such as their size, age, or level of funding. Such studies might help the development of more quantitative theories about the nature of scientific reasoning and consensus formation.

At the moment our "theories" are very vague. At best the "constructivist" view is more like the continental drift theory in the 1920s than the "geopoetry" of Harry Hess. Thus we need more theory construction work, which might be aided by seeking useful models and analogies from other areas of social research. Theories and research on the processes of "persuasion" (e.g., Petty and Cacioppo, 1986), "negotiations" (e.g., Strauss, 1978), and decision-making (e.g, Hammond et al., 1980) have many useful analogies for the behavior of scientists. However, I think that theories about the properties of social organizations will be even more fruitful (e.g., Hargens, 1975). In particular, Randall Collins has analyzed scientific specialties in terms of more general organizational processes and stated a number of fairly specific hypotheses (Collins, 1975).

Reflexive Reflections

One of the theses of the "strong programme" in the sociology of knowledge is "reflexivity": the same processes identified in the construction and acceptance of scientific knowledge also apply to the research produced by those studying science itself. It is apparent that the perspectives presented in this book can be applied to itself and its future influence on others. We have already seen some examples of the reflexive application of these theories. For example, Kuhn's paradigm approach was used to describe the context of its own historical development and the general stages of decision-making were analyzed by March and Simon as being the result of other decisions. It is fitting, then, to conclude with some illustrations and speculations on what this book implies about its own origins and reception.

One of the constructivist theses is that researchers' choice of theories is influenced by their "interests," which raises the issue of my personal interests and how they influenced my choice of topics and theories. These interests might be divided into "nonscientific" and "scientific" categories on the basis of how much these interests would be accepted as "illegitimate" and "legitimate" within a science community. In the nonscientific area I have been influenced greatly by three characteristics of my father. First, he has an overwhelming faith in the advantages of scientific knowledge over other sources of knowledge. Second, he tends to believe that his own beliefs and, in particular, his value system itself are based upon "scientific" analyses, which gives him a tendency to label those who disagree with him as "irrational." Having been accused of behaving in such a manner on numerous occasions, I was motivated to study the nature of scientific knowledge. Finally, I tend to share his distrust of authority and established doctrine, and that distrust encouraged me to examine novel and less established approaches toward the nature of scientific knowledge. These approaches included theories that he would dislike, such as Kuhn's paradigm approach or the related constructivist perspectives, which has created considerable "tension" in my beliefs because I still share my father's faith in science and find it difficult to accept some of the full implications of the "constructivist" approaches. Even though I have left behind my initial functionalist views, I am certain that one could use the history of continental drift and plate tectonics to build a better case for constructivism than I have.

More "scientific" interests are reflected by my professional training. A background in chemistry and sociological training at the University of Wisconsin both encouraged quantitative approaches. Furthermore, my training in theories of social organization encouraged me to adopt this perspective in the study of science. These tendencies were encouraged by the eclectic approach of my graduate advisor, Warren Hagstrom. McCann (1978) opened my eyes to the possibilities available in the analysis of specific articles.

It is more difficult to speculate on the implications of this book's contents for its future reception because that would require not only knowledge of others' interests, but also assessing my own social and cognitive resources. Given the interest in quantitative techniques in American sociology, there should be some interest among American sociologists in this book. However, most constructivists feel that quantitative approaches cannot contribute much to our understanding of the nature of scientific knowledge construction. Even though Chapter 7 illustrates quantitative applications of the relativists and interests perspectives, this book will probably be most interesting to American sociologists, especially since the results can be interpreted from a functionalist point of view.

I have only minimal "social" resources for presenting a "conversation" that will influence others because I do not have an established reputation in sociology or an appointment in one of the elite sociology departments. If this book influences other researchers, it must do so on the basis of such "cognitive" resources as its use of widely shared models and assumptions about the nature of scientific research and its ability to (a) integrate previous research in such a manner that new and interesting problems are identified, (b) propose solutions to methodological problems in the applications of different theories about scientists' behavior, (c) satisfy general criteria, such as quantitative measurements, used to evaluate research results, and (d) produce statements that are analogous to empirical observations of natural phenomena. The last aspect is particularly important and implies that geoscientists should contribute to the sociological discussion about this book. Ideally, I would like to see this book reviewed for sociologists by a geoscientist who did research both before and after this revolution.

It is difficult to evaluate these components in any scientific research, but more so in a discipline lacking an integrating paradigm. Ultimately, one must trust the evaluation of scientific contributions to one's scientific peers. It is their special training and the nature of their social organization that makes the consensus of opinions of scientists more "rational" and "objective" than other means of forming beliefs about the natural world. Their opinions will determine whether this book has described important aspects of the nature of scientific knowledge.

POSTSCRIPT

The manuscript for this book was finished in 1988 without knowledge of H. E. Le Grand's *Drifting Continents and Shifting Theories* (1988, Cambridge University Press).[1] The independent publication of two books with similar titles and themes indicates the great potential that the geoscience revolution offers historians, philosophers, and sociologists of science. These two books have a number of similarities. We both (a) focus on the same time period (1900 to mid-1970s) in the geosciences, (b) compare many of the same events and publications to the perspectives developed by Kuhn, Lakatos, Laudan, and sociologists of science, (c) conclude that each perspective illuminates selected aspects of this history, and (d) even use the history of plate tectonics to suggest how similar theory advances might occur within the study of science itself. Despite these broad similarities, there are some major differences that make our books complementary, rather than redundant.[2]

The titles of our books indicate one of our major differences: I find Kuhn's perspective most useful, whereas Le Grand prefers Laudan's perspective. These preferences reflect not only how well these perspectives fit the historical "facts" (as we each construct them), but also our social and cognitive "interests" and general presuppositions about science. We both try to make these aspects of our analyses fairly explicit. For example, Le Grand's presuppositions include beliefs that "science is normally rife with contending research programs or theories" and that science can be usefully described as a "problem solving activity" (Le Grand, 1988: 268), whereas I find Kuhn's explicit recognition of the key role of social factors more attractive.

Consequently we have different emphases and interpretations of the historical events. As might be expected, Le Grand emphasizes the early debate over continental drift because "theoretical pluralism," degrees of acceptance/rejection, and "problem solving" criteria seem most apparent during this period. However, the current acceptance of plate tectonics theory without a debate over global models creates an "anomaly" for his viewpoint. He tends to minimize this aspect by emphasizing the viability of expansionism as an alternative theory, even though he admits that very few geoscientists even "pursue," much less "accept," expansionism and that all recent textbooks present only plate tectonics as the integrating theory for the geosciences (Le Grand, 1988: 250).

My presuppositions and interests lead me to the opposite choices. The current state of the geosciences provide an excellent example of Kuhn's "normal science," as indicated in Chapter 5. Thus my "anomaly" is the earlier period of debate about continental drift and the transition to the new paradigm. I explain these anomalies by suggesting that prior to plate tectonics the geosciences were in a "multiparadigm" stage of development. That is, I see "*normal science" occurring in different specialties* within a multiparadigm discipline that lacks a useful, global theory, whereas Le Grand sees "*normal sciences" under three different paradigms* that divide the geoscience community (Le Grand, 1988: 34).

The suggestion that specialties marked the major cognitive divisions within the geosciences is consistent with the evidence for "localism" between specialties that Le Grand thoroughly documents, as well as the timing and process of geoscientists' conversions to the new plate tectonics paradigm, and even some of the evidence that Le Grand uses to argue against the usefulness of Kuhn's perspective. For example, he suggests that Kuhn's perspective is inconsistent with the evidence that opponents and proponents could agree on some issues, while those within each group disagreed on other issues (Le Grand, 1988: 269). However, if the geosciences were in a multiparadigm state, then proponents and opponents *within the same specialty* will have consensus on relevant specialty issues, but proponents (or opponents) *in different specialties* will have some dissensus based upon specialty differences.

Even within the same time period, we have different emphases and different assessments of both the typical reasoning processes and the relative merits of specific contributions. He emphasizes the evolving debate among a few geoscientists who actively modified their respective "research traditions" in response to proposed empirical and conceptual problems. I emphasize aspects that he also recognizes (Le Grand, 1988: 76, 92–3): that very few geoscientists entered the debate and those who did so typically had a "parochial" orientation. At best, Le Grand's suggestion that competing paradigms divided the geosciences seems more appropriate for the few who entered the debate, whereas I infer that the vast majority of geoscientists had little concern with how their work related to any global model. This inference should be supported by the examination of typical articles from this time period.

Even if we focus on the few geoscientists who took up the drift debate, we have some differences in our assessments of their reasoning processes. I would emphasize more Le Grand's own statement that the drift debate, ". . . both pro and con, was forensic or rhetorical or persuasive in tone, structure, and content: it did not consist of detached, disinterested comparisons of theory with fact" (Le Grand, 1988: 70). I do so for reasons that we both document and because so few critics of drift changed their minds, even after the contributions of du Toit and Holmes. Thus some of our differences depend upon how we assess these contributions. In particular, I disagree with his assessment of Holmes's later contribution to drift theory. He suggests that Holmes's 1944 model implies that "continents moved through a rigid seafloor" rather than riding with the seafloor as implied by Hess's model (Le Grand, 1988: 199). This is inconsistent with Holmes's 1944 statement: "To sum up: during large-scale convective circulation the basaltic layer becomes a kind of endless travelling belt *on the top of which a continent can be carried along*, until it comes to rest (relative to the belt) when its advancing front reaches the place where the belt turns downwards and disappears into the earth [at trenches]" (my emphasis, from Cox, 1973: 21).

I think that the Holmes's model is close enough to Hess's model that it could have been used as the basis of Vine and Matthews's (1963) explanation of the linear magnetic anomalies in the ocean basins. Hess, of course, built a better case for this model by citing recent oceanographic results, but the basic processes in the two models seem very similar to me.

We also differ in our analytical techniques. I suggest that quantitative analyses may provide useful information, but he is quite hostile about "head-counting" and "number-crunchers" (Le Grand, 1988: 76, 165). Yet the quantitative analysis in the last section of Chapter 7 support his observations on the patterns of who resisted or accepted continental drift before 1950 (Le Grand, 1988: 92, 124, 80, 89, and 219). When he does employ some "head-counting," he tabulates the opinions expressed in textbooks, but does not give significance tests and only uses nationality and publication date to explain the observed differences. Furthermore, I have reservations about how well textbooks reflect what the typical scientist actually believes or uses during research. I think the literature aimed at peers, not students, provides a better data source and this indicates little interest in the debate and probably a tacit assumption of fixed continents.

In addition to our differences in data sources, analytical techniques, and interpretations, we have written our books for somewhat different audiences. With the glaring exception of the quantitative analyses, my book is aimed at the more general reader. I try to give more background and details about the subject of geology, the development and content of the perspectives on science, and the "thinking" processes used by scientists. For example, I use many figures from the scientific literature or create new ones as one means of communicating how scientists "think." This is especially true of the developments in the 1960s, where Le Grand's coverage is quite limited and leaves the naive reader with little understanding of the "cognitive" processes involved in the applications of plate tectonics. I try to explain these "cognitive" processes and then show how "cultural conventions" underlie them. Thus I have sections on (a) the use of models and analogies in science, (b) the typical decisions related to a research project and its influence on others, (c) theories of cognitive processing with illustrations from the geosciences, (d) illustrative quotations from interviews and the literature, and (e) quantitative studies related to how a consensus forms in science.

Finally, we both suggest that the study of science itself is in a state similar to that existing in the geosciences before plate tectonics, but we draw different analogies for future studies of science. He suggests that more systematic use of "multiple working hypotheses" will aid the development of theories about science. Obviously I am sympathetic with this advice because I tried to do this, but I also emphasized the possible utility of more quantitative studies of the "global" properties of science combined with comparisons to similar studies of non-science organizations. If we want to find out what is distinctive about science, then we must systematically compare it to other social "organizations." Thus to continue the analogy I made in Chapter 6 between the history of plate tectonics and future developments in science studies: if we want to understand how (science works)/(the continents were formed), then we should measure (broad quantitative)/(geophysical) properties of (non-science and science "organizations")/(the oceans and continents), rather than focusing on detailed studies of (particular episodes in science)/(continental regions). Since a historian of science, such as Le Grand, would undoubtedly disagree with this particular analogy, this illustrates a point made in Chapter 5: that decisions about what analogies should be drawn from "source models" (the development of plate tectonics in this case) can be the subject of some dispute.

The above comments illustrate how Le Grand and I can look at the "same" phenomena and see different things because we have different "interests," derived in part from differences in our favored "paradigms." These, of course, influence my perceptions of the many positive features of *Drifting Continents and Shifting Theories*. The coverage of the early debate on continental drift theory is the best detailed *description* available: my disagreement is with some of the interpretations. I appreciate his generally sympathetic consideration of sociological perspectives on science, even though he has a tendency to reserve sociological explanations for what the problem solving approach cannot "explain." Finally, in a number of areas where our interpretations are similar, he provides more thorough data and analysis because I wanted to emphasize other subjects. I especially agree with his suggestion that the schematic models or "cartoons" of plate tectonic processes do ". . . not merely *represent* but *constitute* the world for geologists" (his emphasis; Le Grand, 1988: 261).

These few comments should indicate that our books provide complementary perspectives on this exciting revolution in the geosciences. There is clearly room for additional analyses from a wider range of perspectives. Readers are invited to join the "conversation" and help determine what "really" happened in the geosciences.

Notes

1. PERSPECTIVES ON SCIENTIFIC REVOLUTIONS

1. The history of this geoscience revolution is a major subject in numerous books by geoscientists (Hallam, 1973, 1983; Marvin, 1973; Wyllie, 1971, 1976; Uyeda, 1978; Tarling and Tarling, 1971; Takeuchi, Uyeda, and Kanamori, 1967; Van Andel, 1981; Menard, 1969, 1971, 1986; Glen, 1975, 1982; Wood, 1985; Allègre, 1988) or journalists (W. Sullivan, 1974; Wertenbaker, 1974; Horsefield and Stone, 1972; Briggs, 1971; Schlee, 1973; Calder, 1972). Several collected readings exist (Cox, 1973; Wilson, 1972; Bird and Isacks, 1972; Motz, 1975; Bird, 1982). Equally plentiful are historical and philosophical analyses (e.g., Brunnschweiler, 1983; Carozzi, 1970, 1985; Kitts, 1977; R. Laudan, 1980; Giere, 1988; Frankel, 1976, 1978, 1979*a*, 1979*b*, 1980*a*, 1980*b*, 1981*a*, 1981*b*, 1982, 1984, 1987; Le Grand, 1986*a*, 1986*b*; Marvin, 1985; Oreskes, 1988; Ruse, 1981). Comparatively few sociologists have examined aspects of this revolution (Messeri, 1978; Hamilton, 1976; Stewart, 1979, 1983, 1986).

2. Some of the interviewed geoscientists did not respond or are now deceased. In these cases the interview excerpts were cited anonymously if they seemed too controversial.

2. THE RISE AND FALL OF CONTINENTAL DRIFT THEORY

1. Greene (1982) provides a thorough description of changes in geological thought in the 1800s, especially the evolution of theories on mountain building. Hallam (1983) describes several of the controversies in eighteenth century geology.

2. These principles are discussed in most introductory geology textbooks, e.g., Dott and Batten (1981).

3. Burchfield (1975), Faul (1978), and Hallam (1983) provide reviews of the debates about the absolute ages of the earth and the eras and periods in the geological time scale. After their experience with Lord Kelvin, geologists initially distrusted the quantitative evidence based upon the radioactive decay of certain elements, especially since it suggested the earth was billions of years old, instead of the commonly accepted age in the hundreds of millions of years. The current estimated age of the earth is about four and a half billion years.

4. The evolution of the geosyncline model is discussed in Dott (1979), Aubouin (1965), and Greene (1982).

5. The distinction between the two general types of igneous rocks, sial and sima, has undergone considerable refinement since the early 1900s. However, these refinements are not necessary to understand the historical developments and most of this history is written in these terms or the nearly equivalent distinction of granitic and basaltic rocks. The names "sial" and "sima" are derived from the chemical symbols for silicon (Si), aluminum (Al), and manganese (Ma).

6. Exactly who first developed a uniformitarian version of drift theory has been the subject of considerable dispute (cf. Carozzi, 1970; Meyerhoff, 1968; du Toit, 1937; Berkland, 1979).

7. Wegener had a number of other publications, including books on meteorology and paleoclimatology. Marvin (1973: 67) noted the diversity of his interests and his productivity—from two to seven publications every year from 1914 to 1920, a considerable accomplishment since he served in World War I and was wounded twice.

8. The 1929 edition of Wegener's book was organized quite differently from the

1922 edition, which was the basis for the 1924 English edition. The 1929 edition was not translated into English until 1966 (Wegener, 1966).

9. Table 2.1 provides the names and absolute ages for these different geological eras and periods.

10. The figure reproduced in the text is from the 1966 edition (Wegener, 1966) because it provided a better reproduction. The figure in the 1924 edition is identical.

11. Wegener also proposed another mechanism, "polar wandering," to account for the paleoclimatic evidence. Polar wandering holds that the entire crust of the earth can move separately from the interior. Consequently, if the entire crust slipped in the appropriate manner, Pangaea could be moved into the polar region without any relative movement of the continents.

12. In the 1929 edition's presentation of the evidence for drift, Wegener started with the geodetic evidence. Apparently he thought that the most recent measurements of the movement of Greenland away from Europe provided the best evidence for drift, even though they indicated a movement on the order of 36 meters/year. Later, more precise measurements could not detect any movement.

13. Jacoby (1981) has emphasized that Wegener made major changes in his model of the causes of drift between his first presentation of the theory in 1912 (Wegener, 1912*a*) and his first book version in 1915. In his first article Wegener proposed that drift resulted from a form of seafloor spreading, where the "Mid-Atlantic Ridge should be regarded as the zone in the floor of the Atlantic, as it keeps spreading, is continuously tearing open and making space for fresh, relatively fluid and hot sima [rising] from depth." (Jacoby, 1981: 25). Not only did Wegener use this model to explain drift, but he noted that it could explain why the Atlantic was shallower at the Mid-Atlantic Ridge: the newer and hotter sima would be less dense than the older and cooler seafloor further away from the ridge. Both of these suggestions were strongly supported in the 1960s, but Wegener turned to a different explanation for drift in his later writings. This latter explanation will be summarized here because it formed the basis of later discussions of Wegener's theory.

14. The page numbers given by the quotations refer to the pages in *Theory of Continental Drift* (Waterschoot van der Gracht, 1928) without reference to the specific articles in this edited volume.

15. It is interesting to note that the "planetesimal" version of the contraction theory was developed by R. T. Chamberlin's father, T. C. Chamberlin. Although R. T. Chamberlin followed his father in this theory, he failed to follow his father's advice that the best method in geology was one emphasizing "multiple working hypotheses," which was published in a classic paper in 1890 and reprinted a number of times up to 1965 (e.g., T. C. Chamberlin, 1931).

16. Quantitative evidence provided in a later chapter will indicate that more productive geologists expressed stronger opposition to continental drift theory than less productive geologists.

17. Although Holmes's contribution had little effect on the development of drift theory, he did help promote the concept of convection currents. Later in the 1930s David Griggs (1939) produced some experimental models with oil and sawdust to show that convection currents in the oil (mantle) would cause the sawdust (continents) to pile up over the downwelling parts of the convection cells. The Dutch geophysicist F. Vening Meinesz suggested that the downwelling of mantle convection currents explained the ocean trenches, where his gravity surveys showed that the crust was not in isostatic balance. He also endorsed some form of resulting drift. Reviews of these various theories are contained in Marvin (1973), Hallam (1973, 1983), Sullivan (1974), du Toit (1937), and Takeuchi et al. (1967).

3. SPECIALIZATION WITHOUT INTEGRATING THEORY

1. Supplementary information about some of the major researcher is provided in footnotes to help the reader understand their origins, interests, and status. This information will include, when available, the following: name, Ph.D. discipline, year of Ph.D., Ph.D. university, specialty interests, the institutional affiliation of the scientist at the time of the cited publication, and the number of articles cited (for the years in parentheses) and the number of citations to these articles in the *Science Citation Index Cumulative Ten-Year Index: 1955–1964* (Institute for Scientific Information, 1984). A book was counted as equal to 5 articles and citation counts exclude self-citations. The number of cited articles includes only publications in which the scientist was *first author*, so productivity counts may be underestimated for scientists tending to be secondary authors. In addition, occasional tables will give citation counts to specific articles. Citation counts should be used not as absolute measures of importance, but for *relative* comparisons over time and between articles or authors.

2. Menard (1969) and Ericson and Wollin (1964) provide descriptions of these basic oceanographic techniques, the information they give, and the practical problems associated with their use.

3. Bruce Heezen / geology / 1957 / Columbia University / submarine geology, tectonics, and seismicity / Lamont-Doherty Geological Observatory associated with Columbia University /(1957–58) 5 / 9. (See note 1 in this chapter for a description of the contents in these informational notes about specific scientists.)

4. Marie Tharp / geology / 1945 (MA) / University of Michigan / marine geology / Lamont-Doherty / (1957–58) 0 / 0.

5. Maurice Ewing / physics / 1931 / Rice Institute / geophysics, seismology, and submarine geology / Lamont-Doherty and Columbia University / (1955–56) 33 / 60. Wertenbaker (1974) provides an exciting account of Ewing's career and his role in establishing Lamont Geological Observatory.

6. Menard (1986: 64) notes that the "bird's eye view" pictures of the ocean basins were used instead of the preferred contour maps because the Navy classified the exact soundings data. He also noted that soundings were only available along specific courses, so Heezen had used his imagination to fill in the numerous voids (Menard, 1986: 235).

7. H. William Menard / geology / 1949 / Harvard University / marine geology / Scripps Institute of Oceanography / (1955–56) 5 / 5.

8. Walker (1973) and Ericson and Wollin (1964) provide good summaries of the history of the turbidity current hypothesis.

9. Edward Bullard / natural science / 1932 / Cambridge University / geophysics, marine geology, and the earth's magnetic field / University of Toronto / (1955–56) 33 / 72.

10. Ronald Mason / geophysics / 1951 / University of London / geophysics, crustal and upper mantle structure / Imperial College of London and visiting professor at Scripps / (1957–58) 0 / 0.

11. Hugo Benioff / seismology / 1935 / California Institute of Technology / geophysics, seismology, and tectonics / California Institute of Technology / (1955–56) 7 / 11.

12. Mountains can be formed by other mechanisms, such as volcanism, vertical uplift, or "block faulting," where large blocks of the earth's crust are tilted during extension of the crust.

13. The term "polar wandering" had vastly different meanings for the paleomagneticists than for Wegener. Wegener used the term to mean that the whole crust of the earth could slide around on the mantle—see note 11 in Chapter 2. Consequently, a continent embedded in the crust would change its relationship to the magnetic and geographic poles. Wegener suggested that this process and continental drift might ac-

count for the paleoclimatic evidence. For the paleomagneticists polar wandering meant that only the magnetic poles moved; the continents were fixed with respect to the geographic poles.

14. Patrick Blackett / physics / M.A. 1923 / Cambridge University / nuclear-atomic physics and rock magnetism / Imperial College of London / (1955–56) 16 / 35.

15. Stanley Keith Runcorn / physics / 1949 / University of Manchester / geophysics / University of Newcastle / (1955–56) 11 / 27.

16. Polar wandering curves were the standard way to represent the results of paleomagnetic studies. A polar wandering curve for a continent was constructed by studying rocks of different ages from the continent. For each age the apparent position of the magnetic pole was determined from the natural remanent magnetism of the rocks of that age and the calculated pole positions were connected together to form the polar wandering curve for the continent. Despite its label, this data representation method did not imply that polar wandering was the correct interpretation of the data.

17. Paleomagneticists doing directional studies occasionally found rock samples with reversed declinations, but this was simply reversed and plotted with their other data points. Their major interest was in the measured inclinations.

18. Richard Doell / geophysics / 1955 / geophysics and paleomagnetism / Massachusetts Institute of Technology / (1959–60) 5 / 10.

19. Allan Cox / geophysics / 1959 / University of California–Berkeley / geophysics, paleomagnetism, and geomagnetism / U.S. Geological Survey / (1959–60) 3 / 7.

20. S. Warren Carey / geology / 1938 / University of Sidney / ? / University of Tasmania / (1957–58) 4 / 5.

21. Lester King / geology / early 1930s / University of New Zealand / geology and geomorphology / University of Natal, South Africa / (1955–56) 10 / 15.

22. Edward Irving / natural science / 1954 / Cambridge University / paleomagnetism / Australian National University / (1956–57) 5 / 18.

23. J. Tuzo Wilson / geology / 1936 / Princeton University / geophysics and tectonics / University of Toronto / (1955–56) 19 / 27.

4. PLATE TECTONICS

1. Robert Dietz / geology / 1941 / University of Illinois / geology, oceanography, and marine geology / U.S. Coast and Geodetic Survey / (1961–62) 11 / 19. (The organization of the information in these footnotes on specific scientists is described in note 1 in Chapter 3.)

2. Harry Hess / geology / 1932 / Princeton University / geology, mineralogy-petrology, and submarine geology / Princeton University / (1961–62) 24 / 67.

3. In my opinion, the Dietz (1961) article is a good supplement to Hess's (1962) article because it relates more data to the seafloor spreading model and more persuasively leads the reader through the necessary reinterpretations of old data and ideas, which must be done before their support of seafloor spreading is apparent. He also suggested that the linear magnetic anomalies might be due to linear "stress" patterns caused by seafloor spreading. Menard (1986) also notes some of the superior features of Dietz's presentation.

4. It is interesting to compare Holmes's convection current model in Figure 2.3 with Hess's model. They are quite similar and do not support Hess's (1968) later assertion that Holmes had only described a process of "seafloor thinning." Furthermore, Holmes had a non-convecting surface layer in his model, which was added later to Hess's model.

5. G. Brent Dalrymple / geology / 1963 / University of California–Berkeley / geology, isotope geology, geochronology, and origins of volcanic islands / U.S. Geological Survey / (1963–64) 2 / 9.

6. Ian McDougall / geology / 1960 / Australian National University / petrology and geochronology / Australian National University / (1962–63) 6 / 6.

7. Glen (1982) provides very detailed histories of research in the potassium-argon dating method and magnetic reversals, as well as some of the less "friendly" aspects of the competition between the USGS and Australian groups.

8. Fred Vine / marine geology / 1965 / Cambridge University / geology, geophysics, paleomagnetism, and plate tectonics / graduate student at Cambridge / (1962–63) 0 / 0.

9. Drummond Matthews / geology / 1961 / University of Cambridge / marine geology / University of Cambridge / (1962–63) 4 / 4.

10. The Vine-Matthews hypothesis was anticipated by Lawrence Morely. A later chapter will discuss the problems Morely had in getting his ideas published. Glen (1982) and Frankel (1982) describe some of the earlier research that influenced Vine and Matthews.

11. I use "arbitrary" here because these depressions in the seafloor are omitted from Vine's 1966 representation of the Juan de Fuca Ridge. They seemed to have been added to improve the apparent fit of the predicted and observed profiles.

12. James Heirtzler / physics / 1953 / New York University / geophysics and geomagnetism / Lamont-Doherty and Columbia University / (1964–65) 12 / 15.

13. Xavier LePichon / ? / 1966 / University of Paris / ? / Lamont-Doherty Observatory / (1964–65) 1 / 5.

14. Walter Pitman / geophysics / 1967 / Columbia University / marine geophysics / graduate student at Columbia University / (1965–66) 1 / 2.

15. Harrison and Funnell (1964) had reported evidence for reversals in deep sea cores earlier. Menard (1986: 71) mentioned that Ron Mason had found reversals in sediments in the middle 1950s, but had been discouraged from publishing his results. Wertenbaker (1974: 200) and Glen (1982) give accounts of the nearly accidental manner in which Opdyke's group discovered reversals in the sea cores. Glen also describes the conflict between Opdyke and Bruce Heezen when Heezen tried to get into this research area.

16. Neil Opdyke / geology / 1958 / University of Durham / geology, geophysics, and paleomagnetism / (1965–66) 8 / 25.

17. It should be noted that geoscientists could accept continental drift without believing in the seafloor spreading mechanism or they could accept seafloor spreading without accepting the Vine-Matthews hypothesis.

18. Marshall Kay / geology / 1929 / Columbia University / stratigraphy and geosynclines / Columbia University / (1965–66) 10 / 8.

19. John Dewey / geology / 1960 / University of London / structural geology, plate tectonics (continental margins and orogenic belts) / State University of New York–Albany / (1965–66) 6 / 8.

20. Lynn Sykes / geology / 1964 / Columbia University / geophysics, seismology, and tectonics / Lamont-Doherty and Columbia University / (1965–66) 10 / 26.

21. Patrick Hurley / economic geology / 1940 / geochronology / Massachusetts Institute of Technology / (1965–66) 25 / 74.

22. Hurley and Rand's (1968) age dating evidence was not the first age comparison between the southern continents. A similar figure is given by du Toit (1937: 108), but his data were not based upon radiometric dating.

23. Daniel McKenzie / geophysics / 1965 / Cambridge University / geophysics, mantle convection, and plate tectonics / King's College / (1966–67) 2 / 0.

24. Jack Oliver / geophysics / 1953 / geophysics, seismology, and geotectonics / Lamont-Doherty and Columbia University / (1966–67) 18 / 28.

25. Bryan Isacks / seismology / 1965 / Columbia University / seismology / Lamont-Doherty and Columbia University / (1966–67) 3 / 10.

26. Morgan's 1968 article was actually submitted for publication before the 1967 article by McKenzie and Parker and he had presented his ideas earlier at a 1967 conference.

27. W. Jason Morgan / physics / 1964 / Princeton University / geophysics, mantle convection, heat flow, and plate tectonics / Princeton University/ (1967–68) 2 / 10.

28. Tanya Atwater / earth science / 1972 / University of California–San Diego / marine geophysics and paleomagnetism / graduate student at the University of California–San Diego / (1969–70) 1 / 7.

29. John Sclater / geophysics / 1966 / Cambridge University / geophysics and oceanography / University of California–San Diego and Scripps / (1969–70) 8 / 19.

30. Jean Francheteau / geophysics / 1970 / University of California–San Diego / marine geology / Scripps / (1969–70) 4 / 5.

31. William Dickinson / geology / 1958 / Stanford University / geology, petrology, structural geology, and plate tectonics / Stanford University / (1966–67) 8 / 36.

32. Trevor Hatherton / geology / 1954 / University of London / Department of Science and Industrial Research, New Zealand / (1966–67) 19 / 21.

33. John Bird / geology / 1962 / Rensselaer Polytechnic Institute / geology, geotectonics, Appalachian geology / State University of New York–Albany / (1968–69) 4 / 4.

34. Howard Meyerhoff / geology / 1935 / Columbia University / geomorphology and tectonics / University of Pennsylvania / (1969–70) 3 / 4.

35. Arthur Meyerhoff / geology / 1952 / Stanford / geology, structural geology, and geotectonics / Publication Manager with American Association of Petroleum Geologists and Director of Tulsa Science Foundation / (1969–70) 12 / 19.

36. Vladimir Beloussov / geology / about 1937 (unknown degree) / ? / tectonics and tectonophysics / University of Moscow / (1968–69) 23 / 62.

37. Carey (1976) basically argued that the amount of lithosphere consumption in trenches was not adequate for the amount of lithosphere creation at ocean ridges. Consequently, the earth must be expanding. His basic theory was discussed in Chapter 3, but more recent advocates of an expanding earth include Owen (1976) and King (1983).

38. Hall and Robinson (1979) actually measured the magnetic properties of the oceanic crust and find that the simple model of Vine (1966) for these properties appears inconsistent with the observed properties.

5. PHILOSOPHICAL AND HISTORICAL PERSPECTIVES

1. Harré (1986) gives a slightly different discussion of models and analogies than his briefer description in Harré (1976), but the essential points seem similar. It should be mentioned that Harré's "realism" is not a "truth" realism with the implication that scientific theories are "true" descriptions of the real world. Rather he suggests that the *entities or processes* indicated by a theory actually exist, but their properties and behaviors may be less perfectly known. Giere (1988) develops a very similar approach.

2. More recent ocean drilling deep into the basaltic basement rock of the ocean floor has allowed "direct" observation of the magnetic properties of the floor (Hall and Robinson, 1979). The results appear incompatible with the simple seafloor model proposed by Vine and Matthews (1963) or Vine (1966).

3. A number of artificial and natural, mechanical models for plate tectonic processes have appeared in the geoscience literature. Oldenburg and Brune (1972) illustrate the ridge-ridge transform fault concept on the hardening surface of melted wax. Tapponnier *et al.* (1982) use the forced impact of plasticine figures to model the internal fault patterns created by the collision of India with Asia. Finally, Duffield (1972) reported that the floating crust in a naturally occurring lava pool directly models such plate tectonic processes as subduction, spreading at ridges, transform faults, and triple junctions. Wood (1985: 199) notes that over the past hundred years various geoscientists had reported on these characteristics of lava pools without realizing that the crust of the earth could behave in an analogous fashion.

6. SOCIAL PERSEPCTIVES ON DECISION-MAKING IN SCIENCE

1. Hunt (1982) and Mayer (1981) provide a layman's introduction to recent research in artificial intelligence and cognitive psychology. Anderson (1980) and Newell and Simon (1972) provide more technical introductions, whereas Minsky (1985) gives a speculative outline of how the mind is a "community of experts." To those readers influenced by arguments presented by Nobel Prize winners, it may interest them to know that Herbert Simon has won this prize for his research on human decision-making and problem solving.

2. Hargens (1975) and R. Collins (1975) use organizational perspectives in their studies of science.

3. The literature on decision-making is immense and contains many promising analogies that should help us understand the behavior of scientists. The few attempts at applying this literature to scientists generally have a psychological, rather than a sociological, focus (Mahoney, 1976; Grover, 1981; Faust, 1984). Giere (1988) incorporates elements from organizational studies of decision-making.

4. Using a simple count of citations to an article as a measure of its influence on other researchers has a number of problems. First, self-citations should be excluded. Second, "negative" citations do occur, but they are rather infrequent. Chubin and Moitra (1975) find only about 6 percent of the citations in their study are negative. Most articles are simply ignored. For example, Menard reports that of the 3,078 publications in the *Bulletin of the Geological Society of America* from 1888 to 1969, 78 percent were not cited and 2 percent received 65 percent of the citations. Allison and Stewart (1974) show that such inequality is typical of several disciplines.

5. Those advocating an interests approach to science would suggest that the probability that a scientist would read or cite an article is determined partly by the scientist's "interests." This could be incorporated into the iconic model, but we shall use the interests approach in a more simple manner later in this section. A more sophisticated analysis would examine "citation linkages" between articles, so that one could simultaneously examine the effects of resources and interests of both the "speakers" and "listeners." This would require complex statistical procedures, but Schoot (1987) illustrates how this might be done.

6. A multiplicative function for the probabilities is used for several reasons. First, it provides a convenient decompositions when logged. Second, probabilities are constrained to have values between one and zero. This is more easily accomplished with a multiplicative function than with a linear function. Third, a multiplicative function of many independent random variables will produce a lognormal distribution (Aitchison and Brown, 1957), which is consistent with studies indicating that the distribution of citations to articles is lognormal (Stewart, 1979, 1987). Finally, Stewart (1983) reports empirical evidence that citations are modeled best as a multiplicative function of intellectual and social factors.

7. Tannenbaum (1968) and Tannenbaum and Cooke (1979) find that the amount of area *under* the control graph curve, which they call the "amount of control," correlates with other organizational properties. However, they are using only one variable, level in the organization, to predict influence. Since this area is a simple mathematical function of the regression intercept and the slope (centralization score), they might find quite different results when influence is seen as the result of *several* interrelated variables. When only one of such a set of variables is used, its coefficient may be seriously biased (Maddala, 1977).

8. Most studies of the structural dimensions of organizations measure the dimensions in a sample of different organizations and correlate these dimensions across the organizations studied. If the regression approach outlined here is used, one would first do a regression analysis *within* each organization and then study how the within-organization

regression coefficients were correlated *across* organizations. This presents some statistical problems that require the use of the "random-coefficient model" (Swamy, 1971) or other related techniques (Mason, Wong, and Entwisle, 1983).

7. QUANTITATIVE STUDIES

1. Stewart, Hagstrom, and Small (1981) and Stewart (1979) provide more detailed descriptions of the procedures used in this cocitation analysis.

2. Table 7.2 only describes the variables that were significant predictors in the analyses reported in this chapter. Numerous other variables were available predictors, but were not statistically significant. The capitalized variables in the complete list below are described in Table 7.2: INFLUENCE; a measure of how much the article was distributed in preprint form (from the JHU data); RECENT REFS; PUBDELAY; NUM REFS; logged counts of the number of equations and theoretical and empirical figures; LENGTH; dummy variables for journals; subjective assessments of the degree the article emphasizes theory or a review of the literature; ACCEPTS DRIFT; PLATE TECTONICS; PETROL/GEOCHEM; EMPIRICAL; FIELD STUDY; dummy variables for whether the article used laboratory data or reported on instrumental developments; dummy variables for use of geomagnetic data, geophysical data (GEOPHY DATA), geology data, geochemical data, and paleomagnetic data (PALEOMAG DATA); OCEANIC TOPICS; and a dummy variable for studies of the earth's interior. The following are author-related variables and are not listed in any causal order: DEPT QUALITY; DEGREE QUALITY; log of aggregate number of pre-1967 publications cited in the *1965–1969 Science Citation Index* (SCI); log of aggregate number of total citations to these pre-1967 publications; AVE INFLU PRE-1967; the log of aggregate number of 1967–68 publications listed in the *1965–1969 SCI Source Index*; CITES 1967–68 PUBS; log of aggregate average influence of 1967–68 publications (the latter three variables excluded counts for the article in this sample); the number of authors; the average professional age of the authors; NUM CD/PT PUBS; PETROL SPECIALTY; JAPANESE; UNIVERSITY; MARINE SPECIALTY; and the percentage of the authors with listed specialties in geomagnetics or seismology, with present affiliations in North America, the United Kingdom, Europe, or Australia and New Zealand, with industrial or government positions, and with missing data on selected author characteristics.

3. Table 7.2 and Note 2 list variables according to an assumed causal order. Some aspects of this order seem reasonable. For example, citations only occur *after* publication. Similarly, author variables are based upon the characteristics of the scientists *at or before* the time of publication. More problematic is the causal order assumed *within* the set of article-content variables, which constrains the possible predictors of opinions on plate tectonics (ACCEPTS DRIFT). Although the causal ordering among article variables might be questioned, it would not affect the prediction of citations in this section.

4. Stewart (1979, 1983) provides more details about analytical procedures, methodological problems, and their solutions.

5. Stewart (1983) illustrates how to interpret the information provided by the *raw* regression coefficients.

6. Du Toit (1937: 35–36) explicitly ranks over 45 scientists in their degree of support for drift theory. A number of the scientists ranked by du Toit and Wegener were ranked independently by reading the relevant articles. The correlation between these two ranks was about .96.

7. Since article characteristics of these publications were not measured, except for the degree of support for drift theory, this study can be seen as either a sample of articles or a sample of scientists based upon their first stated opinions about drift.

8. The *Bibliography and Index of Geology, Exclusive of North America* did not start until 1934. Corresponding publication counts for the period from 1895 to 1933 came from the *List of Geological Literature Added to the Geological Society Library* (Geological Society

of London, 1895–1935). All counts included coauthored publications and books were counted as equivalent to five articles.

9. Previous analyses with this data set employed slightly different methods and variables. R. Laudan's (1987) critique of the initial analysis (Stewart, 1986) prompted separate analyses for North Americans and "others" by Stewart (1987). However, the latter analysis defined "North Americans" as any scientist with training *or* research experience in North America, which was too broad of a definition. The present analysis limits "North Americans" to those born there, which is a much smaller group. In addition, total publications was logged in this analysis and the SWISS dummy variable was included. Despite these changes, the substantive conclusions in the present analysis are almost identical to those given in both of the earlier analyses.

POSTSCRIPT

1. I might note that in 1986 I sent my manuscript to Cambridge University Press, which published Le Grand's book two years later. Many of their reviewer's comments match unique themes and data used by Le Grand. Although I am convinced that he was the reviewer, I can only speculate on how, if at all, it influenced his own book or his review of the manuscript. Regardless of who wrote the review, I am thankful for the comments and incorporated many of them in the final revisions for this book.

2. For obvious reasons readers should suspect that I have a "vested interest" in this conclusion. After reading my justifications, I encourage the reading of Le Grand's book so that readers can reach their own conclusions.

Bibliography

Aitchison, J., and J. Brown, 1957, *The Lognormal Distribution*, Cambridge: Cambridge University Press.

Allègre, C., 1988, *The Behavior of the Earth*, Cambridge, Massachusetts: Harvard University Press.

Allison, P., 1978, "The Reliability of Variables Measured as the Number of Events in an Interval of Time," in K. Schussler (ed.), *Sociological Methodology 1978*, San Francisco: Jossey-Bass, pp. 238–253.

Allison, P., and J. Stewart, 1974, "Productivity Differences Among Scientists: Evidence for Accumulative Advantage," *American Sociological Review*, 39, pp. 596–606.

Alwin, D., 1988, "Measurement and Interpretation of Effects in Structural Equation Models," in J. Long (ed.), *Common Problems/Proper Solutions: Avoiding Errors in Quantitative Research*, Beverly Hills, California: Sage, pp. 15–45.

Anderson, J., 1980, *Cognitive Psychology and its Implications*, San Francisco: W. H. Freeman.

Atwater, T., 1970, "Implications of Plate Tectonics for the Cenozoic Tectonics Evolution of Western North America," *Geological Society of America Bulletin*, 81, pp. 3513–3536.

Aubouim, J., 1965, *Geosynclines*, Amsterdam: Elsevier.

Barazangi, M., and J. Dorman, 1969, "World Seismicity Map of the E.S.S.A. Coast and Geodectic Survey Epicenter Data for 1961–1967," *Seismological Society of America Bulletin*, 59, pp. 369–380.

Barber, B., 1952, *Science and the Social Order*, New York: Collier.

Barnes, B., 1974, *Scientific Knowledge and Sociological Theory*, London: Routledge and Kegan Paul.

———, 1979, "Vicissitudes of Belief," *Social Studies of Science*, 9, pp. 247–263.

———, 1982, *T. S. Kuhn and Social Science*, New York: Columbia University Press.

Barnes, B., and R. Dolby, 1970, "The Scientific Ethos: A Deviant Viewpoint," *Archives of European Sociology*, 11, pp. 3–25.

Beloussov, V., 1968, "An Open Letter to J. Tuzo Wilson," *Geotimes*, (Dec.), pp. 17–19.

———, 1970, "Against the Hypothesis of Ocean-Floor Spreading," *Tectonophysics*, 9, pp. 489–511.

———, 1974, "Seafloor Spreading and Geologic Reality," in C. Kahle (ed.), *Plate Tectonics: Assessments and Reassessments*, Tulsa, Oklahoma: American Association of Petroleum Geologists Memoir 23, pp. 155–166.

———, 1979, "Why I Do Not Accept Plate Tectonics," *EOS*, (April 24), pp. 207–211.

Ben-Avraham, Z., 1981, "The Movement of Continents," *American Scientist*, 69, pp. 291–299.

Ben-Avraham, Z., A. Nur, D. Jones, and A. Cox, 1981, "Continental Accretion: From Oceanic Plateaus to Allochthonous Terranes," *Science*, 213, pp. 47–54.

Ben-David, J., 1971, *The Scientist's Role in Society: A Comparative Study*, Englewood Cliffs, New Jersey: Prentice-Hall.

Benioff, H., 1954, "Orogenesis and Deep Crustal Structure: Additional Evidence from Seismology," *Geological Society of America Bulletin*, 65, pp. 385–400.

Berkland, J., 1979, "Elisee Reclus—Neglected Geological Pioneer and First (?) Continental Drift Advocate," *Geology*, 7, pp. 189–192.

Bird, J., and B. Isacks (eds.), 1972, *Plate Tectonics: Selected Readings from the Journal of Geophysical Research*, Washington, D.C.: American Geophysical Union.

———, 1982, *Plate Tectonics: Selected Papers from Publications of the American Geophysical Union* (second enlarged edition), Washington, D.C.: American Geophysical Union.

Blackett, P., 1965, "Introduction," in P. Blackett, E. Bullard, and S. Runcorn (eds.), *A Symposium on Continental Drift*, London: The Royal Society, pp. 1–4.

Blackett, P., E. Bullard, and S. Runcorn (eds.), 1965, *A Symposium on Continental Drift*, London: The Royal Society.

Bloor, D., 1976, *Knowledge and Social Imagery*, London: Routledge and Kegan Paul.

———, 1978, "Polyhedra and the Abomination of Leviticus," *The British Journal for the History of Science*, 11, pp. 245–272.

———, 1981, "The Strength of the Strong Programme," *Philosophy of the Social Sciences*, 11, pp. 199–213.

Bonatti, E., and J. Honnorez, 1971, "Nonspreading Crustal Blocks at the Mid-Atlantic Ridge," *Science*, 174, pp. 1329–1331.

Bradshaw, G., P. Langley, and H. Simon, 1983, "Studying Scientific Discovery by Computer Simulation," *Science*, 222, pp. 971–975.

Briggs, P., 1971, *200,000,000 Years Beneath the Sea*, New York: Holt, Reinhart, and Winston.

Brown, H., 1977, *Perception, Theory, and Commitment: The New Philosophy of Science*, Chicago: University of Chicago Press.

Brown, J., 1984, *Scientific Rationality: The Sociological Turn*, Dordrecht: D. Reidel.

Brunnschweiler, R., 1983, "Evolution of Geotectonic Concepts in the Past Century," in S. Carey (ed.), *Expanding Earth Symposium*, Tasmania: University of Tasmania, pp. 9–15.

Bullard, E., 1969, "The Origin of the Oceans," *Scientific American*, 221 (Sept.), pp. 66–75.

———, 1975*a*, "The Emergence of Plate Tectonics: A Personal View," *Annual Review of Earth and Planetary Sciences*, 3, pp. 1–30.

———, 1975*b*, "The Effect of World War II on the Development of Knowledge in the Physical Sciences," *Royal Society of London Proceedings*, A–342, pp. 519–536.

Bullard, E., J. Everett, and A. Smith, 1965, "The Fit of the Continents Around the Atlantic," *Philosophical Transactions of the Royal Society of London*, A–258, pp. 41–51.

Bullard, E., A. Maxwell, and R. Revelle, 1956, "Heat Flow Through the Deep Sea Floor," *Advances in Geophysics*, 3, pp. 153–181.

Burchfield, J., 1975, *Lord Kelvin and the Age of the Earth*, London: MacMillan.

Calder, N., 1972, *The Restless Earth*, New York: Viking.

Callon, M., and J. Law, 1982, "On Interests and Their Transformation: Enrollment and Counter-Enrollment," *Social Studies of Science*, 12, pp. 615–625.

Campbell, A., V. Hollister, R. Duda, and P. Hart, 1982, "Recognition of a Hidden Mineral Deposit by an Artificial Intelligence Program," *Science*, 217, pp. 927–929.

Carey, S. (ed.), 1958*a*, *Continental Drift, A Symposium*, Hobart, Tasmania: University of Tasmania Geology Department.

———, 1958*b*, "A Tectonic Approach to Continental Drift," in S. Carey (ed.), *Continental Drift, A Symposium*. Hobart, Tasmania: University of Tasmania Geology Department, pp. 177–355.

———, 1976, *The Expanding Earth*, New York: Elsevier.

Carozzi, A., 1970, "New Historical Data on the Origin of the Theory of Continental Drift," *Geological Society of America Bulletin*, 81, pp. 283–286.

———, 1985, "The Reaction in Continental Europe to Wegener's Theory of Continental Drift," *Earth Science History*, 4, pp. 122–137.

Chamberlin, T., 1931, "The Method of Multiple Working Hypotheses," *Journal of Geology*, 5, pp. 837–848.

Child, J., 1973, "Predicting and Understanding Organization Structure," *Administrative Science Quarterly*, 18, pp. 168–185.

Chubin, D., and S. Moitra, 1975, "Content Analysis of References: Adjunct or Alternative to Citation Counting," *Social Studies of Science*, 5, pp. 423–441.

Clegg, J., M. Almond, and P. Stubbs, 1954, "The Remanent Magnetism of Sedimentary Rocks in Britain," *Philosophy Magazine*, 45, pp. 583–598.

Cole, J., 1979, *Fair Science*, New York: The Free Press.

Cole, J., and S. Cole, 1973, *Social Stratification in Science*, Chicago: University of Chicago Press.
Cole, S., 1978, "Scientific Reward Systems: A Comparative Analysis," *Research in the Sociology of Knowledge, Sciences, and Art*, 1, pp. 167–190.
Cole, S., J. Cole, and G. Simon, 1981, "Chance and Consensus in Peer Review," *Science*, 214, pp. 881–886.
Coleman, R., 1971, "Plate Tectonic Emplacement of Upper Mantle Peridotites Along Continental Edges," *Journal of Geophysical Research*, 76, pp. 1212–1222.
Collins, H., 1975, "The Seven Sexes: A Study in the Sociology of a Phenomenon or the Replication of an Experiment in Physics," *Sociology*, 9, pp. 205–224.
———, 1981, "Stages in the Empirical Program of Relativism," *Social Studies of Science*, 11, pp. 3–10.
———, 1982, "Knowledge, Norms and Rules in the Sociology of Science," *Social Studies of Science*, 12, pp. 299–309.
———, 1985, *Changing Order*, London: Sage.
Collins, R., 1975, *Conflict Sociology*. New York: Academic.
Coney, P., 1970, "The Geotectonic Cycle and the New Global Tectonics," *Geological Society of America Bulletin*, 81, pp. 739–748.
Cox, A. (ed.), 1973, *Plate Tectonics and Geomagnetic Reversals*, San Francisco: W. H. Freeman.
Cox, A., and R. Doell, 1960, "Review of Paleomagnetism," *Geological Society of America Bulletin*, 71, pp. 645–768.
Craw, R., 1984, "'Conservative Prejudice' in the Debate over Disjunctively Distributed Life Forms," *Studies in History and Philosophy of Science*, 15, pp.131–140.
Crombie, W., 1980, "The new oceanography," *Mosaic*, 11, pp. 2–7.
Dalrymple, G., 1972, "Potassium-Argon Dating of Geomagnetic Reversals and North American Glaciations," in W. Bishop and J. Miller (eds.), *Calibration of Hominid Evolution*, Edinburgh: Scottish Academic Press, pp. 107–134.
Daly, R., 1926, *Our Mobile Earth*, New York: Charles Scribner.
Dean, J., 1979, "Controversy Over Classification: A Case Study from the History of Biology," in B. Barnes and S. Shapin (eds.), *Natural Order*, Beverly Hills, California: Sage, pp. 211–230.
DeMey, M., 1980, "The Interaction Between Theory and Data in Science," in K. Knorr, R. Kohn, and R. Whitley (eds.), *The Social Process of Scientific Investigation*, Dordrecht, Holland: D. Reidel, pp. 3–23.
———, 1982, *The Cognitive Paradigm*, Dordrecht, Holland: D. Reidel.
Dewey, J., 1982, "Plate Tectonics and the Evolution of the British Isles," *Journal of Geological Society of London*, 139, pp. 371–412.
Dewey, J., and J. Bird, 1970, "Mountain Belts and the New Global Tectonics," *Journal of Geophysical Research*, 75, pp. 2625–2647.
———, 1971, "Origin and Emplacement of the Ophiolite Suite: Appalachian Ophiolites in Newfoundland," *Journal of Geophysical Research*, 76, pp. 3179–3206.
Dickinson, W., 1970, "The New Global Tectonics," *Geotimes*, (April), pp. 18–22.
Dickinson, W., and T. Hatherton, 1967, "Andesite Volcanism and Seismicity Around the Pacific Basin," *Science*, 157, pp. 801–803.
Dietz, R., 1961, "Continent and Ocean Basin Evolution by Spreading of the Sea Floor," *Nature*, 190, pp. 854–857.
———, 1968, "Reply." *Journal of Geophysical Research*, 73, p. 6567.
Doppelt, G., 1983, "Relativism and Recent Pragmatic Conceptions of Scientific Rationality," in N. Rescher (ed.), *Scientific Explanation and Understanding: Essays on Reasoning and Rationality in Science*, Boston: University Press of America, pp. 107–142.
Dott, R., 1974, "The Geosynclinal Concept," in R. Dott and R. Shaver (eds.), *Modern*

and Ancient Geosynclinal Sedimentation, Society of Economic Paleontologists and Mineralogists Special Publication No. 19, pp. 1–13.
———, 1979, "The Geosyncline—First Major Geological Concept 'Made in America.' " in C. Schneer (ed.), *Two Hundred Years of Geology in America*, Hanover, New Hampshire: University Press of America, pp. 239–264.
———, 1981," The North American Rejection of Wegener's Drift Theory." Presented at the American Association for the Advancement of Science 1981 annual meeting. Dott. R., and R. Batten, 1981, *Evolution of the Earth*, New York: McGraw-Hill.
Duda, R., and E. Shortliffe, 1983, "Expert Systems Research," *Science*, 220, pp. 261–268.
Duffield, W., 1972, "A Naturally Occurring Model of Global Plate Tectonics," *Journal of Geophysical Research*, 77, pp. 2543–2555.
Duncan, O., 1970, "Partials, Partitions, and Paths," in E. Borgatta and G. Bohrstedt (eds.), *Sociological Methodology 1970*, San Francisco: Jossey-Bass, pp. 38–47.
———, 1975, *Introduction to Structural Equation Models*, New York: Academic.
du Toit, A., 1927, *A Geological Comparison of South America with South Africa*, Washington, D.C.: Carnegie Institute of Washington, Publication No. 381.
———, 1937, *Our Wandering Continents*. London: Oliver and Boyd.
Dziewonski, A., and D. Anderson, 1984, "Seismic Tomography of the Earth's Interior," *American Scientist*, 72, pp. 483–494.
Engel, A., C. Engel, and R. Havens, 1965, "Chemical Characteristics of Oceanic Basalts and the Upper Mantle," *Geological Society of America Bulletin*, 76, pp. 719–734.
Ericson, D., and G. Wollin, 1964, *The Deep and the Past*, New York: Grosset and Dunlap.
Ewing, J., and M. Ewing, 1967, "Sediment Distribution on the Mid-ocean Ridges With Respect to Spreading of the Sea-Floor," *Science*, 156, pp. 1590–1592.
Ewing, J., and F. Press, 1955, "Geophysical Contrasts Between Continents and Ocean Basins," in A. Poldervaart (ed.), *Crust of the Earth*, Geological Society of American Special Paper 62.
Faul, H., 1978, "A History of Geologic Time," *American Scientist*, 66, pp. 159–165.
Faust, D., 1984, *The Limits of Scientific Reasoning*, Minneapolis, Minnesota: University of Minnesota Press.
Fisher, R., and H. Hess, 1963, "Trenches," in M. Hill (ed.), *The Sea*, Vol. 3. New York: Wiley-Interscience, pp. 431–436.
Fowler, P., 1972, "'Logic' vs 'truth.'" *Geotimes*, (July), pp. 11–12.
Frankel, H., 1976, "Alfred Wegener and the Specialists," *Cenaurus*, 20, pp. 305–324.
———, 1979*a*, "The Career of Continental Drift Theory: An Application of Imre Lakatos' Analysis of Scientific Growth to the Rise of Drift Theory," *Studies in the History and Philosophy of Science*, 10, pp. 21–66.
———, 1979*b*, "The Reception and Acceptance of Continental Drift Theory as a Rational Episode in the History of Science," in S. Mauskopf (ed.), *The Reception of Unconventional Science*. Boulder, Colorado: Westview, pp. 51–89.
———, 1980*a*, "Hess's Development of His Seafloor Spreading Hypothesis," in T. Nickles (ed.), *Scientific Discovery: Case Studies*, Boston: D. Reidel, pp. 345–366.
———, 1980*b*, "Problem-solving, Research Traditions, and the Development of Scientific Fields," *PSA 1980*, 1, pp. 29–40.
———, 1981*a*, "The Paleobiogeographical Debate over Disjunctively Distributed Life Forms," *Studies in History and Philosophy of Science*, 12, pp. 221–259.
———, 1981*b*, "The Non-Kuhnian Nature of the Recent Revolution in the Earth Sciences," *PSA 1978*, 2, pp. 197–214.
———, 1982, "The Development, Reception, and Acceptance of the Vine-Matthews-Morley Hypothesis," *Historical Studies in the Physical Sciences*, 13, pp. 1–39.
———, 1983, "Review of *The Road to Jaramillo*," *EOS* (May), pp. 394–396.
———, 1984, "Biogeography, before and after the Rise of Sea Floor Spreading," *Studies in History and Philosophy of Science*, 15, pp. 141–168.

———, 1987, "The Continental Drift Debate," in H. Engelhardt and A. Caplan (eds.), *Scientific Controversies*, Cambridge: Cambridge University Press, pp. 203–248.

Ganascia, J., 1986, "Using an Expert System in Merging Qualitative and Quantitative Data Analysis," in J. Kowalik (ed.), *Knowledge Based Problem Solving*, Englewood Cliffs, New Jersey: Prentice-Hall, pp. 166–182.

Garvey, W., 1973, "Machine-Readable Data Bank on the Communication Behavior of Scientists and Technologists," *Science Studies*, 3, pp. 91–92.

Garvey, W., N. Lin, and C. Nelson, 1970, "Communication in the Physical and Social Sciences," *Science*, 170, pp. 1166–1173.

Gast, P., 1968, "Trace Element Fractionation and the Origin of Tholeiitic and Alkaline Magma Types," *Geochimica et Cosmochimica Acta*, 32, pp. 1057–1068.

Gaston, J., 1978, *The Reward System in British and American Science*, New York: Wiley.

Georgi, J., 1962, "Memories of Alfred Wegener," in S. Runcorn (ed.), *Continental Drift*, New York: Academic, pp. 309–324.

Giere, R., 1988, *Explaining Science: A Cognitive Approach*, Chicago: University of Chicago Press.

Gieryn, T., 1982, "Relativist/Constructivist Programs in the Sociology of Science: Redundance or Retreat," *Social Studies of Science*, 12, pp. 279–297.

Gilbert, G., 1977, "Referencing as Persuasion," *Social Studies of Science*, 7, pp. 113–122.

Gilbert, G., and M. Mulkay, 1980, "Contexts of Scientific Discourse: Social Accounting in Experimental Papers," in K. Knorr et al., (eds.), *The Social Process of Scientific Investigations: Yearbook in the Sociology of the Sciences*, IV. Dordrecht, Holland: D. Reidel, pp. 269–294.

———, 1982, "Warranting Scientific Belief," *Social Studies of Science*, 12, pp. 383–408.

Glen, W., 1975, *Continental Drift and Plate Tectonics*, Columbus, Ohio: Charles E. Merrill.

———, 1982, *The Road to Jaramillo*, Stanford: Stanford University Press.

Gould, S., 1977, *Ever Since Darwin*, New York: W. W. Norton.

Graham, L., 1972, *Science and Philosophy in the Soviet Union*, New York: Alfred Knopf.

Green, A., 1977, "The Evolution of the Earth's Crust and Sedimentary Basin Development," in J. Heacock (ed.), *The Earth's Crust*, Washington, D.C.: American Geophysical Union, Monograph 20, pp. 1–18.

Green, D., and A. Ringwood, 1967, "The Genesis of Basaltic Magmas," *Contributions to Mineralogy and Petrology*, 15, pp. 103–190.

Greenberg, D., 1967, *The Politics of Pure Science*, New York: New American Library.

Greene, M., 1982, *Geology in the Nineteenth Century*, Ithaca, New York: Cornell University Press.

Griffith, B., and H. Small, 1974, "The Structure of Scientific Literatures II: Toward a Macro- and Micro-Structure of Science," *Science Studies*, 4, pp. 339–365.

Griggs, D., 1939, "A Theory of Mountain Building," *American Journal of Science*, 237, pp. 611–650.

Grover, S., 1981, *Toward a Psychology of the Scientist*, Washington, D.C.: University Press of America.

Guttman, L., 1968, "A General Nonmetric Technique for Finding the Smallest Coordinate Space for a Configuration of Points," *Psychometrika*, 33, pp. 469–506.

Hage, J., and M. Aiken, 1967, "Relationship of Centralization to Other Structural Properties," *Administrative Science Quarterly*, 12, pp. 72–92.

———, 1969, "Routine Technology, Social Structure, and Organizational Goals," *Administrative Science Quarterly*, 14, pp. 366–378.

Hagstrom, W., 1965, *The Scientific Community*, New York: Basic Books.

Hall, J., and P. Robinson, 1979, "Deep Crustal Drilling in the North Atlantic Ocean," *Science*, 204, pp. 573–586.

Hall, R., 1963, "The Concept of Bureaucracy: An Empirical Assessment," *American Sociological Review*, 69, pp. 32–40.

———, 1972, *Organizations: Structure and Process*, Englewood Cliffs, New Jersey: Prentice-Hall.
Hall, R., J. Haas, and N. Johnson, 1967, "Organization Size, Complexity, and Formalization," *American Sociological Review*, 32, pp. 904–912.
Hallam, A., 1973, *A Revolution in the Earth Sciences*, Oxford, England: Clarendon Press.
———, 1975, "Alfred Wegener and the Hypothesis of Continental Drift," *Scientific American*, 232, (Feb.), pp. 88–97.
———, 1983, *Great Geological Controversies*, Oxford: Oxford University Press.
Hamilton, L., 1976, "Sociology of the New Tectonics," Mimeographed paper from the Department of Sociology, University of Colorado, Boulder.
Hamilton, W., 1972, "A Rebuttal," *Geotimes*, (June), p. 10.
Hammond, K., G. McClelland, and J. Mumpower, 1980, *Human Judgement and Decision Making*. New York: Praeger.
Hargens, L., 1975, *Patterns of Scientific Research*, ASA Rose Monograph, Washington, D.C.: American Sociological Association.
Harland, W., A. Cox, P. Llewellyn, C. Pickton, A. Smith, and R. Walters, 1982, *A Geologic Time Scale*, Cambridge: Cambridge University Press.
Harré, R., 1970, *The Principles of Scientific Thinking*, London: MacMillan.
———, 1976, "The Constructive Role of Models in the Social Sciences," in L. Collins (ed.), *The Use of Models in the Social Sciences*, Boulder, Colorado: Westview Press, pp. 16–43.
———, 1986, *Varieties of Realism*, Oxford: Basil Blackwell.
Harrison, C., and B. Funnel, 1964, "Relationship of Paleomagnetic Reversals and Micropaleontology in Two Late Caenozoic Cores from the Pacific Ocean," *Nature*, 204, p. 566.
Hatherton, T., and W. Dickinson, 1969, "The Relationship Between Andesitic Volcanism and Seismicity in Indonesia, the Lesser Antilles, and Other Island Arcs," *Journal of Geophysical Research*, 74, pp. 5301–5310.
Heezen, B., M. Tharp, and M. Ewing, 1959, *The Floors of the Oceans: I. The North Atlantic*, Geological Society of America Special Paper 65.
Heirtzler, J., 1968 "Sea-Floor Spreading," *Scientific American*, 219, pp. 60–70.
Heirtzler, J., G. Dickson, E. Herron, W. Pitman, and X. LePichon, 1968, "Marine Magnetic Anomalies, Geomagnetic Field Reversals, and Motions of the Ocean Floor and Continents," *Journal of Geophysical Research*, 73, pp. 2119–2136.
Heirtzler, J., and X. LePichon, 1965, "Crustal Structure of the Mid-Ocean Ridges. III. Magnetic Anomalies Over the Mid-Atlantic Ridge," *Journal of Geophysical Research*, 70, pp. 4013–4033.
Hess, H., 1960, "The Evolution of Ocean Basins: Report to the Office of Naval Research on Research Supported by ONR Contract, Nonr, 1858," (10), 38.
———, 1962, "History of the Ocean Basins," in A. Engel, H. James, and B. Leonard (eds.), *Petrologic Studies: A Volume in Honor of A.F. Buddington*, New York: Geological Society of America, pp. 599–620.
———, 1968, "Reply," *Journal of Geophysical Research*, 73, p. 6569.
Hesse, M., 1966, *Models and Analogies in Science*, New York: University of Notre Dame Press.
———, 1974, *The Structure of Scientific Inference*, London: MacMillan.
———, 1976, "Models Verses Paradigms in the Natural Sciences," in L. Collins (ed.), *The Use of Models in the Social Sciences*, Boulder, Colorado: Westview Press, pp. 6–16.
———, 1980, *Revolution and Reconstruction in the Philosophy of Science*, Bloomington, Indiana: Indiana University Press.
Hickson, D., D. Pugh, and D. Pheysey, 1969, "Operations Technology and Organization Structure: An Empirical Reappraisal," *Administrative Science Quarterly*, 14, pp. 378–397.

Hofstadter, D., 1979, *Gödel, Escher, Bach*, New York: Vintage Books.
Holmes, A., 1913, *The Age of the Earth*, London: Harper Brothers.
———, 1931, "Radioactivity and Earth Movements," *Geological Society of Glasgow Transactions*, 18, pp. 559–606.
———, 1944, *Principles of Physical Geology*, New York: Ronald.
———, 1953, "The South Atlantic: Land Bridges or Continental Drift," *Nature*, 171, pp. 669–671.
Honnorez, J., E. Bonatti, C. Emiliani, P. Bronnirmann, M. Furrer, and A. Meyerhoff, 1975, "Mesozoic Limestone from the Verma Offset Zone, Mid-Atlantic Ridge?" *Earth and Planetary Science Letters*, 26, pp. 8–12.
Horsefield, B., and P. Stone, 1972, *The Great Ocean Business*, New York: Coward, McCann and Geoghegan.
Hunt, M., 1982, *The Universe Within*, New York: Simon and Schuster.
Hurley, P., 1968, "The Confirmation of Continental Drift," *Scientific American*, 218 (April), pp. 52–61.
Hurley, P., and J. Rand, 1968, "Review of Age Data in West Africa and South America Relative to a Test of Continental Drift," in R. Phinney (ed.), *The History of the Earth's Crust*, Princeton, New Jersey: Princeton University Press, pp. 153–160.
Institute for Scientific Information, 1963, *1961 Science Citation Index*, Philadelphia: Institute for Scientific Information.
———, 1972, *Science Citation Index Cumulative Five-Year Index: 1965–1969*. Philadelphia: Institute for Scientific Information.
———, 1973, *1972 Science Citation Index*. Philadelphia: Institute for Scientific Information.
———, 1975, *Science Citation Index Cumulative Five-Year Index: 1970–1974*. Philadelphia: Institute for Scientific Information.
———, 1984, *Science Citation Index Cumulative Ten-Year Index: 1955–1964*. Philadelphia: Institute for Scientific Information.
Irving, E., 1964, *Paleomagnetism and Its Application to Geological and Geophysical Problems*, New York: Wiley.
Isacks, B., J. Oliver, and L. Sykes, 1968, "Seismology and the New Global Tectonics," *Journal of Geophysical Research*, 73, pp. 5855–5899.
Isacks, B., and P. Molnar, 1969, "Mantle Earthquake Mechanisms and the Sinking of the Lithosphere," *Nature*, 223, pp. 1121–1124.
Jacobs, J., R. Russell, and J. Wilson, 1959, *Physics and Geology*, New York: McGraw-Hill.
———, 1974, *Physics and Geology* (second edition), New York: McGraw-Hill.
Jacoby, W., 1981, "Modern Concepts of Earth Dynamics Anticipated by Alfred Wegener in 1912," *Geology*, 9, pp. 25–27.
Jeffreys, H., 1924, *The Earth: Its Origin, History, and Physical Constitution*, Cambridge: Cambridge University Press.
———, 1974, "Theoretical Aspects of Continental Drift," in C. Kahle (ed.), *Plate Tectonics: Assessments and Reassessments*, Tulsa, Oklahoma: American Association of Petroleum Geologists, Memoir 23, pp. 395–405.
Jones, D., A. Cox, P. Coney, and M. Beck, 1982, "The Growth of North America," *Scientific American*, 247, pp. 70–84.
Kahle, C. (ed.), 1974, *Plate Tectonics: Assessments and Reassessments*, Tulsa, Oklahoma: American Association of Petroleum Geologists, Memoir 23.
Karig, D., 1970, "Ridges and Trenches of the Tonga-Kermadec Island Arc System," *Journal of Geophysical Research*, 75, pp. 239–254.
———, 1971*a*, "Origin and Development of Marginal Basins in the Western Pacific," *Journal of Geophysical Research*, 76, pp. 2542–2561.
———, 1971*b*, "Structural History of the Mariana Island Arc System," *Geological Society of America Bulletin*, 82, pp. 323–344.

Kasbeer, T., 1973, *Bibliography of Continental Drift and Plate Tectonics*, Boulder, Colorado: Geological Society of America, Special Paper 14.

Kay, M. (ed.), 1969, *North Atlantic—Geology and Continental Drift*, Tulsa, Oklahoma: American Association of Petroleum Geologists.

Kennett, J., 1982, *Marine Geology*, Englewood Cliffs, New Jersey: Prentice-Hall.

Kerr, R., 1978*a*, "Plate Tectonics: What Force Drives the Plates?" *Science*, 200, pp. 36–38, 90.

———, 1978*b*, "Precambrian Tectonics: Is the Present the Key to the Past?" *Science*, 199, pp. 282–285, 330.

———, 1980, "The Bits and Pieces of Plate Tectonics," *Science*, 207, pp. 1059–1061.

———, 1983, "Ophiolites: Windows on Which Ocean Crust?" *Science*, 219, pp. 1307–1309.

Kessler, M., 1963, "Bibliometric Coupling Between Scientific Papers," *American Documentation*, 14, pp. 10–25.

King, L., 1953, "Necessity for Continental Drift," *Proceedings of the American Association for Petroleum Geologists*, 37, pp. 2163–2177.

———, 1962, *The Morphology of the Earth*, Edinburgh: Oliver and Kirg.

———, 1983, *Wandering Continents and Spreading Sea Floors on an Expanding Earth*, New York: Wiley-Interscience.

Kitts, D., 1977, *The Structure of Geology*, Dallas, Texas: Southern Methodist University Press.

Knorr-Cetina, K., 1981, *The Manufacture of Knowledge*, Oxford: Pergamon Press.

Knorr-Cetina, K., and M. Mulkay (eds.), 1983, *Science Observed*, Beverly Hills, California: Sage Publications.

Koester, D., D. Sullivan, and D. White, 1982, "Theory Selection in Particle Physics: A Quantitative Case Study of the Evolution of Weak-Electromagnetic Unification Theory," *Social Studies of Science*, 12, pp. 73–100.

Köppen, W., and A. Wegener, 1924, *Die Klimate der Geologischen Vorzeit*, Berlin: Gebrüder Borntraeger.

Kosslyn, S., 1983, *Ghosts in the Mind's Machine*, New York: W. W. Norton.

Kuhn, T., 1962, *The Structure of Scientific Revolutions*, Chicago: University of Chicago Press.

———, 1970*a*, *The Structure of Scientific Revolutions* (second edition), Chicago: University of Chicago Press.

———, 1970*b*, "Reflections on My Critics," in I. Lakatos and A. Musgrave (eds.), *Criticism and the Growth of Knowledge*, Cambridge: Cambridge University Press, pp. 231–278.

———, 1970*c*, "Logic of Discovery or Psychology of Research," in I. Lakatos and A. Musgrave (eds.), *Criticism and the Growth of Knowledge*, Cambridge: Cambridge University Press, pp. 1–24.

———, 1977, *The Essential Tension*, Chicago: University of Chicago Press.

Kulp, J., 1961, "Geologic Time Scale," *Science*, 133, pp. 1105–1114.

Lakatos, I., 1970, "Falsificationism and the Methodology of Scientific Research Programmes," in Lakatos, I., and A. Musgrave (eds.), *Criticism and the Growth of Knowledge*, Cambridge: Cambridge University Press, pp. 91–196.

Lakatos, I., and A. Musgrave (eds.), 1970, *Criticism and the Growth of Knowledge*, Cambridge: Cambridge University Press.

LaRoche, M., and S. Pearson, 1985, "Rhetoric and Rational Enterprises," *Written Communication*, 2, 246–268.

Latour, B., 1987, *Science in Action*, Cambridge, Massachusetts: Harvard University Press.

Latour, B., and S. Woolgar, 1979, *Laboratory Life*, Beverly Hills, California: Sage Publications.

Laudan, L., 1977, *Progress and Its Problems*, Berkeley, California: University of California Press.

———, 1981*a*, "A Problem-solving Approach to Scientific Progress," in I. Hacking (ed.), *Scientific Revolutions*, Oxford: Oxford University Press, pp. 144–155.

———, 1981*b*, "The Pseudo-Science of Science?" *Philosophy of the Social Sciences*, 11, pp. 173–198.

Laudan, R., 1980, "The Recent Revolution in Geology and Kuhn's Theory of Scientific Change," in G. Gutting (ed.), *Paradigms and Revolutions*, Notre Dame, Indiana: University of Notre Dame Press, pp. 284–296.

———, 1987, "Drifting Interests and Colliding Continents," *Social Studies of Science*, 17, pp. 317–321.

Law, J., 1975, "Is Epistemology Redundant? A Sociological View," *Philosophy of the Social Sciences*, 5, pp. 317–337.

Lawrence, P., and J. Lorsch, 1967, *Organization and Environment*, Boston: Harvard University Press.

Lear, J., 1967, "Canada's Unappreciated Role as Scientific Innovator," *Saturday Review*, 50, pp. 45–50.

Leatherdale, W., 1974, *The Role of Analogy, Model, and Metaphor in Science*, New York: North-Holland Publishing.

LeGrand, H., 1986*a*, "Specialties, Problems, and Localism: The Reception of Continental Drift in Australia: 1920–1940," *Earth Science History*, 5, pp. 84–95.

———, 1986*b*, "Steady as a Rock: Methodology and Moving Continents," in J. Schuster and R. Yeo (eds.), *The Politics and Rhetoric of Scientific Method*, Dordrecht: D. Reidel, pp. 97–138.

LePichon, X., 1968, "Sea-Floor Spreading and Continental Drift," *Journal of Geophysical Research*, 73, pp. 3611–3697.

Lingoes, J., 1965, "An IBM-7090 Program for Guttman-Lingoes Smallest Space Analysis," *Behavioral Science*, 10, pp. 183–184.

Long, J., 1978, "Productivity and Position in the Scientific Career," *American Sociological Review*, 43, pp. 889–908.

Longwell, C., and R. Flint, 1962, *Introduction to Physical Geology*. New York: John Wiley and Sons.

MacKenzie, D., 1978, "Statistical Theory and Social Interests: A Case Study," *Social Studies of Science*, 8, pp. 35–83.

MacKenzie, D., and B. Barnes, 1979, "Scientific Judgment: The Biometrics-Mendalism Controversy," in B. Barnes and S. Shapin (eds.), *Natural Order*, Beverly Hills, California: Sage Publications, pp. 191–210.

McCann, H., 1978, *Chemistry Transformed*, Norwood, New Jersey: Ablex.

McCulloch, M., and W. Cameron, 1983, "Nd-Sr Isotope Study of Primitive Lavas From the Troodos Ophiolite, Cyprus: Evidence for a Subduction-Related Setting," *Geology*, 11, pp. 727–731.

McKenzie, D., 1969, "Speculations on the Consequences and Causes of Plate Motions," *Geophysics Journal*, 18, pp. 1–32.

McKenzie, D., and J. Morgan, 1969, "Evolution of Triple Junctions," *Nature*, 224, pp. 125–133.

McKenzie, D., and R. Parker, 1967, "The North Pacific: An Example of Tectonics on a Sphere," *Nature*, 216, pp. 1276–1280.

Maddala, G., 1977, *Econometrics*, New York: McGraw-Hill.

Mahoney, M., 1976, *Scientist as Subject: The Psychological Imperative*, Cambridge, MA: Ballinger.

March, J., and H. Simon, 1958, *Organizations*, New York: Wiley.

Marvin, U., 1973, *Continental Drift: The Evolution of a Concept*, Washington, D.C.: Smithsonian Institution Press.

———, 1985, "The British Reception of Alfred Wegener's Continental Drift Hypothesis," *Earth Science History*, 4, pp. 138–159.

Mason, R., 1958, "A Magnetic Survey Off the West Coast of the United States Between

Latitudes 33° and 36° N, Longitudes 121° and 128° W," *Geophysics Journal*, 1, pp. 320–329.
Mason, R., and A. Raff, 1961, "Magnetic Survey Off the West Coast of North America, 32° N Latitude to 42° N Latitude," *Geological Society of America Bulletin*, 72, pp. 1259–1266.
Mason, W., G. Wong, and B. Entwisle, 1983, "Contextual Analysis Through the Multilevel Linear Model," in S. Leinhardt (ed.), *Sociological Methodology 1983–1984*, San Francisco: Jossey-Bass, pp. 72–103.
Masterman, M., 1970, "The Nature of a Paradigm," in I. Lakatos and A. Musgrave (eds.), *Criticism and the Growth of Knowledge*, Cambridge: Cambridge University Press, pp. 59–89.
———, 1980, "Braithwaite and Kuhn: Analogy-Clusters Within and Without Hypothetico-Deductive Systems in Science," in D. Mellor (ed.), *Science, Belief and Behavior*, Cambridge: Cambridge University Press, pp. 61–86.
Maxwell, A., R. Von Herzen, K. Hsü, J. Andrews, T. Saito, S. Percival, E. Milow, and R. Boyce, 1970, "Deep-Sea Drilling in the South Atlantic," *Science*, 168, pp. 1047–1059.
Mayer, R., 1981, *The Promise of Cognitive Psychology*, San Francisco: W. H. Freeman.
Mayr, E. (ed.), 1952, "The Problem of Land Connections Across the South Atlantic," *American Museum of Natural History Bulletin*, 99, pp. 85–258.
Mazur, A., 1987, "Scientific Disputes over Policy," in H. Engelhardt and A. Caplan (eds.), *Scientific Controversies*, Cambridge: Cambridge University Press, pp. 265–283.
Medewar, P., 1964, "Is the Scientific Article Fraudulent? Yes, It Misrepresents Scientific Thought," *Saturday Review*, (Aug. 1), pp. 42–43.
Melson, W., G. Thompson, and T. van Andel, 1968, "Volcanism and Metamorphism in the Mid-Atlantic Ridge 22° N Latitude," *Journal of Geophysical Research*, 73, pp. 5925–5941.
Menard, H., 1964, *Marine Geology of the Pacific*, New York: McGraw-Hill.
———, 1969, *Anatomy of an Expedition*, New York: McGraw-Hill.
———, 1971, *Science: Growth and Change*, Cambridge, Massachusetts: Harvard University Press.
———, 1986, *The Ocean of Truth*, Princeton, New Jersey: Princeton University Press.
Merton, R., 1942, "Science and Technology in a Democratic Order," *Journal of Legal and Political Sociology*, 1, pp. 115–126.
———, 1957, "Priorities in Scientific Discovery: A Chapter in the Sociology of Science," *American Sociological Review*, 22, pp. 635–659.
———, 1963, "Resistance to the Systematic Study of Multiple Discoveries in Science," *European Journal of Sociology*, 4, pp. 250–282.
———, 1968, "The Matthew Effect in Science," *Science*, 159, pp. 56–63.
Messeri, P., 1978, "Obliteration by Incorporation: Toward a Problematics, Theory, and Metric of the Use of Scientific Literature," Presented at the American Sociological Association annual meeting, Sept., 1978.
Meyerhoff, A., 1968, "Arthur Holmes: Originator of Spreading Ocean Floor Hypothesis," *Journal of Geophysical Research*, 73, pp. 6563–6565.
———, 1972, "Review of [Tarling and Tarling, 1971]," *Geotimes*, 17, pp. 34–6.
Meyerhoff, A., and H. Meyerhoff, 1974, "Tests of Plate Tectonics," in C. Kahle (ed.), *Plate Tectonics: Assessments and Reassessments*, Tulsa, Oklahoma: American Association of Petroleum Geologists, Memoir 23, pp. 43–146.
Meyerhoff, A., H. Meyerhoff, and R. Briggs, 1972, "Continental Drift, V: Proposed Hypothesis of Earth Tectonics," *Journal of Geology*, 80, pp. 663–692.
Meyerhoff, H., and A. Meyerhoff, 1978, "Spreading History of the Eastern Indian Ocean and India's Northward Flight from Antarctica and Australia: Discussion," *Geological Society of America Bulletin*, 89, pp. 637–640.
Minsky, M., 1985, *The Society of Mind*, New York: Simon and Schuster.

Mitroff, I., 1974, *The Subjective Side of Science*, Amsterdam: Elsevier Scientific Publishing.

Molnar, P., 1988, "Continental Tectonics in the Aftermath of Plate Tectonics," *Nature*, 335, pp. 131–137.

Molnar, P., and P. Trapponnier, 1975, "Cenozoic Tectonics of Asia: Effects of a Continental Collision," *Science*, 189, pp. 419–426.

———, 1978, "Active Tectonics of Tibet," *Journal of Geophysical Research*, 83, pp. 5361–5375.

Moores, E., 1982, "Origin and Emplacement of Ophiolites," *Reviews of Geophysics and Space Physics*, 20, pp. 735–769.

Morgan, W., 1968, "Rises, Trenches, Great Faults, and Crustal Blocks," *Journal of Geophysical Research*, 73, pp. 1959–1982.

Motz, L. (ed.), 1975, *The Rediscovery of the Earth*, New York: Van Nostrand Reinhold.

Mulkay, M., and G. Gilbert, 1981, "Putting Philosophy to Work: Karl Popper's Influence on Scientific Practice," *Philosophy of the Social Sciences*, 11, pp. 387–407.

Narin, F., 1976, *Evaluative Bibliometrics*, Cherry Hill, New Jersey: Computer Horizons.

Newell, A., and H. Simon, 1972, *Human Problem Solving*, Englewood Cliffs, New Jersey: Prentice-Hall.

Nitecki, M., J. Lemke, H. Pullman, and M. Johnson, 1978, "The Acceptance of Plate Tectonics Theory by Geologists," *Geology*, 6, pp. 661–664.

O'Hara, M., 1965, "Primary Magmas and the Origins of Basalts," *Scottish Journal of Geology*, 1, pp. 19–40.

Oldenburg, D., and J. Brune, 1972, "Ridge Transform Fault Spreading Pattern in Freezing Wax," *Science*, 178, pp. 301–304.

Oliver, J., and B. Isacks, 1967, "Deep Earthquake Zones, Anomalous Structures in the Upper Mantle and the Lithosphere," *Journal of Geophysical Research*, 72, pp. 4259–4275.

Opdyke, N., B. Glass, J. Hayes, and J. Foster, 1966, "Paleomagnetic Study of the Antarctic Deep-Sea Cores," *Science*, 154, pp. 349–357.

Oreskes, N., 1988, "The Rejection of Continental Drift," *Historical Studies in the Physical Sciences*, 18, pp. 311–348.

Owen, H., 1976, "Continental Displacement and Expansion of the Earth During the Mesozoic and Cenozoic," *Philosophical Transactions of the Royal Society of London*, A-281, pp. 223–291.

Pearce, J., and J. Cann, 1973, "Tectonic Setting of Basic Volcanic Rocks Determined Using Trace Element Analysis," *Earth and Planetary Science Letters*, 19, pp. 290–300.

Perrow, C., 1972, *Complex Organizations: A Critical Review*, New York: Scott, Foresman, and Co.

Petty, R., and J. Cacioppo, 1986, *Communication and Persuasion: Central and Peripheral Routes to Attitude Change*, New York: Springer/Verlag.

Phinney, R., 1968*a*, "Introduction," in R. Phinney (ed.), *The History of the Earth's Crust*, Princeton, New Jersey: Princeton University Press, pp. 3–12.

Phinney, R. (ed.), 1968*b*, *The History of the Earth's Crust*, Princeton, New Jersey: Princeton University Press.

Pickering, A., 1980, "The Role of Interests in High Energy Physics," in K. Knorr, R. Krohn, and R. Whitley (eds.), *The Social Process of Scientific Investigation*, 4, pp. 107–138.

Pinch, T., 1981, "The Sun-Set: The Presentation of Certainty in Scientific Life," *Social Studies of Science*, 11, pp. 131–158.

Pitman, W., and D. Hayes, 1968, "Sea-Floor Spreading in the Gulf of Alaska," *Journal of Geophysical Research*, 73, pp. 6571–6580.

Pitman, W., and J. Heirtzler, 1966, "Magnetic Anomalies Over the Pacific-Antarctic Ridge," *Science*, 154, pp. 1164–1171.

Popper, K., 1959, *The Logic of Scientific Discovery*, London: Hutchison.

———, 1968, *Conjectures and Refutations*, New York: Harper and Row.

Pugh, D., D. Hickson, C. Hinings, K. MacDonald, C. Turner, and T. Lupton, 1963, "A Conceptual Scheme for Organizational Analysis," *Administrative Science Quarterly*, 8, pp. 289–315.

Raup, D., 1986, "New Ideas Are 'Guilty until Proved Innocent'," *The Scientist*, (Oct. 20), p. 18.

Reskin, B., 1976, "Sex Differences in Status Attainment in Science," *American Sociological Review*, 41, pp. 597–612.

———, 1977, Scientific Productivity and the Reward System of Science," *American Sociological Review*, 42, pp. 491–504.

Roose, K., and C. Anderson, 1970, *A Rating of Graduate Programs*, Washington, D.C.: American Council on Education.

Runcorn, S., 1956*a*, "Paleomagnetic Surveys in Arizona and Utah: Preliminary Results," *Geological Society of America Bulletin*, 67, pp. 301–316.

———, 1956*b*, "Paleomagnetic Comparisons Between Europe and North America," *Proceedings of Geological Association of Canada*, 8, pp. 77–85.

———, 1962, "Paleomagnetic Evidence for Continental Drift and Its Geophysical Cause," in S. Runcorn (ed.), *Continental Drift*, London: Academic Press, pp. 1–40.

Runcorn, S. (ed.), 1962, *Continental Drift*, London: Academic Press.

Ruse, M., 1981, "What Kind of Revolution Occurred in Geology?" *PSA 1978*, 2, pp. 240–273.

Saleeby, J., 1983, "Accretionary Tectonics of the North American Cordillera," *Annual Review of Earth and Planetary Sciences*, 15, pp. 45–73.

Sarjeant, W., 1980, *Geologists and the History of Geology*, New York: Arno Press.

Saul, V., 1986, "Wanted: Alternatives to Plate Tectonics," *Geology*, 14, p. 536.

Schermer, E., D. Howell, and D. Jones, 1984, "The Origin of Allochthonous Terranes: Perspectives on the Growth and Shaping of Continents," *Annual Review of Earth and Planetary Sciences*, 12, pp. 107–131.

Schlee, S., 1973, *The Edge of an Unfamiliar World*, New York: E. P. Dutton.

Schott, T., 1987, "Interpersonal Influence in Science: Mathematicians in Denmark and Israel," *Social Networks*, 9, pp. 351–374.

Schwarzbach, M., 1986, *Alfred Wegener: The Father of Continental Drift*, Madison, Wisconsin: Science Tech.

Sclater, J., and J. Francheteau, 1970, "The Implications of Terrestrial Heat Flow Observations on Current Tectonics and Geochemical Models of the Crust and Upper Mantle of the Earth," *Geophysical Journal*, 20, pp. 509–542.

Sengör, A., and K. Burke, 1979, "Comments On: Why I Do Not Accept Plate Tectonics," *EOS*, (April 24), pp. 207–211.

Shapin, S., 1982, "History of Science and Its Sociological Reconstructions," *History of Science*, 20, pp. 157–211.

Silver, E., 1971, "Tectonics of the Mendocino Triple Junction," *Geological Society of America Bulletin*, 82, pp. 2965–2978.

Simon, H., 1957, *Models of Man*, New York: Wiley.

———, 1977, *Models of Discovery*, Boston: Reidel Publishing.

Small, H., 1973, "Co-Citation in the Scientific Literature: A New Measure of the Relationship Between Two Documents," *Journal of the American Society for Information Science*, (July-August), pp. 265–269.

———, 1974, "Multiple Citation Patterns in Scientific Literature: The Circle and Hill Models," *Information Storage and Retrieval*, 10, pp. 393–402.

———, 1977, "A Co-Citation Model of a Scientific Specialty: A Longitudinal Study of Collagen Research," *Social Studies of Science*, 7, pp. 139–166.

Small, H., and B. Griffith, 1974, "The Structure of Scientific Literatures I: Identifying and Graphing Specialties," *Science Studies*, 4, pp. 17–40.

Stewart, J., 1979, "Changes in Cognitive and Social Structures During a Scientific Revolution: Plate Tectonics and Geology," Ph.D. Dissertation (Sociology): University of Wisconsin–Madison.

———, 1983, "Achievement and Ascriptive Processes in the Recognition of Scientific Articles," *Social Forces*, 62, pp. 166–189.

———, 1986, "Drifting Continents and Colliding Interests: A Quantitative Application of the Interests Perspective," *Social Studies of Science*, 16, pp. 261–279.

———, 1987*a*, "Drifting or Colliding Interests?: A Reply to Laudan with Some New Results," *Social Studies of Science*, 17, pp. 321–331.

———, 1987*b*, "Discipline Integration, Citation Patterns, and the Plate Tectonics Revolution," paper presented at the 1987 Annual Meeting of the Society for the Social Study of Science.

Stewart, J., W. Hagstrom, and H. Small, 1981, "The Breakup and Dispersion of a New Conceptual 'Continent,'" *Geoscience, Canada*, 8, pp. 8–15.

Storer, N., 1966, *The Social System of Science*, New York: Holt, Rinehart, and Winston.

Strauss, A., 1978, *Negotiations*, San Francisco: Jossey-Bass.

Suess, E., 1904–09, *The Face of the Earth* (translation), 4 vols., Oxford: Clarendon Press.

Sullivan, D., E. Barboni, and D. White, 1981, "Problem Choice and the Sociology of Scientific Competition," *Knowledge and Society: Studies in the Sociology of Culture Past and Present*, 3, pp. 163–197.

Sullivan, W., 1974, *Continents in Motion*, New York: McGraw-Hill.

Suppe, F. (ed.), 1977, *The Structure of Scientific Theories* (second edition), Urbana, Illinois: University of Illinois Press.

Swamy, P., 1971, *Statistical Inference in Random Coefficient Regression Models*, New York: Springer-Verlag.

Sykes, L., 1966, "The Seismicity and Deep Structure of Island Arcs," *Journal of Geophysical Research*, 71, pp. 2981–3006.

———, 1967, "Mechanisms of Earthquakes and Nature of Faulting on Mid-Ocean Ridges," *Journal of Geophysical Research*, 72, pp. 2131–2153.

Takeuchi, H., S. Uyeda, and H. Kanamori, 1967, *Debate about the Earth*, San Francisco: Freeman, Cooper, and Co.

Tannenbaum, A., 1968, *Control in Organizations*, New York: McGraw-Hill.

Tannenbaum, A., and R. Cooke, 1979, "Organizational Control: A Review of Studies Employing the Control Graph Method," in C. Lammers and D. Hickson (eds.), *Organizations Alike and Unlike*, London: Routledge and Kegan Paul, pp. 183–210.

Tapponnier, P., G. Peltzer, A. Le Dain, R. Armijo, and P. Lobbold, 1982, "Propagating Extrusion Tectonics in Asia: New Insights from Simple Experiments in Plasticine," *Geology*, 10, pp. 611–616.

Tarling, D., and M. Tarling, 1971, *Continental Drift: A Study of the Earth's Moving Surface*, London: G. Bell and Sons.

Taylor, F., 1910, "Bearing of the Tertiary Mountain Belt on the Origin of the Earth's Plan," *Geological Society of America Bulletin*, 21, pp. 176–226.

Travis, G., 1981, "Replicating Replication?: Aspects of the Social Construction of Learning in Planarian Worms," *Social Studies of Science*, 11, pp. 11–32.

Uyeda, S., 1978, *The New View of the Earth*, San Francisco: W. H. Freeman.

Vacquier, V., 1959, "Measurement of Horizontal Displacement Along Faults in Ocean Floor," *Nature*, 183, pp. 452–453.

van Andel, T., 1981, *Science at Sea: Tales of an Old Ocean*, San Francisco: W.H. Freeman.

van Andel, T., and C. Bowin, 1968, "Mid-Atlantic Ridge Between 22° and 23° North Latitude and the Tectonics of Mid-Ocean Rises," *Journal of Geophysical Research*, 73, pp. 1279–1298.

Verhoogen, J., 1983, "Personal Notes and Sundry Comments," *Annual Review of Earth and Planetary Sciences*, 11, pp. 1–9.

Vine, F., 1966, "Spreading of the Ocean Floor: New Evidence," *Science*, 154, pp. 1405–1415.

Vine, F., and D. Matthews, 1963, "Magnetic Anomalies Over Ocean Ridges," *Nature*, 199, pp. 947–949.

Vine, F., and J. Wilson, 1965, "Magnetic Anomalies Over a Young Oceanic Ridge off Vancouver Island," *Science*, 150, pp. 485–489.

Walker, R., 1973, "Mopping Up the Turbidite Mess," in R. Ginsburg (ed.), *Evolving Concepts in Sedimentology*, Baltimore: Johns Hopkins University Press, pp. 1–37.

Waltz, D., 1982, "Artificial Intelligence," *Scientific American*, 247 (Oct.), pp. 118–133.

Waterschoot van der Gracht, W. (ed.), 1928, *Theory of Continental Drift*, Tulsa, Oklahoma: American Association of Petroleum Geologists.

Weber, M., 1947, *Theory of Social and Economic Organization*, New York: Free Press.

Wegener, A., 1912*a*, "Die Entstehung der Kontinente," *Petermann's Geographische Mitteilungen*, 58, pp. 185–195, 253–256, 305–308.

———, 1912*b*, "Die Entstehung der Kontinente," *Geologische Rundschau*, 3, pp. 276–292.

———, 1915, *Die Entstehung der Kontinente und Ozeane*, Braunschweiz: Vieweg.

———, 1924, *The Origin of Continents and Oceans*, London: Methuen.

———, 1966 [1929], *The Origin of Continents and Oceans*, New York: Dover.

Wegmann, C., 1963, "Tectonics Patterns at Different Levels," *Geological Society of South Africa*, 66 (annex), pp. 1–78.

Weimer, W., 1979, *Notes on the Methodology of Scientific Research*, Hillsdale, New Jersey: Lawrence Erlbaum.

Werner, O., 1970, "Cultural Knowledge, Language, and World View," in P. Garvin (ed.), *Cognition: A Multiple View*, New York: Spartan Books, pp. 155–175.

Wertenbaker, W., 1974, *The Floor of the Sea*, Boston: Little, Brown and Co.

West, S., 1960, "The Ideology of American Scientists," *IRE Transactions in Engineering and Management*, EM–1, pp. 54–62.

———, 1982, "A Patchwork Earth," *Science 82*, 3, pp. 46–52.

Williams, R., and J. Law, 1982, "Putting Facts Together: A Study in Scientific Persuasion," *Social Studies of Science*, 12, pp. 535–558.

Willis, B., 1944, "Continental Drift, ein Marchen," *American Journal of Science*, 242, pp. 509–515.

Wilson, J., 1959, "Geophysics and Continental Growth," *American Scientist*, 47, pp. 1–25.

———, 1961, "Continental and Oceanic Differentiation—Discussion of 'Continents and Ocean Basin Evolution by Spreading of the Sea Floor,' " *Nature*, 192, pp. 125–128.

———, 1963*a*, "Evidence from Islands of the Spreading of the Sea Floor," *Nature*, 197, pp. 536–538.

———, 1963*b*, "Hypothesis of the Earth's Behavior," *Nature*, 198, pp. 925–929.

———, 1963*c*, "A Possible Origin of the Hawaiian Islands," *Canadian Journal of Physics*, 41, pp. 863–870.

———, 1963*d*, "Continental Drift," reproduced in J. Wilson (ed.), *Continents Adrift*, San Francisco: W. H. Freeman, pp. 41–55.

———, 1965*a*, "A New Class of Faults and Their Bearing on Continental Drift," *Nature*, 207, pp. 343–347.

———, 1965*b*, "Evidence from Ocean Islands Suggesting Movement in the Earth," in P. Blackett, E. Bullard, and S. Runcorn (eds.), *A Symposium on Continental Drift*, London: Royal Society of London, pp. 145–167.

———, 1968*a*, "A Revolution in Earth Science," *Geotimes*, 13, pp. 10–16.

———, 1968*b*, "A Reply to V.V. Beloussov," *Geotimes*, 13, pp. 20–22.

———(ed.), 1972, *Continents Adrift*, San Francisco: W. H. Freeman.

———, 1974, "The Limits to Science," Radio program for the Public Broadcast System.

Wood, R., 1985, *The Dark Side of the Earth: The Battle for the Earth Sciences*, London: Allen and Unwin.

Wyllie, P., 1971, *The Dynamic Earth*, New York: John Wiley.

———, 1974, "Plate Tectonics, Seafloor Spreading, and Continental Drift: An Introduction," in C. Kahle (ed.), *Plate Tectonics: Assessments and Reassessments*, Tulsa, Oklahoma: American Association of Petroleum Geologists, Memoir 23, pp. 5–15.

———, 1976, *The Way the Earth Works*, New York: John Wiley.

Yearley, S., 1981, "Textual Persuasion: The Role of Social Accounting in the Construction of Scientific Arguments," *Philosophy of the Social Sciences*, 11, pp. 409–435.

Yoder, H., and C. Tilley, 1962, "Origin of Basaltic Magmas: An Experimental Study of Natural and Synthetic Rock Systems," *Journal of Petrology*, 3, pp. 342–532.

Ziman, J., 1968, *Public Knowledge*, Cambridge: Cambridge University Press.

———, 1978, *Reliable Knowledge*, Cambridge: Cambridge University Press.

Zuckerman, H., and R. Merton, 1971, "Patterns of Evaluation in Science," *Minerva*, 9, pp. 66–100.

Index